Frying

Related titles from Woodhead's food science, technology and nutrition list:

Antioxidants in food (ISBN: 1 85573 463 X)

Antioxidants are a major ingredient in food processing, both in controlling oxidation and in influencing other aspects of food quality as well as providing potential health benefits. This collection reviews antioxidant use, particularly the increasing role of natural antioxidants in food processing.

Thermal technologies in food processing (ISBN: 1 85573 558 X)

Thermal technologies are traditionally a compromise between their benefits in the enhancement of sensory characteristics and preservation, and their shortcomings, for example in reducing nutritional properties. The need to maximise process efficiency and final product quality has led to a number of new developments including refinements in existing technologies and the emergence of new thermal techniques. This collection reviews all these key developments and looks at future trends, providing an invaluable resource for all food processors.

Food processing technology: Principles and practice (ISBN: 1 85573 533 4)

The first edition of *Food processing technology* was quickly adopted as the standard text by many food science and technology courses. The publication of this completely revised new edition is set to confirm the position of this textbook as the best single-volume introduction to food manufacturing technologies available. New chapters include computer control of processing, novel 'minimal' technologies including processing using high pressures or pulsed electric fields, ohmic heating and an extended chapter on modified atmosphere packaging.

Details of these books and a complete list of Woodhead's food science, technology and nutrition titles can be obtained by:

- visiting our web site at www.woodhead-publishing.com
- contacting Customer services (e-mail: sales@woodhead-publishing.com; fax: +44 (0) 1223 893694; tel.: +44 (0) 1223 891358 ext.30; address: Woodhead Publishing Limited, Abington Hall, Abington, Cambridge CB1 6AH, England)

If you would like to receive information on forthcoming titles in this area, please send your address details to: Francis Dodds (address, tel. and fax as above; e-mail: francisd@woodhead-publishing.com). Please confirm which subject areas you are interested in.

Frying

Improving quality

Edited by
J. B. Rossell

CRC Press
Boca Raton Boston New York Washington, DC

WOODHEAD PUBLISHING LIMITED
Cambridge England

Published by Woodhead Publishing Limited
Abington Hall, Abington
Cambridge CB1 6AH
England

Published in North and South America by CRC Press LLC
2000 Corporate Blvd, NW
Boca Raton FL 33431
USA

First published 2001, Woodhead Publishing Limited and CRC Press LLC

British Library Cataloguing in Publication Data
A catalogue record for this book is available from the British Library.

Library of Congress Cataloging-in-Publication Data
A catalog record for this book is available from the Library of Congress.

Woodhead Publishing Limited ISBN 1 85573 556 3
CRC Press ISBN 0-8493-1208-6
CRC Press order number: WP1208

Cover design by The ColourStudio
Project managed by Macfarlane Production Services, Markyate, Hertfordshire (e-mail: macfarl@aol.com)
Typeset by MHL Typesetting Limited, Coventry, Warwickshire
Printed by T J International, Padstow, Cornwall, England

Contents

Preface ix
List of contributors xi

1 Introduction 1
J. B. Rossell, Leatherhead Food Research Association

Part I General issues **5**

2 The market for fried food 7
R. Piper, Europanel, London
2.1 Introduction 7
2.2 The range of fried foods available 8
2.3 Factors influencing the British and other European food markets 9
2.4 The market for fried food in the UK 10
2.5 The frozen food market in other European countries 12
2.6 Future trends 15
2.7 Sources of further information and advice 16

3 Regulation in the European Union 19
R. Fox, Pura Foods Limited, Belvedere
3.1 Introduction – the legal context 19
3.2 The structure of the frying industries 22
3.3 The sale of food 22
3.4 The life of frying oils 28
3.5 Environmental protection 34
3.6 Future trends 36

3.7 Sources of information........ 38
3.8 References........ 43

4 Regulation in the United States........ 49
D. Firestone, Food and Drug Administration, Washington DC
4.1 Introduction........ 49
4.2 FDA regulations and guidelines........ 49
4.3 USDA/FSIS guidelines and directives........ 52
4.4 State and city regulations........ 53
4.5 Sources of further information and advice........ 54
4.6 References........ 57

5 Health issues........ 59
B. Ruiz-Roso and G. Varela, Universidad Complutense de Madrid
5.1 Introduction........ 59
5.2 Dietary lipids: structure and function........ 60
5.3 Sources of dietary lipids........ 61
5.4 Digestion and absorption........ 62
5.5 Transport and metabolism........ 65
5.6 Health issues relating to fat and oil intake........ 68
5.7 The role of deep-frying in the fat intake........ 71
5.8 The impact of repeated frying........ 74
5.9 Measuring the impact of frying on fat intake........ 75
5.10 Conclusions........ 78
5.11 References........ 80

Part II Frying oils........ 85
6 The composition of frying oils........ 87
S. P. Kochhar, Good-Fry International NV, Rotterdam
6.1 Introduction........ 87
6.2 Types of frying oils and fats........ 88
6.3 Minor components and frying oil stability........ 91
6.4 Combined effects of natural products on stabilisation of frying oils........ 105
6.5 Future trends........ 108
6.6 Conclusions........ 109
6.7 References........ 110

7 Factors affecting the quality of frying oils and fats........ 115
J. B. Rossell, Leatherhead Food Research Association
7.1 Introduction........ 115
7.2 Properties and composition of oils and the relationship betweeen oil composition and its suitability as a frying oil... 116
7.3 Oil authenticity........ 127
7.4 Minor components........ 142

7.5 Quality limits for a fresh (unused) frying oil 148
7.6 Transport, delivery and storage 149
7.7 The frying process 152
7.8 Future trends 158
7.9 References 159

8 The measurement of frying oil quality and authenticity 165
R. F. Stier, Consultant
8.1 Introduction 165
8.2 Maintaining quality during frying 166
8.3 Regulatory issues 169
8.4 Quality measurements for refining operations 171
8.5 Developing purchasing specification and certifying vendors .. 173
8.6 Quality control during frying 175
8.7 Adulteration of fats and oils 177
8.8 Tests for frying fats and oils 178
8.9 The future for monitoring oil quality 189
8.10 References 190

Part III Improving product quality 195
9 The manufacture of pre-fried potato products 197
M. J. H. Keijbets, Aviko BV, Steenderen
9.1 Introduction 197
9.2 What are pre-fried potato products? 198
9.3 Range of pre-fried potato products and their use 198
9.4 Key requirements for pre-fried potato products 199
9.5 Key manufacturing processes for pre-fried French fries 200
9.6 Key manufacturing processes for 'formed' pre-fried potato products 208
9.7 Storage and distribution 210
9.8 Major quality-determining factors during manufacture of pre-fried French fries 211
9.9 Future trends 212
9.10 Sources of further information and advice 213
9.11 References 213

10 Managing potato crisp processing 215
R. M. Bennett, Consultant
10.1 Introduction 215
10.2 Oil and fat management 216
10.3 Raw material management 220
10.4 Managing the processing operation 222
10.5 Packaging 233
10.6 Future trends 234
10.7 References 235

11 Effective process control in frying 236
G. B. Quaglia and F. M. Bucarelli, Istituto Nazionale della Nutrizione, Rome
11.1 Introduction 236
11.2 The HACCP approach 237
11.3 Flow diagrams examination 239
11.4 Hazard evaluation and preventative measures 241
11.5 Monitoring critical control points in the frying process 252
11.6 Future trends 259
11.7 Bibliography 259

12 Flavour and aroma development in frying and fried food 266
P. Gillatt
12.1 Introduction 266
12.2 Flavour of raw potatoes 267
12.3 Degradation reactions occurring in edible oils and fats during frying 268
12.4 The Maillard and Strecker degradation reactions 272
12.5 Flavour development in foods 277
12.6 The importance of fatty acid composition on the flavour production
in the frying process 285
12.7 The effect of antioxidants in frying oils 296
12.8 Oil uptake by fried food 303
12.9 Effect of frying techniques, frying regime, cooking method, and additives on flavour of fried food 312
12.10 The influence of the food being fried 314
12.11 Sensory issues 315
12.12 Application of flavours 318
12.13 The future 325
12.14 Acknowledgement 327
12.15 References 327
Appendix: Flavours and aromas derived from lipid oxidation 335

13 Improving the texture and colour of fried products 337
C-S. Chen, C-Y. Chang and C-J. Hsieh, Da-Yeh University, Taiwan
13.1 Instrumentation for measuring the texture and colour of fried products 337
13.2 Influences on the texture and colour of fried products 340
13.3 Using response surface methodology (RSM) 343
13.4 A case study: fried gluten balls 348
13.5 Conclusions 355
13.6 References 356

Index 359

Preface

Pan-frying and deep-frying have been very popular and ancient methods of food preparations for more than 4000 years. Pre-fried and fried food products like potato crisps, fish fingers or French fries have become a main component of our diet. It is estimated that the total usage of frying fats and oils in restaurants, commercial frying and households is about 20 million tons a year.

During the last international symposium on Deep Fat Frying in Germany in March 2000, experts reaffirmed that there are no health concerns associated with consumption of frying fats and oils that have not been abused during normal frying conditions. In contrast, used cooking oils, containing high levels of degradation products, lead to a loss of organoleptic quality and a decrease in the nutrition value in fried foods, and cause strong foaming. Consumers also sometimes suffer from gastrointestinal distress after consuming food fried with such oils. Attempts by some frying establishments to bring about savings in frying oil costs have resulted in various improper practices such as the over use of frying oils, frying for too long a time, and recovery and reprocessing of spent frying oil for use in animal and poultry feeds which led indirectly to the Belgian catastrophe in which animal feed based on spent frying oil became accidentally contaminated with industrial transformer oil containing PCBs and dioxin.

The chemistry of oils and fats at frying temperatures is rather complex. More than 500 different chemical compounds have been detected as a result of oxidation, pyrolysis, polymerisation and hydrolysis. The kind and quantity of these reaction products also vary from one frying process to the other. It is impossible to identify any one compound or a group of compounds as a key indicator of the deterioriation process. Only a combination of different physical parameters may offer a solution of this problem. The March symposium *recommended* the combination of *two* tests like the determination of polar materials and the determination of the polymer triglycerides as the best way of analysing suspect frying fats and oils. Many countries have official or informal guidelines or legal regulations for assessing the quality of frying fats and oils as a way of protecting the consumer. However, these regulations are usually based only on one method and need correction.

In the past, research on deep fat frying has concentrated mostly on health, regulation and analytical aspects of the subject. The theory of deep frying and its application to industrial frying processing have been relatively neglected. The March symposium identified a number of new areas for research such as the use of natural ingredients with stabilising properties and new ideas about the acid catalysed polymerisation of triglycerides which acts in tandem with the acid catalysed dehydration of sterols, squalenes and others natural compounds. The symposium therefore identified the need to:

> Encourage and support basic research focused on understanding the dynamics of deep fat frying and the frying process. Research should be cross-disciplinary encompassing oil chemistry, food engineering, sensory science, food chemistry and nutritional sciences.

The present book will help us to get a better understanding both of current research on frying oil quality and of the nature of deep frying itself. Such an understanding is necessary to produce good fried products more economically with an optimum flavour and a better shelf-life.

Dr Christian Gertz
Chemiedirektor
Chemisches Untersuchungsamt Hagen

Contributors

Chapters 1 and 7

Dr J. B. Rossell
Leatherhead Food Research Association
Randalls Road
Leatherhead
Surrey KT22 7RY
England

Tel: +44 (0)1372 822286
Fax: +44 (0)1372 386228
E-mail: *brossell@lfra.co.uk*

Chapter 2

Mr Richard Piper
Europanel Raw Database GIE
Taylor Nelson AGB House
Westgate
London W5 1UA
England

Tel: +44 (0)208 967 4559
Fax: +44 (0)208 967 4002
E-mail: Richard.Piper@tnsofres.com

Chapter 3

Mr Richard Fox
Pura Foods Limited
Technical Development Centre
Crabtree Manorway South
Belvedere
Kent DA17 6BB
England

Tel: +44 (0) 208 320 9000
Fax: +44 (0) 208 320 9003
E-mail: *richard.fox@pura.co.uk*

Chapter 4

Dr David Firestone
Food and Drug Administration
Contaminants Chemical Division
200 C St. S.W. HFS-336
Washington DC 20204
USA

Fax: +1 202 205 4422

Chapter 5

Professor Baltasar Ruiz-Roso and
Professor Gregorio Varela
Departamento de Nutricion
Universidad Complutense de Madrid
28040 Madrid
Spain

Fax: +34 91 394 17 32
E-mail: *ruizrojo@eucmos.sim.ucm.es*

Chapter 6

Dr S. P. Kochhar
48 Chiltern Crescent
Earley
Reading RG6 1AN
England

Tel: +44 (0)118 962 16 11
Fax: +44 (0)118 962 60 79
E-mail: spkochhar_99@yahoo.ac.uk

Chapter 8

Mr Richard F. Stier
c/o E. T. Stier
1309 Avenida Sebastiani
Sonoma
CA 95476
USA

Fax: +1 925 484 9788
E-mail: *Rickstier4@aol.com*

Chapter 9

Dr Martin J. H. Keijbets
Head of Research and Development
Aviko BV
PO Box 8
7220 AA Steenderen
The Netherlands

Tel: +31 (0)575 458200
Fax: +31 (0)575 458380
E-mail: *m.keijbets@aviko.nl*

Chapter 10

Dr Reid M. Bennett
5200 Keller Springs Road
Suite 511
Dallas
Texas 75248
USA

Tel: +1 972 490 5727
Fax: +1 972 960 8507

Chapter 11

Professor G. B. Quaglia and
Professor F. M. Bucarelli
Istituto Nazionale della Nutrizione
Via Ardeatina
No. 546 00179 Roma
Italy

Fax: +39 06 503 1592
E-mail: *quaglia@inn.ingrm.it*

Chapter 12

Dr Peter Gillatt
32 Lowestoft Road
Reydon
Southwold
Suffolk IP19 6RJ
England

E-mail: *gillattp@kettlefoods.co.uk*

Chapter 13

Dr C-S. Chen, Dr C-Y. Chang and
Dr C-J. Hsieh
Department of Food Engineering
Da-Yeh University
Chang-Hwa
Taiwan 515

Fax: +886 4 2376 9738

1

Introduction

J.B. Rossell, Leatherhead Food Research Association

Frying is one of the fastest, oldest and simplest methods of food cooking, since it involves heating an edible oil or fat and simply using the hot oil to cook the food. It was probably invented by the ancient Chinese but became so popular that it is now used throughout the world in domestic, restaurant and industrial establishments. The popularity relates to the speed with which the food can be cooked as well as the pleasant attractive properties of the cooked food. Frying cooks the food through to the middle and while so doing generates a 'crust' on the surface of the food as well as a distinctive fried food flavour.

Frying is useful in cooking all types of food, viz. meat, fish and vegetables. In fact, a single vegetable, the potato, is probably the food most closely associated with frying, since potatoes are used to generate both French fries and crisps (chips in US parlance). Banks (1996) relates that the introduction of crisps can be traced back to an event in 1853, when Commodore Cornelius Vanderbilt was vacationing in Saratoga Springs, where he ordered French fries for his evening meal. He complained that the potatoes were sliced too thick and sent them back to the kitchen. The chef, George Crum, was angered and prepared some paper thin slices of potato and fried these until they were golden brown, fully crisp and dehydrated. This was no doubt intended as a riposte to the unreasonable demands of the self-opinionated restaurant guest, but, to his surprise, the Commodore found the crisp slices to his liking. This led to a new food, initially called 'Saratoga Chips', afterwards chips in the USA and crisps in the UK. They were initially available only as a restaurant item, but, just before the turn of the century, crisps were introduced as a snack food by frying in open kettles and serving loose over the counter in paper bags from a bulk stock. As popularity increased, small factories began to produce pre-packaged crisps, which could be sold in garages and other retail outlets.

In 1929, the J.D. Ferry Company introduced a continuous fryer, which provided a commercial boost to the development of industrial frying as we now know it. In today's world, huge installations are totally dedicated to crisps production, manufacturing different types, such as traditional, salted, and a variety of flavoured types. The crisps industry has now become a major part of the food manufacturing industry and it is not unusual for a crisps-making factory to consume over 400 tonnes of frying oil per week.

The French fries industry has also progressed, and several factories worldwide now manufacture pre-fried products in a variety of forms. These include not only 'pre-fried chips' which just need finish frying, but also 'oven ready' chips, which need simple re-warming, and low-fat chips, all produced to a variety of thicknesses and shapes to suit the palate of the consumer or the fashion of the day.

These developments with two distinct forms of fried potato have run alongside the development of the ready meals trade. Many types of fried food are now produced and sold retail. Frying is a very attractive way of 'setting' a batter on the surface of a food such as fish fingers or battered and breaded turkey legs. The coating generates added value to the food, and provides a pleasing appearance. It has, for instance, been claimed that children find golden fish fingers far more visually attractive than boiled white fish, and are then more easily persuaded to eat this nutritious food. In addition, frying also helps protect the food from microbial attack. The initial frying effectively sterilises the surface of the food and, provided the food is initially of good quality, a sterilisation of the surface is sufficient to ensure a good shelf life, especially if the fried product is subsequently frozen.

A contrast arises, however, between the initially prepared Saratoga chips and the crisps, etc., that we now consume. This is because the crisps, pre-fried frozen foods and ready meals that we now eat are cooked in a large-scale industrial plant and then transported to retail outlets, where they may be displayed for several weeks before they are purchased and consumed. They are not eaten hot straight out of the kitchen. This entails a long period of time between production and consumption, during which oxidation and deterioration reactions can continue to take place. A consequence of this is that frying oils for the industrial frying sector must withstand not only the stress of frying at 180°C but also the subsequent storage, and still be of good flavour.

Although frying is one of the simplest cooking methods for the chef, it is, in contrast, one of the least well understood for the food scientist. This is due to the fact that both oxidation and hydrolysis take place during the frying operation. Above all, an understanding of oxidation is confounded by the fact that a variety of different frying oils and fats is used, each having a profusion of different constituent fatty acids. Even experiments in which oil is heated on its own to the frying temperature of about 180°C involve a complex series of oxidation reactions, but this becomes even more complicated when food is introduced into the hot oil. The constituent fatty acids are oxidised initially to hydroperoxides, but these are unstable at the frying temperature, breaking down quickly to secondary oxidation products, such as aldehydes and ketones. Some of these are 'steam distilled' out of the oil by the steam liberated from the food as it cooks, but sufficient remain to form pro-oxidants, assisting further oxidation of the oil. On the other hand, some of the oxidised components react with protein in the food to generate the flavours that we find so attractive. Other decomposition products act as surface-active agents, breaking down the interfacial tension between the oil and the food, assisting heat transfer between the hot oil and the food and thus in turn assisting the cooking process. Furthermore, some components escape from the food, catalysing, inhibiting, or otherwise participating in the reactions in the oil. It is for these reasons that a fresh oil needs to be 'broken in' before the optimum fried products can be produced. Although these aspects can be explained in general terms, the devil is in the detail, preventing a full understanding and full optimisation of the frying process. This book sets out to correct this lack of understanding of the frying process.

The book therefore covers the market for fried foods, in which the range of different fried foods is reviewed together with the size of the market in the UK and other European countries. This general section of the book also covers regulatory issues in the EU and the USA, two of the main markets for industrially fried foods. Health issues are next discussed, since there are conflicting issues. One the one hand, consumers want a diet containing polyunsaturated fatty acids free from food additives such as antioxidants, but on the other hand they also want foods that are free of oxidised fats and the rancid and perhaps deleterious oxidation products that result when polyunsaturated oils are used with insufficient care and attention.

The oil or fat used in the frying operation becomes part of the food we eat and is, of course, the major factor in the quality and nutritional value of the food we eat. A large section of the book is therefore devoted to the properties and use of this important raw material. There are therefore chapters on the composition of frying oils, factors influencing the quality of frying oils and measurement of fat quality during and at the end of frying. The actual frying process is, of course, also important, and the book therefore concludes with chapters on effective process control, measures that need to be taken in order to maximise flavour, texture and colour, as well as the production of pre-fried foods and potato crisps.

Reference

BANKS, D. (1996). In *Deep Frying* edited by E.G. Perkins & M.D. Erickson, AOCS Press, Champaign Illinois, USA, pp. 1–2.

Part I

General issues

2

The market for fried food

R. Piper, Europanel, London

2.1 Introduction

2.1.1 Europanel

Europanel was founded in 1964 and is jointly owned by two of the World's biggest market information suppliers, Taylor Nelson Sofres and GfK. Taylor Nelson Sofres operates consumer panels in Great Britain, France, Spain, Portugal and Ireland, as well as a number of countries overseas. GfK is based in Germany but has panels in most European countries not covered by TNS. The Europanel companies provide continuous market research information based (primarily) on consumer panels which is a major data source for the majority of companies in the packaged food business.

2.1.2 Consumer panels

A consumer panel is a representative sample of private households (or individuals) from whom information about their purchasing is collected on a regular basis. Each of Europanel's consumer panels is nationally representative of the country in which it operates and the data from panel members are collected either via wanding of bar codes using a hand-held terminal (electronic data capture, EDC) or from their completion of paper diaries (Diary). Panel details by country are shown in Table 2.1.

The resulting rich database can provide not only quantified market structures and trends but also insights into the behaviour underlying them (for instance demographics of buyers, numbers of buyers, weight of purchase and loyalty, switching between sectors and brands). Because they are consumer based, panels can offer greater comparability between countries than that which can be

Table 2.1

Country	Number of households	Methodology
Germany	12,000	EDC
Great Britain	1,0000	EDC
France	8,000	EDC
Italy	5,000	EDC
Spain	5,000	EDC*
The Netherlands	4,400	EDC
Belgium	3,000	EDC*
Switzerland	3,000	Diary
Austria	2,800	Diary
Portugal	1,826	Diary
Ireland	1,500	EDC*

* Since 1999

obtained from retailer sourced information, which suffers from dramatically differing retail structures and levels of co-operation.

2.1.3 The market for fried food

This chapter will look briefly at the factors influencing the food market in Great Britain and other European countries, examine the large and dynamic frozen food market in some detail and comment on the trends in impulse savoury snacks.

2.2 The range of fried foods available

Both the frozen foods and impulse savoury snacks markets are driven by innovation, particularly to satisfy the whims of children and young adults (though this emphasis will need to change as the population ages).

2.2.1 Frozen foods

The main sectors of interest from the fried foods angle are fish, poultry and potatoes. In fish, the main emphasis is on the traditional battered and breaded ranges, including the ever popular fish finger, but chicken has seen growth from the products familiar from fast-food chains, such as nuggets, dippers and shapes (e.g. dinosaurs). Potatoes have also seen growth in shaped products, such as alphabet letters, and innovation and new product development continue to be of importance.

2.2.2 Impulse savoury snacks

This category is also distinguished by high levels of product innovation and fragmentation. Although children tend to be key, there has been significant growth in premium-priced, adult-orientated snacks for sharing.

2.3 Factors influencing the British and other European food markets

There is a wide range of factors at work, including:

- *Smaller households* A decline in birth rates and household sizes is taking place across Europe. Because the pace of change is faster in the Catholic south, we are seeing a more homogeneous pattern developing with clear implications for types of food products and the way they are packaged.
- *More working women* The increased importance of women in the labour force has led to more affluence and, of course:
- *The demand for convenience* This has been apparent for many years but the joy of it from the point of view of the innovative manufacturer or retailer is that it is self-perpetuating, as new generations of consumers grow up without the skills or desire to cook conventional foods on a regular basis. Coupled with this is a change in lifestyles leading to:
- *Eating 'on the hoof'* To some degree this reflects the growth of fast food outlets, itself a sign of increased:
- *Europeanisation and globalisation*

And:

- *The increased importance of lifestyle factors*

On a different track we see:

- *Increasing health consciousness* fuelled by food scares such as BSE and by concerns about GM crops and, last but not least
- *Global warming*

Clearly these factors are varied and may be at odds with each other. For example, the demands for convenience and health may be in conflict and, where they are, convenience is likely to win because of the weight of influences behind it. The growth in the number of working women and the fragmentation both of families and meal occasions is well documented and tend to be to the benefit of the packaged foods industry in general and of many pre-fried products in particular. Health consciousness tends to have a major impact only where the healthy alternative presents few or no disadvantages of taste, price or utility (for example half-fat milk). Global warming may be having some impact, for instance in the trend from hot drinks to cold, but it is difficult to disentangle from more basic influences such as growth of central heating and improved insulation.

Table 2.2

Winners	Losers
Chicken and fish	High-fat red meats
Fruit	High-calorie, high-starch desserts
Fresh vegetables	Rich cake and pastry dishes
Ethnic dishes	Hot drinks/thicker drinks
Cold drinks	
Convenience, near-fresh foods	

There is undoubtedly an increased importance for 'lifestyle' and a tendency towards 'globalisation', particularly amongst the young. As this is often primarily 'Coca-Colonisation' it is to the benefit of some key types of pre-fried foods.

It is possible to try to combine the impact of these developments and group products into 'winners' and 'losers' as shown in Table 2.2. Again, this suggests continued opportunities for pre-fried food and the areas suggested as losers are not of too much relevance.

2.4 The market for fried food in the UK

2.4.1. Frozen foods – overall

The market is huge (£3.4 bn.) and in 1999 was still showing growth in value of over 6%, reflecting stronger price movements, product innovation and consumers trading up the quality spectrum. Amongst major grocery sectors only soft drinks and confectionery showed higher growth. However, in volume terms growth was less than 1% and even this disguised declines in the traditional meat and fish categories offset by growth in the more modish ready-cooked meals and pizzas. Vegetables were flat but within this potatoes were up by 19%. Frozen foods were used in around 25% of domestic preparation occasions involving frying.

2.4.2 Fish

This market is worth £520m. and bought by 88% of the population. In volume terms it has averaged only 1–2% volume growth in the last ten years. The dominant sectors continue to include fingers (17%), battered (15%) and breaded fillets (13%) but recent growth has come from breaded steaks and fishcakes as manufacturers try to target adults to a greater extent.

2.4.3 Vegetables

Frozen green vegetables are worth £322m. and potato products a larger, and growing £388m., with each of them being bought by over 80% of the population. Within potato products, chips, notably oven chips, are dominant and growing but there is also growth from the new products in the children's sector, mentioned earlier.

2.4.4 Meat

This sector was fairly sluggish, even before BSE, but Poultry has shown growth, notably coated and flavoured portions and segments – reflecting consumer desire for value added.

2.4.5 Ready-cooked meals

These products, of course, benefit from changing consumer lifestyles and needs and the market is worth almost £400m., though volume growth has slowed in recent years with the competition from the chilled sector. However, frozen products are purchased by 75% of the population and there is scope to move further up market.

2.4.6 Pizzas

This sector is now worth £275m., sought by 66% of households and showed renewed growth in 1999 with success for both deep pan and thin and crispy. Multipacks and the commodity end of the market in general lost out.

2.4.7 Crisps and snacks

These may be considered part of the broader impulse purchasing market which looked like Fig. 2.1 in 1999.

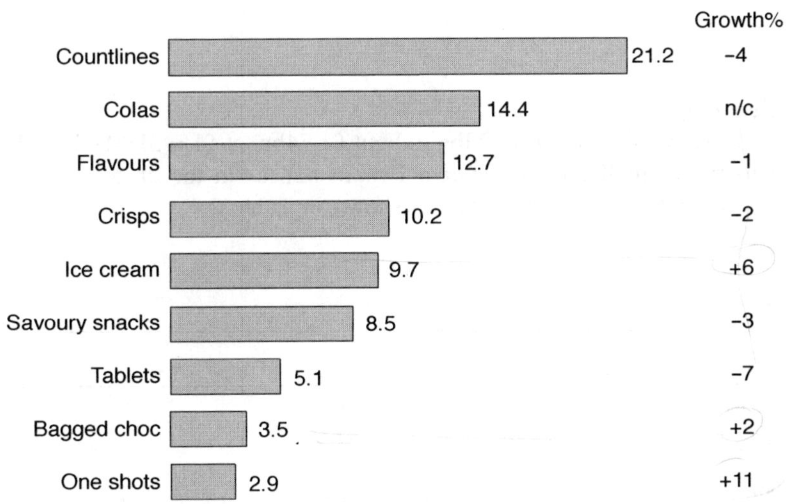

Fig. 2.1 % share of the value of the wider impulse market (by sector – latest year).

Both crisps and savoury snacks have major shares of this market but both are struggling to retain volume. This has been the case for crisps for some years but is a newer challenge to savoury snacks. The pressure on snacks is illustrated by reduced likelihood to be present in children's lunchboxes, possibly due to competition from 'healthier' sectors, such as dairy products.

2.5 The frozen food market in other European countries

2.5.1 France

The frozen food market in France is less highly developed than in Great Britain, with none of the market sectors being purchased by more than 50% of households during the course of a year. This is to be expected granted the more traditional French attitudes to food and cookery, but things are changing and most sectors have showed volume growth in the last three years, averaging as shown in Table 2.3. These growth rates are all ahead of Great Britain, except those for pizza (perhaps not a surprise).

Table 2.3

	%
Fish	+0.2
Vegetables	+5.4
Meat	+3.8
Ready-cooked meals	+4.2
Pizzas	−1.9

2.5.2 Germany

The total market was worth DM5.23bn. (about £1.74bn. at £1 = 3DM). It is thus also considerably smaller than in Great Britain but, as in the case of France, volume growth rates tend to be better as shown in Table 2.4.

Table 2.4

	%
Fish	−0.6
Vegetables	+1.6
Meat	+13.0
Ready-cooked meals	+4.3
Pizzas	+4.3

2.5.3 Italy

The most notable feature of the Italian market in recent years has been the phenomenal growth in the frozen ready meals market. As in France, this reflects a decline in traditional family meal patterns but the growth rate is of a completely different order (34% per annum rather than 4%). Frozen pizzas are a relatively small market (!) but growing at around 6% a year. The picture by sector is shown in Table 2.5.

Table 2.5

	%
Fish	+3.7
Vegetables	+0.3
Meat	+1.0
Ready-cooked meals	+34.0
Pizzas	+5.7

2.5.4 Spain

The market is highly developed for fish and vegetables, much less so for meat and pizzas. Domestic and retailing patterns are changing fast and overall this is by far the fastest growing market in the 'big five' countries. Percentages are shown in Table 2.6. However, this is a three-year picture and that for 1999 was depressed for meats, ready-cooked meals and (to an extent) pizzas.

Table 2.6

	%
Fish	+15.3
Vegetables	+22.6
Meat	+3.3
Ready-cooked meals	+10.0
Pizzas	+4.6

2.5.5 The Netherlands

The Dutch market is well developed, except for ready-cooked meals, and fairly stable, again except for ready-cooked meals, which are falling by 15% a year – a very different picture from most countries (Table 2.7).

Table 2.7

	%
Fish	−0.6
Vegetables	+0.3
Meat	+2.3
Ready-cooked meals	−15.0
Pizzas	+4.0

2.5.6 Belgium

Unfortunately we do not have trend figures for 1999 but in 1998 and 1997 all sectors (including the relatively underdeveloped ones of ready-cooked meals and pizzas) were buoyant.

2.5.7 Switzerland

By contrast to Belgium, all sectors have been depressed. The last three years' picture is shown in Table 2.8, however, 1999 saw a recovery for vegetables (+7.9%) and ready-cooked meals (+4.1%).

Table 2.8

	%
Fish	−3.8
Vegetables	N/C
Meat	−2.2
Ready-cooked meals	−3.6
Pizzas	−5.7

2.5.8 Austria

The market overall is fairly flat, with growth due only to the launch of discount retailers into frozen food and a continued increase in home delivery. However, ready-cooked meals are an exception – increasing by 19% in 1999.

2.5.9 Portugal

The state of development of the market tends to be similar to Spain but the growth rate (Table 2.9) is somewhat less good.

Table 2.9

	%
Fish	+8.0
Vegetables	+5.0
Meat	−1.3
Ready-cooked meals	+11.7
Pizzas	+0.7

2.5.10 Ireland

We do not have trend figures for 1999 but the market is as highly developed as Great Britain and, on the basis of 1997 and 1998 figures, much more buoyant.

2.5.11 Continental Europe overview

Most of the frozen food markets are less developed than Great Britain (and Ireland!) but have scope for continued growth, particularly in perceived value-added sectors. The southern countries in particular are growing, including in sectors of interest to the fried-food industry, such as chips. The main winner in meat products has been coated, flavoured poultry. Savoury snacks are much less developed in most countries than in Great Britain and offer considerable growth potential.

2.6 Future trends

For frozen foods in general there is unlikely to be a return to rapid growth rates, due to the increasing maturity of the market in most countries and the threat from chilled. Also specifically from the fried angle the (at least claimed) consumer interest in healthy eating will continue to be a factor. However, there are better growth rates in the southern countries, and there are the other positive factors highlighted earlier. This can be illustrated by some findings from a study we carried out in Great Britain into the attitudinal differences between heavy and light frozen food buyers (Table 2.10). Except possibly in relation to health, it seems likely that we will see more of the heavy buyer types as time goes on and less of the light.

Table 2.10

Typical statements of light buyers	Typical statements of heavy buyers
'I enjoy cooking'	'I don't have time to prepare food'
'My family sits down to meals together'	'I don't enjoy cooking'
'I try to eat a range of healthy foods'	'I like to try new products'

We believe that the following areas should continue to be considered as key success factors for new product development to help ensure this growth:

- Convenience
- Innovation
- Healthy
- Advertising
- Kids
- Adults
- Taste
- Quality
- Promotions.

Continuing strategic opportunities could include:

- Ethnic
- Organic
- Freezer to mouth
- Fresh market
- Sandwiches/lunchbox
- Kids/infant ranges
- For savoury snacks, some of the same factors apply – above all *innovation* and *market stimulation* via advertising and promotions.

2.7 Sources of further information and advice

This chapter is based on information collected by the member countries of the Europanel network including the following:

Austria
Fessel-GfK
Institut für Marktforschung
Ges.m.b.H.
Hainburger Straße 33
1030 Wien
Tel: 43 (1) 717 10-0
Fax: 43 (1) 717 10-314
43 (1) 717 10-194
Peter Damisch
Tobias Schediwy
peter.damisch@gfk.at
tobias.schediwy@gfk.at

Belgium
GfK Belgium NV
Thames House - Riverside Park
Internationalelaan 55-b11
B-1070 Brussels
Te: 32 (2) 558 0558
Fax: 32 (2) 558 0559
Paul Belckx
Paul.belckx@gfk.be

France
TN Sofres/Secodip
2 rue Francis-Pédron
F-78241 Chambourcy, Cedex
Tel: 33 (1) 3074 8080
Fax: 33 (1) 3074 8029
Jean-Loup Guyot
Valérie Tillon
jean-loup.guyot@tnsofres.com
valerie.tillon@tnsofres.com
http://www.secodip.com

Germany
GfK Panel Services Consumer Research GmbH
Nordwestring 101
D-90319 Nürnberg
Tel: 49 (911) 395 0
Fax: 49 (911) 395 4013
Thomas Bachl
Wolfgang Twardawa
thomas.bachl@gfk.de
wolfgang.twardawa@gfk.de

Great Britain
Taylor Nelson Sofres (UK)
Westgate
London W5 1UA
Tel: 44 (20) 8967 0007
Fax: 44 (20) 8967 4887
Mike Penford
David White
mike.penford@tnsofres.com
david.white@tnsofres.com
http://www.tnsofres.co.uk

Ireland
TN Sofres Ireland
Temple House
Temple Road
Blackrock Co. Dublin
Tel: 353 (1) 278 1011
Fax: 353 (1) 278 1022
Finn Raben
finn.raben@tnsofres.com

Italy
IHA Italia SpA
Via Vittor Pisani 31
20124 Milano
Tel: 39 (02) 6708 0208
Fax: 39 (02) 6707 1249
Carlo Pescetti
c.pescetti@ihaitalia.it
http://www.ihagfm.ch

Netherlands
GfK Nederland bv
Mgr. Schaepmanlaan 55
NL-5103 BB Dongen
Tel: 31 (162) 384 000
Fax: 31 (162) 322 337
Dick Valstar
dick.valstar@gfk.nl
http://www.gfk.nl

Portugal
TN Sofres Euroteste
Avenida Engenheiro Arantes e Oliveira
N5 S/L, 1900 Lisboa
Tel: 351 (21) 843 7050
Fax: 351 (21) 840 7995
Nelson Piteira Furtado
Alfredo Hasslocher
euroteste@ip.pt

Spain
TN Sofres/Dympanel SA
Cami de Can Calders 4
08190 Sant Cugat del Vallés
Barcelona
Tel: 34 (93) 581 9400
Fax: 34 (93) 581 9401
Pedro Ros
Josep Montserrat
pedro.ros@dympanel.es.tnsofres.com
josep.montserrat@dympanel.es.tnsofres.com

Switzerland
IHA Institut für Marktanalysen AG
Obermattweg 9
CH-6052 Hergiswil
Tel: 41 (41) 632 9111
Fax: 41 (41) 632 9465
Rebekah Bruhwiller
Walter Quakernack
Thomas Hochreutener
bruhwillerr@ihagfm.ch
quakernack@ihagfm.ch
hochreut@ihagfm.ch
http://www.ihagfm.ch/

3

Regulation in the European Union

R. Fox, Pura Foods Limited, Belvedere

3.1 Introduction – the legal context

3.1.1 The basis of EU law

The law of the European Union on food issues looks satisfyingly similar at every level but the one where it ultimately matters – in the detail. In matters of principle, most countries in Europe legislate in the same way, which ought to be of much comfort to the food technologist. After all, food poisoning bacteria do not respect border patrols, and good nutrition should be the same the world over. Fraud, most countries agree, is a 'bad thing'.

In the national law of most European countries, there exists 'primary legislation', often known as 'laws', and 'secondary legislation' or 'regulations'. The European Union (EU), and the European Economic Community (EEC) before it, overlay that structure with 'directives'. No explanation of food law can go far without defining these terms, and this one will be no exception.

A law is passed by the legislature of a country. The UK's Food Safety Act (1990) sets out the framework for all subsequent legislation. Germany's 'Law on Foods and Commodities' (Lebensmittel- und Bedarfsgegendständgesetz – LMBG) of 1997 does the same, as does Finland's Elintarvikelaki (Food Law) 361/1995, and so on. What the primary law does is to grant the Executive the right to make regulations. It also imposes duties, creates offences, sets up enforcement agencies and defines penalties. The Food Safety Act says that food shall

- not be injurious to health
- be of a nature or substance or quality demanded by the purchaser
- not have a false description or misleading label.

So it sets out the principles of safety and fraud.

A Regulation (or a Decree, an Ordinance, an Order or a Statutory Instrument) is made by the government of the country, not its legislature. Who signs it (minister or monarch) varies, but the effect is the same. This is where the food manufacturer must look for hard data. For example, it is the appendices of the Dutch margarine, fats and oils order (M.V.O. verordening 1975) on edible oils and fats, that define permitted antioxidants, colours, antifoams and flavours.

The European Council generated Directives. Although directives appear to have the force of law, each 'is addressed to the member states', and instructs them to amend their laws if necessary to conform. In most cases, the member states then enact a regulation to implement the EC directive. Nowadays, the EU generates Regulations itself, which come into force on specific dates throughout the EU. The national governments then generate a regulation which states hardly more than 'EU Directive number ... enters our law'. Some countries, for example Greece, Spain and Austria, have a Food Code with the force of law. For definitive information by country, see the tables in sections 3.3 to 3.5, and the sources of information in section 3.7.

3.1.2 The international context – Codex Alimentarius

Codex Alimentarius comprises the consensus of committees from around the world on issues which may be covered by law in some of the member countries. It is a conscious attempt to harmonise regulation across the world. Codex consists of a series of committees, that deliberate on proposals brought by their members, and generate standards. Where legislation exists, Codex standards have little force. Nevertheless, governments do take notice of Codex when revising regulations, so it may be seen as a pointer to the future. Since legislation is framed in some countries with an imprecise duty on the manufacturer, Codex can be used in prosecutions, in the absence of strict regulations. The courts then decide how much weight to place on the Codex standards.

With additives having no fixed limit, Codex uses a principle of Good Manufacturing Practice (GMP), implying that the minimum quantity consistent with technological need should be used. European legislation prefers the term *quantum satis* (QS), which means the amount needed, with no implication of minimising usage (see Table 3.1 on page 26).

The Codex Alimentarius Commission has a standard for the composition of vegetable oils,[1] which is the principal reference standard in disputes over the authenticity of oils. Some of the limits are very wide, and more detailed studies taking geographical source into account may be of more use for commercial decisions.

3.1.3 Areas covered by the law

Many countries have compositional standards for certain foods (often staples or those found in past times to have been subject to fraud). The EU has a major preoccupation about olive oil, a commodity closely linked to the economies of several of its members' influential farming constituencies, and one historically subject to systematic attempts at fraud. Thus Regulation 356/92 specifies seven grades of olive oil. Edible oils and fats are subject to such standards in most European countries. In some, there is a distinction between 'seasoning' oils (for salads, sauces, etc.) and those for frying as well. France has a distinction of this nature, revolving on the content of linolenic acid (maximally 2% for frying, although 70% linoleic acid is considered perfectly satisfactory).

Additives (q.v.) are regulated by EU and national law in a consistent and systematic way. The criteria for an additive to be permitted (q.v.) are safety, functionality and need. Safety is obvious, but implies that some poor animal has been tested to destruction to discover the dose at which it dies. Hence the 'need' criterion; legislators do not want to have too many additives tested. Functionality means that the additive has to do something useful. Very few additives are permitted in frying oils, since there are only two agreed functions: antioxidants and antifoaming agents (see section 3.3.5).

Processes and processing aids (q.v.) are controlled in some countries. France is attempting to impose limits on processing aids in edible oil refining at the time of writing. Several countries define the refining operations deemed appropriate for edible oil, and the solvents which can be used.

3.1.4 Non-legal pressures

Most countries have bodies so influential that their pronouncements have almost the force of law. The Food Advisory Committee (FAC) in the UK publishes guidelines that Trading Standards Officers (the local government officers charged with enforcing retail sale laws) attempt to impose on manufacturers as though they were law. One such guideline of interest to oil suppliers is the one discouraging claims on the absence of cholesterol. Austria has in its food code indications of 'an established change in the frying oil' which are guidelines. In Germany, the working group of the Food Chemistry Expert Representatives of the Länder (states) and the BgVV published a position paper on frying fats, which most companies follow as though it was law. This is where the much-quoted polar compounds limit comes from. Denmark has a draft Order on *trans* fatty acids, which although unfinalised owing to lack of scientific consensus, is followed by many manufacturers.

Below this level of influence are respected institutions like the Institut Pasteur in France or commercially powerful groupings like the Institute of Grocery Distribution (IGD) or the British Retail Consortium (BRC), which can effectively impose their guidelines on suppliers. Finally, there are the consumers and no-one should be in any doubt of their ability to impose their collective will, especially after the practical disappearance of soya oil in much of Europe because of its genetic modification.

3.2 The structure of the frying industries

3.2.1 The supply chain

The principle of due diligence (see section 3.3) means that each member of the supply chain has to take 'reasonable' precautions to ensure his suppliers are complying with the law and good practice. The members of the chain are the manufacturers of the foods to be fried, the makers of the frying media, processors who include frying in the preparation of their food, caterers and fast-food outlets, and lastly the retailer of bottled oil. Each has a different interest in the law. An important distinction here is between the nature of the oil as traded, and its condition in the fryer. The user of an oil (caterer, food manufacturer, or consumer) is legally protected as to the composition of the oil. He or she also has a right to 'fitness for purpose', so the oil should not deteriorate rapidly when used in accordance with instructions. The user then has the duty not to use the oil beyond its reasonable life, and a different set of rules apply.

Food manufacturers will often have very different requirements from caterers. So, on the one hand, a caterer may reuse the oil over a period of many weeks, topping up where necessary. He needs a stable long-life oil in order keep his costs down, and legislation may well restrict his options there (e.g. on *trans* fatty acid levels). As the oil deteriorates, polar compounds, free fatty acids and polymers will build-up, and the law often has something to say about that. On the other hand, a manufacturer of pre-fried chips may never have to dispose of the oil, which may have an average residence time of only a few hours. This is because the throughput of chips is so large that even if they take up only 5% of their weight in oil, the contents of the frier are consumed rapidly and need to be constantly replenished. The chip manufacturer can use a relatively unstable but 'healthy' oil such as sunflower. Because the oil never gets old, most of the legislation on end-of-fry life is just not relevant.

3.2.2 Transport

Transport of frying media is covered by all legislation relating to food, but when handled in bulk, there are additional regulations and non-legislative codes of practice. Internationally, a Codex committee is looking into a code of practice for transport. Across the EU, Council Directive 93/43/EEC specifies lorries for bulk road transport. There is a corresponding Commission Directive 96/3/EC covering transport by sea. In the UK, the Seed Crushers and Oil Processors Association (SCOPA)[2] has developed a code of practice covering road tanker construction, tanker registration and identification, operator training, cleaning, and recording of at least the three previous loads.

3.3 The sale of food

The sale of food constitutes a legal act, and a battery of laws applies. 'Sale' usually means offering for sale, or having foods on the premises with the reasonable supposition that they will be offered for sale. Retail sale ('sale to the ultimate consumer') invokes a further set of laws, such as weights and measures and labelling.

3.3.1 Safety and durability

All countries require that food is safe. The expression in Britain is that it should not be 'injurious to health'. Germany has a similar provision, but extends it to any substance that children might mistake for foods. But what does 'safe' mean? Food has a habit of deteriorating into an unsafe condition. Therefore, there is usually a legal requirement for a statement of durability on packaged food. For foods liable to become unsafe, there is a 'use by' date printed on the pack. The 'display until' date is not legally binding, just a guide for the retailer. Foods that deteriorate in quality but not normally into an unsafe condition, have a 'best before' date, which is a legal provision.

Frying oils carry a 'best before' warning, because they contain no water, so do not support microbial growth. Unless stored under very extreme conditions, they do not change chemically in an unsafe way, and even then taste disgusting long before they cause medical conditions. Frying media may be unsafe by virtue of chemicals present before sale. One ought to be able to assume that the manufacturer or trader has done nothing grossly stupid or negligent. However, two examples in 1999 prove that untrue. In Belgium, someone dumped transformer oil into recycled vegetable oil. That was compounded into feed, some of which was fed to pigs, and their fat was rendered into lard. The result was frying lard potentially contaminated with polycyclic biphenyls (PCBs) and dioxins, rather nasty carcinogens.

The second example took place in Indonesia, where palm oil was apparently diluted with cheaper diesel oil. The contamination of around 1% was picked up by a superintendent at Rotterdam, but not before contaminated palm oil had entered the food chain. By chance (or perhaps not, see below), the deodorisation step of physically refining palm oil almost completely removes diesel mineral hydrocarbons. Hydrocarbons are not particularly hazardous to health, and the nauseous smell of diesel may have been the greatest risk to which the consumer could have been exposed. These examples illustrate that even the most diligent food manufacturers can be caught out by unforeseen events in their supply chain.

3.3.2 The due diligence defence

These examples illustrate well the problem of who is to blame in law for unsafe food. The person or company selling the food to the consumer bears the initial responsibility. Clearly, if he has committed an unsafe act (for example a butcher contaminating cooked meat with uncooked), he alone probably carries the burden of guilt. If not, he may be able to shift the responsibility back to his supplier, if he has exhibited 'due diligence', and so on back up the supply chain.

Due diligence requires that the company in question has taken all 'reasonable' steps to ensure the safety of the product supplied to him. What is reasonable can only be settled in a court of law. A large company will probably be required to take more precautions than a small one.

In the case of lard containing PCBs, a legal prosecution would turn on whether the seller might reasonably have foreseen the contamination. The answer is probably no, up to the date upon which the story broke in the press. From that date, any supplier might have been held liable if any lard still in the supply chain reached the consumer containing dangerous levels of PCBs. For he might be expected to have realised the significance of the contaminated animal feed to his supply chain. In practice, this is what happened, and lard was held up in the supply chain while laboratories worked long hours doing very sophisticated GCMS analyses. They showed the lard had only 'safe' levels of PCBs.

3.3.3 Safe levels of contaminants

Contaminants such as pesticides, mycotoxins, polycyclic aromatic hydrocarbons (PAHs) and PCBs may be the subject of regulations. Contaminants are controlled at the EU level under Regulation 315/93, with detailed requirements on nitrates, aflatoxin and various metals in subsequent Directives. The way in which safe levels are calculated is complex, and depends on the amount of the food typically consumed, and hence the amount of contaminant ingested, compared against a 'no effect' limit in animal studies, usually with a margin of error. The Acceptable Daily Intake (ADI) is the amount of the contaminant judged not to be dangerous, expressed in micrograms of contaminant per kilogram of body weight per day. Given average and peak consumption data, then a maximum recommended limit (MRL) in an individual food can be defined.

In the case of pesticides, a World Health Organisation committee known as JMPR has been establishing MRLs since 1966, and they are the authoritative source. Dioxins and PCBs have been the subject of EU legislative attention since Decision 94/652/EC set up the risk assessment. Apart from in animal tissues (which are prompted by the Belgian dioxin incident), no acceptable levels have been set by the time of writing. The UK's agriculture ministry (MAFF) has published considerable surveillance data in reports on its web site, and these can form the basis of judgements on 'safe' levels. The surveillance shows that animal fats are a greater risk to human health than vegetable sources. The consumer seldom understands that such a thing as a no-effect level of a contaminant can exist.

3.3.4 HACCP applied to fried food

The principle of food safety required or recommended in much food safety legislation is that of Hazard Analysis by Critical Control Points (HACCP). A critical control point is a stage in processing which, if not done to specification, compromises the consumer's safety. There are entire books devoted to HACCP, so suffice it to say here that frying will usually be a critical control point, since it may very well be the last or even the only heat treatment the food will undergo. In this case, food safety demands that cooking in all parts of the food is sufficient to kill any pathogenic bacteria that might be present. Frying is a high-temperature, short-time cooking method, with heat penetrating from the outside. So the coolest part of the food will always be the centre. Let us consider a few examples:

- Flash frying of fish. Even for double coating of batter with an intermediate frying, the fish remains raw and frozen on the inside. This food is designed for cooking by the consumer. Centre temperature is irrelevant to safety in this case.
- Cooking of bhajees (deep fried Indian vegetable balls). Here we have an assembled ball, which is then deep fried to cook it. The centre may well have been contaminated by the hands that prepared it, and the ball may be 50 mm in diameter. Clearly, the rate of heat conduction into the centre determines safety, and the bhajee is not safe until the centre has experienced pasteurisation conditions (e.g. at least 72°C for 2 minutes).
- Frying chips (French fries). Provided the chips have not been mistreated before frying, the centre is essentially uncontaminated, even though the surface may well be. The frying sterilises the surface. In the centre, it is only a matter of cooking the starch to make it taste good.

From these examples, it emerges that measuring the centre temperature of the fried article would be the precaution most likely to ensure safety or eating quality. The food factory may well do so. However, much frying goes on in catering establishments, often by staff with little food safety knowledge. Here, it is better to rely on rigid frying temperatures and times. One usually finds that oil manufacturers print recommended cooking temperatures and times on tins or buckets of oil. That is their contribution to the safety of the food cooked in their product.

3.3.5 Regulation of additives

An additive is an ingredient of a food not normally of itself consumed as a food, and having a function useful in the food. Thus, an antioxidant such as tocopherol is an additive, but salt is not. The law distinguishes a processing aid as an additive which, because of its level or form in the food as sold, has no functionality. Citric acid is used in refining edible oils, and residues remain in the finished oil. It is a processing aid in cooking oil, because it does not perform its normal function in food as an acidity regulator (the function for which it is permitted). As such, it does not need to be declared as an additive.

The legal principle for food additives is that unless they are permitted in the food in question, then they may not be used. The legislation of all countries contains tables of additives listed against various foods. Table 3.1 lists the

Table 3.1 Additives permitted in frying

Additive	Class	Units	Codex Alimentarius CCFAC proposals, 2000	EU
E900 DMPS	Antifoam	mg/kg		10
E310,311,312 gallates	Antioxidant	mg/kg	100	200
E320 BHA	Antioxidant	mg/kg	200	200
E321 BHT	Antioxidant	mg/kg	75	100
319 TBHQ	Antioxidant	mg/kg	Ban (currently 120)	ban
E306,7,8,9 tocopherols	Antioxidant		GMP	QS
E304 ascorbyl esters	Antioxidant	mg/kg	500	QS
Colours in oils & fats generally	Colour			no
E160b annatto in solid fats	Colour	mg/kg	20	10
E160a carotenes in solid fats	Colour	mg/kg	GMP	QS
E100 curcumin in solid fats	Colour	mg/kg	GMP	QS

QS = Quantum satis (i.e. as much as needed) GMP = Good Manufacturing Practice (i.e. minimum needed)
The appropriate EU legislation is implemented in each of the member states.

permitted additives in cooking oils in Europe. The relevant EC Directive 95/2/EC (as amended) is implemented in all EU countries. For American readers, the notable absentee is tertiary butyl hydro-quinone (TBHQ), which is permitted in the USA and in some other countries. Some palm oil is dosed with TBHQ before shipment, but the refiner then strips it out during refining to comply with local regulation. Even if residues were found, they would be classed as a processing aid, since the concentration would be too low to be technologically significant. Nevertheless, port authorities have been known to take a strict view about the legality of the practice.

The Codex Alimentarius figures given in Table 3.1 are the proposals put forward to the March 2000 meeting of the Codex Committee on Food Additives and Contaminants (CCFAC).[3] There is some doubt about the permitted level of TBHQ. The Codex web site quotes 200 ppm, but the Codex standard 19-1981 (revision 2 – 1999) allows only 120 ppm. Any national legislation will of course take precedence, but for the purposes of international trade it would be wise to conform to 120 ppm.

3.3.6 Packaging declarations: nutrition and ingredients

In most countries, there is regulation of what can and must be stated on packaged food for retail sale. In Germany, according to the Food Labelling Ordinance, you must state:

- the sales description
- the name of the manufacturer, packer or seller
- a list of ingredients
- the date of minimum durability or the use-by date
- an indication of quantity/amount.

There are similar requirements in other countries, being based on EC directive (79/119 and its amendments).

The list of ingredients has to be in descending order of quantity. A new aspect of labelling is the Quantitative Ingredient Declaration (QUID). Although required by EC Directive 79/119 of 1978, the effective date is as recent as February 2000. At the time of writing, there is some disagreement as to its applicability. A QUID is required if an ingredient is named or implied by the pack graphics, but the directive is strictly not applicable to categories having their own vertical directives. Thus, the countries of the EU have interpreted the directive differently. France has been keen on QUID for some years. Partly, this seems to be because of a consumer preference for multi-component oils, and it is not unusual to find a product claiming it contains 'five oils'.

In Britain, by contrast, enforcement officers have been advised to go softly with manufacturers who appear not to be complying with QUID. One oddity is the requirement to label a flavoured frying product as containing 99% sunflower oil. The remainder is salt, flavour and vitamins. The law states in Britain that packaged food may contain a declaration of nutritional content, but if it does so, then the form is prescribed. I refer the reader to national regulations on this topic.

All the labelling law relates to retail sale. It is intended to inform the consumer. There is no obligation to print any such information on packs for further processing. Business-to-business trade is still governed by any law that uses the word 'sell', but labelling law is typically not one of these. In practice, of course, the buying power of large processors and caterers is such that they require their suppliers' specifications to carry infinitely more information than any retail pack. They may very well be using the specification to work out the declarations on a large number of their own products, which may make a variety of claims, all of which will need to be backed up with specified nutritional and ingredient declarations.

Another point to bear in mind is that the industrial buyer may not have a legal right, but has a due diligence duty to receive accurate and sufficient information from his supplier. A good example is in the labelling of fat components. 'Mono-unsaturates' are defined in labelling regulations as 'fatty acids containing one cis- double bond'. The manufacturer could hide behind a statement of 'total mono-unsaturates', in which he includes *trans* monoenes. He would be unwise to do so, for his customer may transcribe this loose category into the tighter legal definition, and may take his supplier to court if he is himself prosecuted. He would certainly withdraw his business!

Finally, there is the grey area of catering sale. The same design of 20-litre drum of frying oil sold to an industrial customer may be offered in a catering supplies outlet to which the consumer has access for retail purchase. In practice, most manufacturers will play safe and comply with all the legal requirements of retail sale.

3.3.7 Weights and measures

A pack for retail sale must bear an accurate declaration of the weight or volume of its contents. The most frequent form of that declaration is in accordance with average weight legislation. National legislation follows the EC model. Essentially, if a packer opts to use average weight declaration, then:

- the average weight of packs must be not less than the declared weight
- no more than 2.5% of packs can fall more than one 'tolerable negative error'* below the declared weight, and
- effectively none should be more than twice the tolerable negative error below the declaration.

The tolerable negative error (TNE) is defined on a sliding scale, so that it is nine grams between 200 and 300 grams, and 1.5% between one litre and 10 litres. Packs that comply can be marked with the 'e' symbol. Oils, as liquids, are generally sold by volume. There are no prescribed pack sizes for liquid oils, unlike those for edible solid fats (sold by weight). However, the average weight regulations based on EC Directive 76/211/EEC do specifically mention edible oil in liquid or gel form.

Fats and oils for industrial use lie beyond the scope of prescribed pack size legislation, though still fall within the limits of average weight provisions (which apply up to 20 kg or 20 litres). Regardless of whether the regulations strictly apply to industrial products, the provisions constitute a reasonable guideline for the tolerable negative error; a figure of 150 grams should be used between 10 kg and 15 kg, and 150 ml between 10 litres and 15 litres.

3.4 The life of frying oils

3.4.1 Why frying oil has to be discarded

The reader of this book will have gathered that frying oil has to be discarded at the end of its useful life. Intuitively, you can see why; the oil darkens, it thickens, it may contain deposits, and it may acquire an acrid flavour. Fried food will look and taste poor. What is not so clear is why the law should take an interest in frying oil life. Let us examine the arguments.

The law often concerns itself with unfair practices. A frying oil that is used until the food is acrid and blackened could be construed as unfair, inasmuch as

* The 'tolerable negative error' is defined in legislation for various pack sizes, falling as a percentage of the declared size as that size increases. For example, for 500 gram packs it is 15 grams (3%), and for 15 litres it is 150 ml (1%).

the enterprise doing the frying can reduce its costs over a competitor using good practice. Yet it is unusual for the law to intervene when the quality deterioration is so obvious to the consumer. If a food looks and tastes bad, the customer will not buy it. However, poor quality of food is an imprecise legal concept. Therefore, some countries, but by no means all, have enacted regulations on specific chemical analyses that measure oil deterioration (see section 3.4.3).

There is only thin evidence that poor oil quality is a health risk. These issues are covered in more detail in Chapters 4 and 5. The quantities of the chemicals of concern are enormous by most contamination standards; several percent for free fatty acids and for polymerised oils, up to a quarter of the oil for total polar compounds. Humans have been consuming these compounds for centuries. Two classes of compound present in abused heated oils have attracted the attention of toxicology researchers: cyclic compounds and polymers.

Cyclic compounds arise by cyclisation between C_{15} and either C_{10} or C_{11}, to produce either a 6-member or 5-member ring respectively. Purified cyclic fatty acid monomers (CFAM) have been isolated from heated linseed oil (a mixture of 5- and 6-membered rings) and sunflower oil (mainly 5-membered rings). It appears that all CFAM are easily absorbed and incorporated into fatty tissues. There have been many metabolic studies. Fatty acids are broken down in cells in 2-carbon chunks (β-oxidation), which for CFAM ceases when the ring is encountered. The 6-membered rings are excreted rapidly, so have very low toxicity. The 5-membered rings are preferentially absorbed, and can lead to decreased liver lipogenesis. The levels of CFAM in these studies was from 0.0075% to 0.15% of diet.[4] In another study,[5] liver enzyme activity was reduced in rats fed purified CFAM derived from used hardened soya oil. The levels fed in both these studies are several fold greater than those normally found in food.[6]

Polymeric compounds are the gums and thickening compounds of used frying oils, and can reach 10% of the oil. They may be neutral or oxidised, and hence polar. Both groups are included in the generic class of 'polar compounds'.[7] In one study,[8] dimers at 0.1%, 1% or 5% of diet were fed to rats. No effect on weight gain was observed. Other studies have used up to 20%, when the rats suffered diarrhoea from what is a highly unbalanced diet.

The above detailed studies are supplemented by many in which used frying oils have been fed. Only mild, if any effects have been reported.[9,10] One extreme study[11] fed rats for 18 months at 20% of the diet with fresh oil, used oil and its polar or non-polar fractions. Only the diet of 20% polar compounds caused a small reduction in growth and increased liver and kidney weights.

In conclusion, then, the case for legislation is weak on both fraud and safety grounds.

3.4.2 Regulation of fresh frying oils

Ever since oilseed rape was stripped of its high erucic acid content (more than 40% originally), European law has distinguished between the old varieties,

Table 3.2 Regulations on unused frying oils

	EU	Austria	Belgium	Denmark	Finland	France	Germany	Greece	Ireland	Italy	Luxembourg	Netherlands	Portugal	Spain	Sweden	United Kingdom
Erucic acid (max %)	5	5	5	5	5	5	5	5	5	5	5	5	5	5	5	5
Free fatty acids (%)		0.4														
Smoke point (°C)		205														
Lauric fats		no														
Linolenic acid (max %)						2										
Oils specified														yes		
Used oils														ban		
Trans fatty acids (max %)				15												

which it relegates to industrial use, and edible varieties. The EC Directive 76/621 limits erucic acid to 5% of the fat. All EU countries have incorporated the directive. In practice, no refinery now expects to see more than 1%.

France is somewhat isolated in having a limit on linolenic acid content. The limit could nowadays be argued as a barrier to trade, since there is no safety justification for it, and most European countries have happily used rapeseed oil for deep frying, either in its native form, or hydrogenated to increase its stability. Likewise, much of America fries in soyabean oil, which also exceeds the French 2% linolenic threshold.

Denmark restricts *trans* fatty acids in a draft Order. The proposed limit is 15% at the time of writing, reducing to 10% eventually. Despite the order not having reached the statute book, many Danish oils and fats businesses already follow the provision.

Austria has regulation of the quality of fresh frying oil in terms of free fatty acids and smoke point. Chapter B30 of the Austrian Food Code goes further in describing as unsuitable: oils with a significant medium or short chain fatty acid content (though it is unclear whether palm oil would be caught by this recommendation); polyunsaturated fatty acids (particularly linoleic acid, and therefore sunflower oil); and animal fats and oils. The United Kingdom, by contrast, uses large amounts of palm oil and lard for frying traditional fish and chips, even though the Draft Regulations and Guidance of Nutritional Standards for School Lunches recommend sunflower oil as the best option if fried food is served in school meals There is an apparent discrepancy here. Olive oil would typically have a free fatty acid content sufficient to render it ineligible as a frying oil in Austria (see Table 3.2). The important distinction is that while it would be illegal to describe olive oil as a frying oil in Austria, it would not be illegal to deep fry using it.

Spain's 1983 'Reglamentación Técnico-Sanitaria de aceites vegetales comestibles' (Technical and sanitary regulation on edible vegetable oils) adopts a very different approach. It lays down the composition of a whole series of oils, their Lovibond colour scores, iodine values, saponification values, and saturates in the 2-position. Only these oils are permitted in frying. Unusually, this vertical regulation also governs the pack sizes and the labelling, subjects that in other countries and in EU directives are covered in horizontal legislation (i.e. covering a single subject for all foods).

3.4.3 End of frying life

Section 3.4.1 argued that the case for regulation of the end of frying life is weak. There is no EU legislation on the subject. I cannot claim that this is because Europe's legislators agree with me. Far more likely is that the commissioners have seen no need to develop Europe-wide rules on the subject. Much of the EU's legislation is designed to establish a free market across the Union. It is in the nature of frying that its products do not travel well. Fast food cooked in France cannot compete with fast food cooked even in Luxembourg, so the EC

Table 3.3 Regulations on end of frying life

	EU	Austria	Belgium	Denmark	Finland	France	Germany	Greece	Ireland	Italy	Luxembourg	Netherlands	Portugal	Spain	Sweden	United Kingdom
Max frying temp. (°C)		180	180			180*				180			180			
Smoke point (min °C)		170					170									
Free fatty acids (max %)			2.5													
Acid value (max)		2.5					2									
Polar compounds (max %)		27				25	24			25			25			
Oxidised fatty acids (max %)		1					0.7									
Dimers and polymers (max %)			25									16				
Viscosity 50°C (max mPa.s)			37													
of liquid oils			27													

* Relates to automatic chip vending machines. 200°C for highly saturated oils.

has not attempted to force the issue. Food (such as potato crisps) that has been fried legally as part of its processing in one country, can be sold in another country of the EU, even if it would have been illegal to fry in that way in the selling country. The reason is that it would be a restraint of free trade to ban its sale. The only grounds for such a ban are safety issues, and the country where a manufacturer tried to sell the food would have to take legal action to prevent sale. The furore over beef suspected of BSE contamination demonstrates just how much burden of proof is required of a threat to safety.

Individual countries have legislation (or guidelines with the effective force of legislation) on the limit of acceptable quality of frying oil, and hence the point at which it needs to be discarded (Table 3.3). Several countries specify a maximum frying temperature, although France makes an exception in respect of especially stable oils, when 200°C is permitted. Ironically, British manufacturers of long-life oils often recommend a frying temperature of around 190°C, which would be illegal in some European lands. I also find it anomalous that Austria permits frying at 180°C, but sets the smoke point limit at 170°C. It would be quite legal to fry with a continuously smoking oil! The acidity of the oil can be regulated by acid value (grams of potassium hydroxide per 100 grams of oil) or the free fatty acids grams oleic acid per 100g of oil). The latter is approximately twice the former.

The consensus when it comes to polar compounds is weak, as befits such a contentious measure. There is also some confusion, because of the varying measures. The data in Table 3.4 from Hamilton & Perkins[12] for sunflower oil after six minutes of frying, illustrate the typical relationship between some of the measures. Diglycerides and free fatty acids did not change from the values in the fresh oil.

The most common confusion is between polar compounds and dimers and polymers (the measure in Belgium and the Netherlands). As can be seen, dimers and polymers are only one class of polar compounds (as defined by AOCS method Cd 20-91 or IUPAC method 2.507). The confusion is beginning to be a barrier to free trade in frying oils across the EU, because frying trials in one country are not accepted as evidence in another, solely on the grounds that the correct analytical measure was not used. Germany and Austria have another

Table 3.4

	After 6 mins.	Fresh oil
Free fatty acids	0.5%	0.6%
Diglycerides	1.0%	1.2%
Oxidised monomers of triglycerides	3.9%	0.9%
Triglyceride dimers	5.1%	0.6%
Triglyceride polymers	1.1%	0
Total polar compounds	11.7%	3.2%

measure in addition to polar compounds, that of oxidised fatty acids insoluble in petroleum ether.

The various countries cannot agree on an acceptable maximum for polar compounds, Austria quoting 27%, and Germany 24%. However, if you are taken to task over an assay of polar compounds, you probably need not worry about these discrepancies: the reproducibility (at 95% confidence) of polar compound analyses across laboratories is quoted as 2.17% in AOCS method Cd 20-91.

The United Kingdom has no rules on the end of frying life. The absence of criteria in Table 3.3 against other countries should not necessarily be taken as an absence of regulation, and I advise the reader to check with users in the country in question; the rules are not necessarily laid down in formal legislation. It seems likely that guidelines developed at an international conference at Hagen in April 2000 will have considerable influence in the future (see section 3.6.2).

3.5 Environmental protection

3.5.1 Why the law addresses discharges from frying operations

The principle of all legal systems is that the actions of one person should not damage another person, and any action that causes harm is likely to be forbidden. The environment has recently come to be seen as something that needs protecting in the same way. The earliest environmental legislation worried only about acts that immediately harmed other people. Nowadays, such outcomes as global warming and de-oxygenation of rivers are taken into account.

Frying involves no really nasty materials, no radioactivity or potent chemicals likely to poison or cause cancer. Its hazards are mostly those of excess nutrient. Environmental law covers liquid effluents and waste packaging, which are discussed below. The law usually also deals with smells and gaseous effluents. I shall make no attempt to discuss these, other than to say they are normally based on the principle of nuisance.

3.5.2 Liquid effluents

Fat is the most energy-packed of all foods. Its energy yield in nutrition is 37 kJ/g. In effluent terminology, we refer to its biological and chemical oxygen demand (BOD and COD). These are the weights of oxygen needed to convert the fat to carbon dioxide and water, and hence a measure of the intensity of treatment needed at the sewage treatment works before a relatively clean stream of water can be discharged into a river or the sea. The theoretical COD and BOD of pure fat can be calculated as 3,800,000 ppm. It is this enormous effluent loading that is the principal reason for intercepting fat in fat traps. The other reason is that even a liquid oil will solidify in cold weather (or when it starts thickening in use) and can block drains.

Premises are granted an 'effluent consent' by the authority responsible for such matters. A large food factory will often be expected to pre-treat its effluents to meet the consent. In some cases, the consent applies across a group of factories. A classic case is the big complex of food factories on Grimsby Docks in the UK, where there is a single landlord (the original port authority) and a complex sewer system. Small units, such as a fast-food outlet on a high street, may appear to be less heavily regulated. In practice, assumptions are being made that assess the demands they are placing on the sewage system, without actually measuring them. Issues such as this are dealt with by departments of government concerned with the environment in most cases, not food authorities.

3.5.3 Waste oils

What is the fate of used oils? They wax up the drains, and create an expensive effluent problem if discharged to the sewer. In solid waste, they generate gas problems if sent to landfill. The usual means of disposal is to recycle them via specialist recovery firms.

Spain, Luxembourg and France either implicitly or explicitly ban the use of spent oils in frying. I have been unable to find any other specific ban on the use of spent oil in human foods. The reply from Eire is specific that there is no ban, while Austria and Finland state that it is not covered by food legislation. The Belgian position is typical: 'Article 1 of Royal Order of 3rd January 1975 defines as harmful foods or foodstuffs prepared from raw materials unfit for human consumption'. The implication is that if a spent oil is still within the tolerance of a usable oil, then it can legally be used. In practice, once an oil has been recovered, its traceability is compromised, and good manufacturing practice demands that it should not be used for human consumption. In the light of Belgium's ban on use of recovered edible oil in animal feed (see below), a court might well decide that it was not suitable for human use.

As to spent oils being used in animal feed, my reply from Eire is that 'recovered vegetable oil is not prohibited as an ingredient in animal feeds'. It is allowed in Britain also. Belgium's Advisor General of DG4 has stated that 'a merchant can no longer recycle frying fats to deliver them to an animal feed factory', although a food factory can compound its waste oil into animal feed. Used frying oils can be used for industrial purposes, although it is probably uneconomic to do so. One likely destination in the future is biodiesel.

The EU is discussing introducing controls on the sourcing, collection, storage and distribution of used edible oil for animal feed. The trigger is the incident at the Belgian Verkest animal feed plant, where transformer oil containing large quantities of dioxins and PCBs found its way into the waste edible oil tanks. The animal feed compounded from this stock caused the death of chickens fed with it, and farm animals across Belgium, Netherlands and parts of Germany had to be removed from the food chain. The bill for lost livestock, recall of food products, not to mention analysis to demonstrate products were free of the contamination, was colossal. Emergency controls on movement of animals and their products were in place for a year.

3.5.4 Packaging

The packaging materials in contact with foodstuffs are controlled. There is a series of eleven EC Directives in place, generally incorporated into national legislation. Most significant of these are those concerned with migration from plastics into the food, and the methods for testing. Since fatty foods are the most at risk of dissolving toxins from plastics, the bottles, pails and plastic cans used for delivering frying oils in packs from ½ litre to 20 litres must comply with the directives. Packaging waste is now controlled, with varying vigour according to the environmental credentials of the national governments. Although the principles are governed by EC Directive 94/62 EC, national schemes vary widely.

3.6 Future trends

3.6.1 The supremacy of EU law

The 'ever closer union' of the states of Europe may be resisted by some governments, and perhaps even by a majority of the population in some nations. Nevertheless, there is a historical inevitability about the convergence of its laws. As far as food law is concerned, it is easy to predict that common features will progressively outnumber differences. Table 3.2 demonstrates that for fresh frying oils there is not much commonality yet. Yet, this is the area where, in my opinion, we are likely to see most progress. The French 2% limit on linolenic acid can be seen as a restraint on trade, especially when so much of Europe already uses rapeseed oil for frying. The Danish *trans* fatty acid limit is more difficult to predict, because it could go either way, depending on public opinion and the emerging scientific evidence; either a Europe-wide limit, or dropping of the Danish position. As I have argued, there is less of a case for legislation on end of frying life, so it would take a major upset for the EU to get involved. Expect the wonderful diversity of legislation to continue.

3.6.2 Pseudo-legislative pressures

Much of the pressure in Europe is in non-legislative but nonetheless effective regulation. The German situation is the most obvious case, with an influential opinion taken by all involved as having quasi-legal status. An international conference at Hagen[13] in Germany in April 2000 extends the German philosophy of influential opinion. A round-table discussion at the conference led to the internet publication of draft recommendations. They included:

- There should be no health concerns associated with consumption of frying fats and oils that have not been abused at normal frying conditions.
- Analyses of suspect frying fats and oils to confirm abuse should comprise:
  - total polar compounds (max. 24%)
  - polymeric materials (max. 12%).

- The use of rapid tests correlating with internationally recognised standard tests is recommended.

3.6.3 Codex Alimentarius

Codex teams are working on international standards for additives. Limits for colours annatto, curcumin and β-carotene appear in the latest draft lists. There is a Codex committee working on fat spreads,[14] and another on storage and transport.[15] While Codex has limited legal significance, many legislators take its expert deliberations into account when setting standards, therefore in some respects Codex is a pointer to the future.

3.6.4 Consumer pressure

In many countries of the EU, consumer pressure goes far beyond legislation. The Unilever submission for novel food clearance on stanol esters was effectively held up by pressure on national governments not to clear 'technological' products. This stems from the shattering experience of the GM issue. American readers may need some explanation.

The European attitude to genetic modification was, for most of the 1990s, benign. The FoodFuture campaign built the case around extensive ethical and consumer reaction surveys, and especially the principle of choice. Innovations in tomato paste and cheese rennet quietly outsold their conventional rivals. The bombshell was the arrival of unsegregated American soya and maize. The press and consumer response was alarm at the extent that these products had permeated the food chain. Most manufactured foods, it seemed, contained soya lecithin, or maize starch, or soya oil, if only as a carrier of minor ingredients and additives.

Consumer pressure came to define 'genetically modified' as 'having an ingredient or additive originating from a genetically modified crop'. Supermarket chains in the UK vied with each other to advertise to their customers that no trace of GM product was present in their own-label food. By contrast, EU Directive 1139/98, and national legislation based on it, defined 'genetically modified' as containing new DNA or protein. Additives were not within its scope, and the legislation envisaged a 'negative list' and a *de minimis* threshold. Subsequently, the EC bowed to the pressure, so that Directive 50/2000 extended the law to additives and flavourings, while 49/2000 established the *de minimis* threshold at 1% of adventitious contamination. This threshold does not allow 1% of GM material, it merely states that for food from a non-GM source, a 1% non-deliberate contamination is not illegal. Some retailers have estimated that this 1% limit translates into 0.05% to 0.1% in finished foods on average. The threshold has legal but little commercial significance; the polymerase chain reaction (PCR) test can detect levels below 0.1%, and any supplier whose products tested PCR-positive within the legal threshold is likely to come under pressure to examine his supply chain.

The consequence for frying is the effective disappearance of soya and maize oils, unless 'identity preserved'. This means that the supplier can show by record-keeping that only crops not genetically modified are included in his product. The British Retail Consortium and Food & Drink Federation have published a standard for how the audit trail can be demonstrated. There are two levels: a minimum compliant standard and a 'recommended best practice' standard, which is exceedingly onerous.

It is interesting to note that the standards of exclusion for GM material are more stringent than in the organic standard. The organic standard was introduced as a thoughtful answer to defining what could be described as organic in selling food. The GM standard is an emotional reaction to a technology that the consumer fears.

What will be the next food scare? It would be a brave man who predicted it. However, given that food legislation is driven by the two principles of safety and the prevention of fraud, it is reasonable to expect that any major shift in the law will result from an emerging safety scare.

The long-term concerns with fried food relate to heart disease, and one can see the effect of the link with saturated fat and *trans* fats in guidelines at present. Danish draft legislation sets a maximum *trans* level. The Pasteur Institute in France recommends a target for frying oils of 5% *trans* maximum. Here we see health pressures (to reducing saturates and *trans*) in direct opposition to economic constraints (stability by hydrogenation or using saturated oils). How can the conflict be resolved? The ideal long-life oil is a mono-unsaturated oil laden with natural antioxidants such as tocopherols and other phenols, and with negligible saturated and *trans* fatty acids. Rapeseed oil can approach the ideal only by hydrogenation, which builds the *trans* level. Olive oil would be a good candidate if its price could be cut by a factor of six.

The animal feed industry saw a similar issue in the 1970s. Technology in the form of single-cell protein seemed the answer until agriculture cut off its economic legs by developing high-yielding soya. On the principle that nature is always cheaper than chemical industry in the long run, we can look to biotechnology to breed the perfect frying oil. Whether that is by recombinant genetics or by conventional breeding depends on the public climate. The recombinant technology is faster and more specific, but public fears could yet force the conventional route. It is perhaps worth noting that 'conventional' breeding can involve inducing mutations by means that would scare the consumer if he or she knew about them.

3.7 Sources of information

I heartily recommend anyone entering the market in Europe for the first time to check with the relevant national regulatory body, or better still with a company or consultant familiar with the country. The EU law is a good starting point, but significant differences still exist between nation states. There follows a list of contact points with regulatory bodies, publishers of legislation (not normally the same organisation), and Table 3.5 shows the applicable laws and regulations.

My experience is that national authorities appreciate an approach in their own language, but that in many cases it is not necessary. The Scandinavian countries and the Netherlands are very comfortable communicating in English. Germany's Bundesministerium für Gesundheit provided me with a booklet in English on Consumer Protection in Food Legislation, in addition to an extensive package in German. Most national government web sites have an English version, as well as each of their national languages.

3.7.1 EU and national regulatory bodies

EU

The executive of the European Union consists of the European Commission, organised into 23 Directorates General and 12 Services. There are 17 Commissioners, each responsible for one or more DGs. Depending on the objective of Directives (harmonisation, safety, monetary), it is not possible to generalise about which DG is responsible for food matters, and in any case the national authorities implementing EU law are better points of contact. Each country also has an Office of Representation of the European Commission on its own soil. The European Commission itself is at Rue de la Loi 200, 1049 Brussels, Belgium, tel: +322 299 1111, fax: +322 295 0122, URL: http://www.europa.eu.int.

Austria

Bundeskanzleramt (Bundesministerium für Frauenangelegenheiten und Verbraucherschutz), Gruppe VI/B (Lebensmittelangelegenheiten), Radetzkystraße 2, A-1031 Wien, Tel: +43 1 711 72/0, Fax: +43 1 713 79 52, DVR: 0000019

Belgium

Ministère Fédéral des Affaires Sociales de la Santé Publique et de l'Environnement, Inspection générale des Denrées alimentaires, Boulevard Pachéco 19, bte 5, B-1010 Bruxelles, Tel: +32-2-210 48 43, Fax: +32-2-210 48 16, email: ewida@health.fgov.be, URL: http://www.minsoc.fgov.be/en/index.htm or http://belgium.fgov.be/pa.

Denmark

Fødevaredirektoratet, Mørkhøj Bygade 19, 2860 Søborg, Denmark, tel: +45 33 95 60 00, fax: +45 33 95 66 96, email: vfd@vfd.dk, URL: http://www.vfd.dk

Finland

National Food Administration, PO Box 5, 00531 Helsinki, Finland, tel: +358 9 77261, fax: +358 9 7726 7666, URL: http://www.elintarvikevirasto.fi.

France
Direction Générale de la Concurrence, de la Consommation et de la Répression des Fraudes, Bureau D3, 59 boulevard Vincent Auriol, 75703 Paris cedex 13, Tel. (+33) 1 44 97 04 65, Fax. (+33) 1 44 97 05 27

Germany
Bundesministerium für Gesundheit, Am Propsthof 78a, D-53108 Bonn, Germany, Tel: +49 228 941 4230, Fax: +49 228 941 4989, URL: http://www.bmgesundheit.de/gesetze.

Greece
Higher Chemical Council, Food Directorate, 16 Anast. Tsocha Street, GR-115 21 Ampelokipi, Athens, Greece, Tel: +301 64 28 211, Fax: +301 64 65 123, Telex: 218311 GCSL GR, email: gk-foodiv@ath.forthnet.gr.

Ireland (Eire)
Department of Agriculture and Food, Agriculture House, Kildare Street, Dublin 2, Ireland, tel: +353 1 607 2000, fax: +353 1 661 6263, URL: http://www.irlgov.ie/daff. Food Safety Authority of Ireland, Abbey Court, Lower Abbey Street, Dublin 1, tel: +353 1 672 4711.

Italy
Ministero della Sanità, Dipartimento degli alimenti e nutrizione e della sanità pubblica veterinaria, Piazza Marconi 25, I-00144 Roma, Italy, Tel: +39 6 5994 1, Fax: +39 6 5994 3676, Telex: 613169, URL: http://www.sanita.it/sanita/servizi.htm.

Luxembourg
Ministère de la Santé, 57 boulevard de la Pétrusse, L-2935 Luxembourg, Tel: +35 2 478 5527, Fax: +35 2 491 337

Netherlands
Ministerie van Volksgezondheid, Welzijn en Sport, Postbus 20350, 2500 EJ Den Haag, The Netherlands, Tel: +31 70 340 6884, Fax: +31 70 340 5177, URL: http://www.minvws.nl/international. Ministry of Agriculture Nature Management and Fisheries, Bzuidenhoutsweg 73, Postbus 20401, NL-2500 EK Den Haag, The Netherlands, tel: +31 70 378 4062, URL: http://www.minlnv.nl/international.

Portugal
Ministerio da Agricultura, Instituto da Qualidade Alimentar, Av. Conde Valbom 98, 1100 Lisboa, Portugal, Tel: +351 1 796 2161, Fax: +351 1 797 1750. Direccão-Geral de Saúde, Divisão de Saúde Ambiental, Ministério de Saúde, Alameda D. Afonso Henriques 45, 1056 Lisboa Codex, Portugal, tel: +351 1 847 5515, fax: +351 1 795 9211.

Spain
Subdirección General de Higiene de los Alimentos, Ministerio de Sanidad y Consumo, Paseo del Prado 18, 28071 Madrid, Spain, Tel: +34 91 596 1000 /596 1608, Fax: +34 91 596 1547 /596 1548. Ministerio de Agricultura Pesca y Alimentación, Paseo Infanta Isabel 1, 28014 Madrid, Spain, tel: +34 91 347 5403, fax: +34 91 347 5006

Sweden
Statens livsmedelsverks, Box 622, S-75126 Uppsala, Sweden, Tel: +46 18 17 5500, Fax: +46 18 10 5848, email: livsmedelsverket@slv.se, URL: http://www.slv.se.

United Kingdom
Ministry of Agriculture, Fisheries and Food, Ergon House, 17 Smith Square, London SW1P 3JR, tel: +44 20 7238 3000, fax: +44 20 7238 6591, email: consumer@info.maff.gov.uk, URL: http://www.maff.gov.uk. Food Standards Agency, PO Box 31037, Ergon House, 17 Smith Square, London SW1P 3WG, tel: +44 20 7238 6480, fax +44 20 7238 6763, email: helpline@foodstandards.gsi.gov.uk, URL: http://www.foodstandards.gov.uk.

Codex Alimentarius Commission
Vialle delle Terme di Caracalla, 00100 Roma, Italy, tel: +39 06 57051, fax: +39 06 5705 4593, email: codex@fao.org, URL: http://www.codexalimentarius.net.

3.7.2 Publishers of legislation

EU
The Official Journal of the European Communities is the official source of all EU directives, decisions and regulations. Copies of relevant issues are available in each of the relevant community languages through the official seller in each community state. This is mostly the seller of national legislation documents. It is also now published on the internet at http://www.europa.eu.int/eur-lex/en/oj/index.html. Each issue is accessible for 20 days following the date of publication.

Austria
Official journal: Bundesgesetzblatt der Republik Österreich, from Österreichische Staatsdruckerei, Rennweg 16, A-1037 Wien, Austria, tel: +43 179 789294, fax: +43 179 789419, available on the internet to subscribers: http://www.verlagoesterreich.at/gbbl/. Austrian Food Code published by Brüder Hollinek (projektsitz), Luisenstrasse 20, 3002 Purkersdorf, Austria, tel/fax: +43 223 167 365.

Belgium
Moniteur Belge (in French, or Belgisch Staatsblad in Flemish), la Direction du Moniteur Belge, rue de Louvain 40-42, 1000 Bruxelles, Belgium, tel: +32 2 552 2211, fax: +32 2 511 0184, URL: http://www.moniteur.be for issues since June 1997.

Denmark
Decrees published by: Schultz Information, Herstedveg 10-12, 2620 Albertslund, Denmark, tel: +45 43 63 23 00

Finland
Suomen Säädöskoskoelma (in Finnish, or Finlands Författningssamling in Swedish) from: OY Edita AB, FIN-00043 Edita, Finland, journals freely available on http://www.edita.fi/fs.

France
Journal Officiel de la République Française, from: Journaux Officiels, Service Information Diffusion, rue Desaix 26, 75727 Paris Cedex 15, France, tel: +33 1 40 58 79 79, fax: +33 1 45 79 17 84, URL for issues since January 1998: http://www.journal-officiel.gouv.fr.

Germany
Bundesgesetzblatt, Bundesanzeiger. Verlagsgesellschaft mbH, Südstraße 119, 53175 Bonn, tel: +49 228 382 080, fax: +49 228 382 0836; Bundesanzeiger, tel: +49 221 976 680, fax: +49 221 976 68115, URL: http://www.bundesanzeiger.de.

Greece
Government Gazette available from the National Printing Press, fax: +30 1 523 4312. Greek Food Code published by GS Alysandratos and Associates, Colokoltroni 13, 15772 A Ilisia, Greece, tel: +30 1 775 6767, fax: +30 1 959 2322.

Ireland (Eire)
The Government Publications Office, Molesworth Street, Dublin 2, tel: +353 1 671 0309.

Italy
Official Gazette published by: Istituto Poligranco e Zecca della Stato, Direzione Editonale, Settore vendite e abbonamenti, Via Marciana Marina, 00199 Roma, Italy, tel: +39 06 8508 2307, fax: +39 06 8508 4117.

Luxembourg
Mémorial Journal Officiel du Grand-Duché de Luxembourg, Imprimérie de la Cour Victor Buck, BP 1341, Luxembourg 1013, tel: +35 24 99 86 61, fax: +35 24 99 41 64.

Netherlands
PBO publications from: SER, Bezuidenhoutsweg 60, postbus 90405, 2509 LK Den Haag, Netherlands, tel: +31 70 3 499 499, fax: +31 70 3 832 535. Nederlandse Staatscourant and Staatsblad van het Koninkrijk der Nederlanden available from tel: +31 70 3 789 880, fax: +31 70 3 789 783, email: adv.staatscourant@sdu.nl.

Portugal
URL for electronic version of the official journal: Diário da República Electrónico http://www.dr.incm.pt. email: dre@incm.pt.

Spain
Boletín Oficial published by: La Librería del BOE, Trafalgar 27, 28010 Madrid, Spain, tel: +34 91 538 2121, email: clients@com.boa.es.

Sweden
Ordinances and Guidelines published by the Statens livsmedelsverk (see §3.7.1 above).

United Kingdom
All legislation published by: Stationery Office, PO Box 276, London SW8 5DT

3.7.3 Legislation – the laws and regulations

Table 3.5 lists the laws and regulations by country. The absence of an entry does not necessarily mean that no legislation exists, merely that I, and those I have consulted, have not identified a specific law or regulation covering that topic.

3.8 References

1. Codex Alimentarius Commission Alinorm 99/17 Appendix II, see Codex web site at *http://www.fao.org/es/esn/codex*
2. Seed Crushers and Oil Processors Association, 6 Catherine Street, London WC2B 5JJ, tel: +44 20 7836 2460, fax: +44 20 7379 5735
3. Codex Committee on Food Additives and Contaminants: Codex General Standard for Food Additives available on *ftp://ftp.fao.org/codex/ccfac32/fa9915be.pdf.*
4. IWAOKA W T, PERKINS E G: Metabolism and lipogenic effects of the cyclic monomers of linolenate in the rat. *JAOCS* **55**, 734–738 (1978)
5. LAMBONI C, SEBEDIO JL, PERKINS EG: Cyclic fatty acid monomers from dietary heated fats affect rat enzyme liver activity. *Lipids* **33**, 675–691 (1998).

(References continued on page 48).

Table 3.5 The laws and regulations by country

	EU	Austria	Belgium	Denmark	Finland	France	Germany	Greece	Ireland	Italy	Luxembourg	Netherlands	Portugal	Spain	Sweden	United Kingdom
Primary law on food safety and avoidance of fraud		A1	B1	DK1	SF1	F1	D1	G1	EI1	I1	L1	NL1	no	E1	S1	UK1
Composition of oils and fats		A2	B2			F2	D2	G2			L2	NL2	P1	E2	No	no
Erucic acid content	EU1	A3	B3	DK2			D3	G3	EI2	I2	L3				S2	UK2
Permitted colours	EU2	A4	B4	DK3	SF2	F3	D4	G4	EI3	I3	L4	NL3	P2	E3	S4	UK3
Permitted flavours	EU3	A5	B5				D4	G5	EI4	I4	L5	NL4	P3	E4	S3	UK4
Permitted miscellaneous additives (e.g. antioxidants)	EU4	A6	B6	DK3	SF3	F3	D4	G6	EI5	I3	L6	NL5	P4	E5	S4	UK5
Labelling	EU5	A7	B7	DK4	SF4	F4	D5	G7	EI6	I5	L7	NL6	P5	E6	S5	UK6
Weights and measures in general		A8				F5		G8		I6	L8		P6	E7		UK7
Average weights	EU6								EI7		L9				S6	UK8
Control of used edible oil			B8	no		F6			no		L10			E8		no
Packaging recycling or disposal	EU7								EI8	I7	L11	NL7	P7	E9	S7	UK9

Key: An entry 'no' means that controls do not exist. The absence of an entry should not be taken to imply the same.

European Union
EU1 erucic: Directive 76/161
EU2 colours: Directive 94/36
EU3 flavours: Directive 88/388, completed by Directive 91/71
EU4 miscellaneous additives: Directive 95/2
EU5 labelling: Directive 79/112, amended by 93/102, 95/42
EU6 average weights: Directive 76/211, amended by 78/891
EU7 packaging: Directive 94/62

Austria
A1 safety: Lebensmittelgesetz 1975, BGBI nr. 86
A2 oils: Österreichische Lebensmittelbuch, 3. Auflage, Kapitel B30, section 1.6
A3 erucic: Erucasäureverordnung BGBI nr. 468/1994
A4 colours: Farbstoffverordnung BGBI nr. 541/1996
A5 flavours: Aromenverordnung BGBI nr. 42/1998
A6 miscellaneous additives: Zusatzstoffverordnung BGBI II nr. 383/

1998
A7 labelling: Lebensmittelkennzeichnungsverordnung, 1993
A8 weights: Fertigpackungsverordnung BGBI nr. 867/1993, BGBI nr. 132/1995, BGBI II nr. 139/1997

Belgium
B1 safety: Loi du 24/1/77
B2 oils: Arrêté royal du 23/4/74 (edible oils), arrêté royal du 22/1/88, amended 3/5/99 (frying), arrêté royal du 2/10/80 (human consumption)
B3 erucic: Arrêté royal du 26/2/76
B4 colours: Arrêté royal du 9/10/96
B5 flavours: Arrêté royal du 24/1/90
B6 miscellaneous additives: Arrêté royal du 1/3/98
B7 labelling: Arrêté royal du 13/11/86
B8 used oils: opinion of Advisor General, DG4 Agribex, Brussels of 7/2/2000.

Denmark
DK1 safety: Fødevareloven nr. 471 af 1/2/98
DK2 erucic: Bekendtgørelse nr. 57 af 22/1/99
DK3 additives: Bekendtgørelse nr. 942 af 11/6/97,
DK4 labelling: Bekendtgørelse nr. 598 af 14/8/93 (general), 198 af 20/3/92 (nutrition)

Finland
SF1 safety: Elintarvikelaki (Food Law) 361/1995, 1/4/95
SF2 colours: ruling 1756 of 1/1/96
SF3 additives including emulsifiers and antioxidants: ruling 811/1997 0f 2/8/99
SF4 labelling: regulation 794/1991 of 10/5/91 and ruling 795/1991 of 1/6/91

France
F1 safety: Code de la Consommation (loi no. 93-949 of 26/7/93)
F2 oils: décret du 11/3/1908, décret no. 73-139 du 12/2/73, arrêté du 19/11/90
F3 additives: arrêté du 2/10/97
F4 labelling: arrêté du 7/12/84, décret no. 03-1130 du 27/9/93, arrêté du 3/12/93
F5 weights: arrêté du 21/3/85
F6 used oil: loi du 15/7/75

Germany
D1 safety: Lebensmittel- und Bedarfsgegendständegesetz (LMBG 1/1/97)
D2 oils: Guidelines on edible oils and fats of 17/4/97
D3 erucic: Verordnung vom 24/5/77 (BGBl I p782), last amended 26/10/82 (BGBl I p1945)
D4 additives (including colours, flavours, others): Verordnung vom 29/1/98 (Bundesgesetzblatt 1998 Teil 1, nr. 8)
D5 labelling: Lebensmittel-Kennzeichnungsverordnung vom 6/9/84 (BGBl I p1221), last amended by BGBl I p460; Nährwert-Kennzeichnungsverordnung vom 25/11/94 (BGBl I p3526)

Greece
G1 safety: Greek Food Code (GFC)
G2 oils: articles 70 to 78 of GFC
G3 erucic: articles 70 to 78 of GFC
G4 colours: article 33 of GFC
G5 flavours: article 44 of GFC
G6 miscellaneous additives: article 35 of GFC
G7 labelling: article 11of GFC
G8 weights: EC directives have been implemented

Table 3.5 Continued

Ireland

EI1 safety and fraud: Sale of Food and Drugs Act 1879, amended 1879 and 1899. See also Health Acts 1947 (no.28), 1953 (no.26), 1970 (no.1) and Statutory Instrument (SI) no.333 (1991)

EI2 erucic: Health (Erucic Acid in Food) Regulations 1978 (SI no.123), amended by SI no. 67 (1992) and SI no. 271 (1982)

EI3 colours: SI no. 344 of 1995

E14 flavours: SI no. 22 of 1992

E15 miscellaneous additives: SI no. 128 of 1997

EI6 labelling: SI no. 205 of 1982 as amended

EI7 average weight: Packaged Goods (Quantity Control) Act 1980 (no.11), Regulation SI no.39 of 1981, as amended by Metrology Act 1996 (no.27)

EI8 Waste Management Act 1996 (no. 10), Regulations SI no. 242 of 1997

Italy

I1 safety: Law no. 283 of 30/4/62

I2 erucic: Law no. 659 of 9/10/80

I3 all additives: Ministerial decree no. 209 of 27/2/96

I4 flavours: Decree no. 107 of 25/1/92

I5 labelling: Legislative decree no. 109 of 27/1/92 (labelling); Legislative decree no.77 of 16/2/93 (nutrition)

I6 weights: Law no. 690 of 25/10/78; Presidential decree no.391 of 26/5/80; Decree-law no. 450 of 3/7/76 as amended

I7 packaging: Legislative decree no. 22 of 5/2/97

Luxembourg

L1 safety: Act of 25/9/53 as amended

L2 oils: Regulation of 4/8/75

L3 erucic: Regulation of 29/12/77

L4 colours: Regulation of 19/3/97

L5 flavours: Regulation of 20/12/90

L6 miscellaneous additives: Regulation of 10/4/97

L7 labelling: Regulation of 16/4/92 as amended (labelling); Regulation of 22/6/92 (nutrition)

L8 weights: Regulation of 26/11/81 as amended

L9 averafe weights: Regulation of 19/10/77 as amended

L10 used oil: Ministerial order of 30/6/99

L11 packaging: Regulation of 31/10/98

Netherlands

NL1 safety: Warenwet (Dutch Commodities Act); Dutch Food Law of 10/12/91

NL2 oils: Decree of 10/4/75

NL3 colours: Decree of 27/9/95

NL4 flavours: Decree of 24/1/80

NL5 miscellaneous additives: Decree of 23/9/96

NL6 labelling: Decree of 10/12/91 as amended (labelling); Decree of 7/9/93 (nutrition)

NL7 packaging: Regulation of 30/12/97

Portugal

safety: no basic food law

P1 oils: Decree-law no. 32/94 of 5/2/94; Order no. 928/98 of 23/10/98

erucic: EC directive applies

P2 colours: Order no. 759/96 of 26/12/96

P3 flavours: Order no. 620/90 of 3/8/90

P4 miscellaneous additives: Decree-law no. 363/98 of 19/11/98

P5 labelling: Decree-law no.560/99 of 18/12/99 (labelling); Order no.

751/93 of 23/8/03 (nutrition)
P6 weights: Order no. 359 of 7/6/94; Order no. 1198/91 of 18/12/91
P7 packaging: Decree-law no. 366-A/97 of 20/12/97 and Order no. 29-B/98 of 15/1/98

Spain
E1 safety: Spanish Food Code, as approved by decree no.2484/1967 of 21/9/67 as amended
E2 oils: Royal decree no. 1011/1981 of 10/4/81 as amended; Royal decree no.308/1983 of 25/1/83 as amended
erucic: EC restrictions apply
E3 colours: Royal decree no. 2001/1995 of 7/12/95
E4 flavours: Royal decree no. 1477 of 2/11/90 as amended
E5 miscellaneous additives: Royal decree no. 145/1977 of 31/12/97
E6 labelling: Royal decree no.1334/1999 of 31/7/99 (labelling); Royal decree no. 930/1992 of 17/7/92 (nutrition)
E7 weights: Royal decree no.723 of 24/6/88; Royal decree no. 1472 of 1/12/89
E8 used oil: Order of 26/1/89 as amended
E9 packaging: Law no. 11/1997 of 24/4/97

Sweden
S1 safety: Food Act (SFS 1971:511 as amended)
oils: none
S2 erucic: Ordinance SLV FS 1993:15
S3 flavours: Ordinance SLV FS 1996:1 as amended
S4 additives including colours: Ordinance SLV FS 1999:22
S5 labelling: Ordinance SLV FS 1993:19 as amended (labelling); Ordinance SLV FS 1993:21 (nutrition)
S6 average weights: Ordinance STAFS 1993:18
S7 packaging: Ordinance SFS 1994:1235 as amended

United Kingdom (England & Wales – different statutory instrument numbers relate to Scotland and to Northern Ireland)
UK1 safety: Food Safety Act 1990
UK2 erucic: 1977/691, amended by 1982/264
UK3 colours: 1995/3124
UK4 flavours: 1992/1971, amended by 1994/1486 and 1996/1499
UK5 miscellaneous additives: 1995/3187
UK6 labelling: 1996/1499
UK7 weights: Weights and Measures Act 1963, Order 1988/2040, amended by 1990/1550, 1994/2868
UK8 average weights: 1986/2049, amended by 1987/1538, 1992/1580, 1994/1852
UK9 packaging: 1997/648

6. MAHUNGU S M, ARTZ W E, PERKINS E G: Oxidation products and metabolic processes. Chapter 2 in *Frying of Food* ed. Boskou D and Elmadfa I, Technomic, 1999.
7. KOCHHAR S P (ed.) *New Developments in Industrial Frying*, SCI, London, 1997.
8. PERKINS E G, TAUBOLD R: Nutritional and metabolic studies of noncyclic dimeric fatty acid methyl esters in the rat. *JAOCS* **55**, 632–634 (1978).
9. KEANE K W, JACOBSON G A, KRIEGER G H: Biological and chemical studies on commercial frying oils. *J Nutr* **68**, 57–74 (1959).
10. MARQUEZ-RUIZ G, DOBARGANES M C: Nutritional and physiological effects of used frying fats. In *Deep Frying, Chemistry, Nutrition and Practical Applications*, ed. PERKINS E G, ERICKSON M D, AOCS, 1996.
11. BILLEK G, GUHR G, WAIBEL J: Quality assessment of used frying oils: a comparison of four methods. *JAOCS* **55**, 728–733 (1978).
12. HAMILTON R J, PERKINS E G: Chemistry of Deep Fat Frying, in KOCHAR S P (ed.): New Developments in Industrial Frying, publ. SCI, London ISBN 0-9526542-8-8.
13. Third International Symposium on Deep Fat Frying, Hagen/Westphalia (Germany) March 20–21, 2000, organised by Deutsche Gesellschaft für Fettwissenschaft e.V., conclusions to be found on http://www.gdch.de/dgf/recomm.htm, reported in *Inform* **11**, 630–631 (2000).
14. Codex draft standard for fat spreads and blended spreads, Alinorm 99/17, Appendix VI.
15. Codex recommended code of practice for the storage and transport of edible oils and fats in bulk, Alinorm 99/37, para 165 and Appendix VII.

4

Regulation in the United States

D. Firestone, Food and Drug Administration, Washington DC

4.1 Introduction

It has been recognized for some time that improper frying operations result in degradation of frying fat and reduce the quality and wholesomeness of fried foods. Although there are no worldwide regulations and guidelines for control of frying fat and manufacture of fried foods, a number of European countries, concerned with possible health risks to consumers, have issued regulations and guidelines for control of frying fats.[1,2] However, the US Food and Drug Administration (FDA) has not established specific regulations or guidelines to control the quality of frying oils or fried foods and the US Department of Agriculture (USDA) Food Safety and Inspection Service (FSIS) Meat and Poultry Inspection Manual contains some general guidelines for frying meat and poultry products[3] and the agency has issued a plant sanitation directive requiring cleaning of frying equipment at regular intervals.[4] State and local regulatory agencies have no specific regulations for control of frying fats or frying operations other than the general provisions of Title 21 of the Code of Federal Regulations and the FDA Food Code.[5]

4.2 FDA regulations and guidelines

Federal food laws and regulations in the US are intended to protect the consumer by assuring the integrity, wholesomeness and proper labeling of food products in interstate commerce. Section 402 of the Federal Food, Drug, and Cosmetic Act, section 402, states that a food is deemed adulterated if it 'bears or contains any poisonous or deleterious substance which may render it injurious to health'

[402(a)(1)]; if it 'bears or contains any added poisonous or added deleterious substance' (other than specified substances such as pesticides) [402(a)(2)]; if it 'consists in whole or in part of any filthy, putrid, or decomposed substance, or if it is otherwise unfit for food' [402(a)(3)]; or if it 'has been prepared, packed or held under insanitary conditions whereby it may have become contaminated with filth, or whereby it may have been rendered injurious to health' [402(a)(4)]. Section 404 of the Act mandates the Secretary of Health and Human Services to promulgate regulations to control contamination of food with micro-organisms during the manufacture, processing, or packing of food products distributed in interstate commerce. The FDA, concerned about food safety, has been leading an effort in the US to improve coordination among public health and food regulatory officials to improve food safety programs to minimize outbreaks of foodborne illness.[6]

FDA has not established specific regulations to control the quality of frying fats since it has not been determined that frying fats used in deep-frying operations are injurious to health. However, frying fats are subject to the general provisions of the Federal Food, Drug, and Cosmetic Act and to specific food safety programs such as proposed Hazard Analysis and Critical Control Point (HACCP) systems applied to food manufacture.[7] The Food Code[5] is a general reference document intended for use and adoption by state and local government agencies responsible for overseeing food safety in retail establishments (restaurants, child care centers, health care institutions, etc.). While neither federal law nor federal regulation, the Food Code provisions are consistent with federal food laws and regulations and are written for ease of legal adoption at all levels of government.

The various sections of the Food Code, intended primarily to protect consumers from foodborne diseases, cover employee health; personal cleanliness and hygiene practices; food handling, preparation and presentation; equipment installation, use and sanitation; water, plumbing and waste handling; physical facilities; storage and use of toxic materials (sanitizers, drying agents, pesticides, fruit and vegetable washing chemicals, etc.); and compliance and enforcement procedures including approval of HACCP plans. Section 4-301.14 requires ventilation hood systems to be sufficient in number and be able to prevent grease or condensation from collecting on walls and ceilings. An Annex sets forth a series of enforcement mechanisms and references including management and personnel guidelines for assuring food safety; food establishment inspection and preparation of inspection reports; HACCP guidelines including procedures to assure that HACCP systems are working plus typical flow diagrams; food processing requirements; and a set of model forms and guides including HACCP guidelines and HACCP Food Establishment Inspection Report forms. Page 1 of a HACCP Inspection Data form is shown in Fig. 4.1. The Food Code does not specifically address optimum frying temperatures since it is mainly concerned with destruction of micro-organisms of public health concern. Accordingly, the Code specifies that all parts of a food be heated to a temperature of 63°C for 3 minutes (minimum) and for longer holding times at lower temperatures (121 minutes at 54°C).

DEPARTMENT OF HEALTH AND HUMAN SERVICES
PUBLIC HEALTH SERVICE
FOOD AND DRUG ADMINISTRATION

HACCP INSPECTION DATA

EST. NAME:	**PERMIT NO.**	**INSPECTOR:**	
DATE:	**TIME IN:** :AM / PM	**TIME OUT:** :AM/ PM	

Record all observations below - transfer violations to Inspection Report

FOOD TEMPERATURES / TIMES / OTHER CRITICAL LIMITS
Use Additional Forms If Necessary

FOOD STEP	1.	CRITICAL LIMIT	2	CRITICAL LIMIT	3	CRITICAL LIMIT	4	CRITICAL LIMIT
A. SOURCE								
B. STORAGE								
C. PREP BEFORE COOK								
D. COOK								
E. PREP AFTER COOK								
F. HOT/COLD HOLD								
G. DISPLAY/ SERVICE								
H. COOL								
I. REHEAT								

OTHER FOOD TEMPERATURES OBSERVED — Use steps from above for location

FOOD	TEMP. °C/°F	STEP	FOOD	TEMP. °C/°F	STEP	FOOD	TEMP. °C/°F	STEP

Fig. 4.1 Page 1 of FDA HACCP Inspection Data form.

In 1998, FDA drafted a document 'Managing Food Safety: A HACCP Principles Guide for Operators of Food Establishments at the Retail Level',[8] based on input from industry, academia and consumers as well as state and local food regulators, in order to assist food establishment employees in their efforts to prepare safe food. The document is intended to serve as a guide in preparation of a simple plan based on HACCP principles. It includes sections on identifying critical control points, developing corrective actions, carrying out verification procedures (checking monitoring and corrective action records, etc.) and maintaining facilities equipment.

4.3 USDA/FSIS guidelines and directives

The Meat and Poultry Inspection Manual of the US Department of Agriculture, Food Safety and Inspection Service (USDA/FSIS)[3] contains general guidelines for frying meat and poultry. Noting that deep fat frying times vary with temperature, amount of replacement fat added periodically and fat treatment during use, the guidelines state that 'excessive foaming, darkened color and objectionable odor or flavor are evidence of unsuitability and require fat rejection.' The guidelines also note that frying fat should be discarded when it 'foams over the vessel's side during cooking, or when its color becomes almost black as viewed through a colorless glass container. Serviceable life of fat can be extended by holding frying temperature below 204°C (400°F), daily replacing one third or more, filtering as needed, and cleaning the system at least weekly. Adding an antifoam agent (methyl-polysiloxane) to new fat is helpful.'

For poultry, the FSIS guidelines advise that

> to completely fry poultry parts, time and temperature required depends upon product type and weight, and upon equipment. Acceptable frying operation should be carried out at approximately 190°C (375°F) or higher for 10 to 13 minutes when parts are not precooked ... commercially prepared fats may contain antioxidants or antifoaming agents ... used fat may be made satisfactory by filtering, adding fresh fat, and regularly cleaning the equipment. Acceptable frying operation should be carried out at approximately 190°C (375°F) or higher for 10 to 13 minutes when parts are not precooked ... commercially prepared fats may contain antioxidants or antifoaming agents ... used fat may be made satisfactory by filtering, adding fresh fat, and regularly cleaning the equipment. Large amounts of sediment and free fatty acid content in excess of 2% are usual indicators that frying fats are unwholesome and require reconditioning or replacement. Sediment is usually removed by filtering. Adding fresh fat or new fat reduces the free fatty acid to acceptable levels.

The guidelines point out that fat used for fish products is not satisfactory for frying poultry. Solid frying fat may be kept liquid providing the holding temperature does not fall below 54°C (130°F) in order to prevent localized excess heating and fat breakdown during melting.

FSIS directive 11,000.2[4] requires cleaning of frying equipment at regular intervals and allows continuous filtering or flushing with clean fat for limited periods of time.

> Complete drainage, followed by dismantling and scouring or otherwise thorough cleaning, is necessary for acceptable sanitizing. Traces of water and detergents increase rate of fat breakdown. They must be completely removed from pipelines, valves, filters, pumps, etc., must be of sanitary construction, readily accessible to cleaning, and preferably constructed of stainless steel. Rubber and some types of plastic connecting lines are not acceptable.

4.4 State and city regulations

Inquiries were made at several intervals during 1989–2000 to 35 US cities and all 50 US state health departments and food control agencies to determine whether specific laws and regulations were available for control of frying fats and frying operations in processing plants and restaurants. The replies frequently emphasized that there were no specific laws and regulations other than general requirements that fats used in food service establishments are obtained from approved sources and are not adulterated. Many health departments pointed out that there were no specific regulations for frying fats, frying operations and fried foods in industrial and food service operations other than general regulations for sanitation in these facilities. Many regulatory agencies specifically noted that the 1999 or earlier version of the Food Code was adopted to regulate the preparation of fried food.

The City of Philadelphia Department of Health reported that examination and evaluation of frying fats and oils are 'generally limited to organoleptic comparison of used cooking oils with fresh oils. The nature of foods fried, volume of frying, and the establishment of filtering and replacement regimen are also evaluated.' The San Francisco Department of Health noted that 'as a routine practice inspectors check for color, sediments, excessive smoke and odors of oils used in frying ... corrections are made through replacement or filtration of cooking oils.'

The Chicago Department of Health regulates fats and oils under the general provisions of chapter 4-344 of the Municipal Code of Chicago which basically addresses sanitation practices in food establishments. Food products require approved labels and should be free of rancidity. During routine inspections, frying oils are checked for color, sediments and foreign objects as well as excessive smoke. If necessary, oil samples are collected for determination of rancidity by the Kreis test. The State of Wisconsin Department of Health stated that frying fats and oils are not considered a health hazard from bacterial contamination because of the high cooking temperatures used in deep-fat frying operations. Concerns are related to proper exhausting of frying fumes and controlling 'off' odors and flavors. The Fulton County (Atlanta, Georgia) Food Service Sanitation Regulations do not address the control of frying fats except for the cleaning of frying equipment and proper disposal of spent cooking fats.

FDA's Division of Federal-State Relations advised the Connecticut Department of Health in 1990 (in response to an inquiry by the state agency): (a) there is no standard frequency for filtering fat used in a deep-frying operation (the filtering material should be clean and the oil should be clear and properly stored); (b) any presence of 'off' odors or visible evidence of foreign material, filth, or other adulterants would warrant discarding the fat; and (c) fat must be adequately protected from contamination during use, storage, or filtering.

The State of Montana's Food and Consumer Safety Section regulations specify that ventilation hoods in food service establishments shall be

> installed at or above all commercial type deep fryers, broilers, fry grills, steam-jacketed kettles, hot-top ranges, ovens, barbecues, rotisseries, dishwashing machines, and similar equipment which produce comparable amounts of steam, smoke, grease, or heat ... ventilation hoods and devices shall be designed to prevent grease or condensation from collecting on walls and ceilings, and from dropping into foods or onto food-contact surfaces ... filters or other grease extracting equipment shall be removable for cleaning and replacement if not designed to be cleaned in place.

Denver's Department of Health requires in addition to adequate ventilating hoods in food establishments, use of a velometer to test the equipment and confirm that hoods maintain suitable air velocities. The City of St. Louis Department of Health also reported that its Food Service Establishment Ordinances require ventilating hoods 'designed to prevent grease or condensation from collecting on walls and ceilings, and from dripping into food contact surfaces. Filters or other grease extracting equipment shall be readily removable for cleaning and replacement if not designed to be cleaned in place.'

The City of New Orleans Department of Health provided a copy of its regulations concerning fats, oil and grease disposal in food service establishments. These facilities are required to have grease control devices for separating and retaining water-borne fats, oil and grease prior to the wastewater exiting the trap and entering the sanitary sewer collection and treatment system. Discharged wastewater should be free of oil or grease exceeding 250 mg/l. Specifications and instructions are provided for grease interceptors and operation of oil and grease waste disposal systems.

4.5 Sources of further information and advice

Federal state and local agencies in the US are primarily concerned with improving the safety of the nation's food supply by enhancing surveillance of foods to prevent or improve the response to outbreaks of food-borne disease, as well as developing a strategy for greater control or elimination of food-borne pathogens from the food supply. Nevertheless, availability in food codes and regulations of a uniform set of guidelines for frying fats and frying operations in food establishments would help provide better quality as well as safe fried foods.

Several European countries have issued guidelines and advice for handling frying fats which could provide the basis of a generally accepted set of rules for preparation of fried foods. The General Advice on Handling Frying Fats, issued by the Swedish National Food Administration[9] is shown in Table 4.1. The Environmental and Food Agency of Iceland issued several years ago the following set of guidelines:

- Use fats and oils intended for deep frying. Many types of salad oil do not maintain their quality at the temperatures used for deep frying.

Table 4.1 Swedish National Food Administration's guidelines for deep-fat frying

1. All the fat in the deep-fat fryer must be changed before it starts smoking or foaming. Use e.g., Food Oil Sensor or Oxifrit Test to indicate when it is time to change.
2. Strain the fat and clean the fryer once a day. Rinse carefully after cleaning. Solid material in the fat and detergent residues accelerate breakdown of the fat. Store strained fats at room temperature or at lower temperatures in a covered stainless-steel vessel. If iron pots are used, they should be rinsed only with hot water. Detergents remove the protective film of polymerized fat that builds up during use.
3. The frying temperature should be 160–180ºC (320–356ºF). At lower temperatures, the product absorbs more fat. At higher temperatures, the fat deteriorates quicker.
4. Use fat that is specially intended for frying.
5. Avoid salting or seasoning the fried food over the fryer. Salt or seasoning can accelerate breakdown of the fat.
6. Lower the temperature when not frying and protect the fat from light.
7. The fryer should have no iron, copper or brass parts that come in contact with the heated fat.
8. Keep a constant level of fat in the fryer. Fry a little at a time to keep the temperature as even as possible. Prefry when large amounts are to be prepared.
9. Use a separate fryer, if possible, for frying potatoes. The fat deteriorates more rapidly when meat or fish is fried than when only potatoes are fried.

Caution: Do not overheat. If the fat temperature rises above 300ºC (572ºF), the fat may start to burn.

- Do not mix used fats or oils with new ones, as that would accelerate deterioration.
- Clean all frying equipment regularly and filter the fat. All dirt and residue of detergents and cleaning products adversely affect the quality of fats and oils. Avoid contact of copper or copper compounds with fat. Do not apply salt to foodstuffs above the frying pan, since metal compounds in salt could result in deterioration of the fat.
- The appropriate frying temperature is 165–190ºC. Higher temperatures result in dark color, oxidation, hydrolysis, and polymerization. If the temperature is too low, the frying time is too long, affecting the quality of foodstuffs. To minimize the drop in temperature, it is important not to overload the frying pan.
- When the fat is heated, the temperature should not be set higher than the temperature to be used for frying.
- Durability of fat can be prolonged by keeping the temperature between 90 and 120ºC when the fat is not in use.
- As the heat transfer in solid fat is low, it should be melted at low temperatures to avoid overheating certain parts of the fat. Slight burning or overheating of fat can accelerate deterioration and spoil all of the fat in the pan.
- Remember that spoiled frying fat can have adverse health effects.

The 3rd International Symposium on Deep-Fat Frying held 20–21 March 2000 in Hagen, Westphalia, Germany was concluded with issuance of eight recommendations for frying oil, as follows:[10]

INSPECTION FORM — FINLAND NATIONAL FOOD ADMINISTRATION
(For The Food Laboratory, To Be Filled Out By The Health Inspector)

No. and Date ....

Amount of fat in kettle .... Added daily ....

Fat totally renewed, date ....

How often will the fat be renewed totally ....

Criteria for total removal (except time) ....

How long daily will the fat be kept hot ....

At the time of sampling, how long kept hot ....

Fat temperature, reported .... measured ....

Filtering and storage of the fat ....

....

What will be fried just now ....

What else will be fried ....

Data of the kettle ....

Cleaning of the kettle, last (when) ....

How was kettle cleaned ....

....

Sensory evaluation of the fried food:

Taste ....Smell ....Color ....

Observations on the spot:

Overall cleanliness..........

Smoke..........Ventilation..........

Fig. 4.2 Finnish inspection form.

1. The principal quality index for deep-fat frying should be the sensory parameters of the fried food.
2. Analysis of suspect frying fats and oils should utilize two tests to confirm abuse. Recommended analyses should include (references are added by this author):
   - total polar materials (11) (maximum, 24%)
   - polymeric materials (12) (maximum, 12%)
3. Use of rapid tests (quick tests) (13) are recommended. Rapid tests should:
   - correlate with internationally recognized standard methods
   - provide an objective index
   - be easy to use
   - be safe for use in food processing and preparation areas
   - quantify with oil degradation
   - be rugged enough for field use.
4. Affirming previous work, no health concerns are associated with consumption of frying fats and oils that have not been abused at normal frying temperatures.
5. Encourage development of new and improved methods that give chemists and the food industry the tools to conduct work more quickly and easily. Methods should be environmentally friendly and use less hazardous and lower quantities of solvents.
6. Encourage and support basic research focused on the dynamics of deep-fat frying. Research should be cross-disciplinary, encompassing oil chemistry, food engineering, sensory science, food chemistry, and nutritional sciences.
7. Use filtration to maintain oil quality.
8. Used, but not abused, frying oils may be diluted with fresh oil with no adverse effects on oil quality.

The National Food Administration of Finland issued a circular letter in 1991 to inspection entities outlining suggested procedures for sampling and analyzing frying fat. Test criteria included sensory evaluation; total polar materials, maximum of 25%; acid value of vegetable oil and solid fat, 2.0 and 2.5, respectively; smoke point of vegetable oil and solid fat, minimum 180 and 170ºC, respectively; Fritest, vegetable oil, maximum 2 (scale 1–3); Oxifrit Test, below 3 (scale 1–4); and food oil sensor, below 4 (scale 0–6). An inspection form was also made available (Fig. 4.2), to be completed by the inspector and submitted to the food laboratory.

4.6 References

1 FIRESTONE D, STIER R F and BLUMENTHAL M M, 'Regulation of frying fats and oils', *Food Technol.,* 1991 **45**(2) 90–94.

2 FIRESTONE D, 'Worldwide regulation of frying fats and oils', *INFORM*, 1993 **4**(12) 1366–71.

3 *Meat and Poultry Inspection Manual*, Food Safety and Inspection Service, U.S. Department of Agriculture, Washington, D.C., December 1990, section 1840, page 125.

4 *Combined Compilation of Meat and Poultry Inspection Issuances for 1984–1990*, Food Safety and Inspection Service, U.S. Department of Agriculture, Washington, D.C., FSIS Directive 11,000.2, 4-28-87, section cc, page 14.

5 *Food Code*, 1999 Recommendations of the United States Public Health Service Food and Drug Administration, Washington, D.C., PB 99-115925.

6 *Food Safety Initiative Update*, Center for Food Safety and Applied Nutrition, FDA, Washington, D.C., January 28, 2000 (see *http://www.cfsan.fda.gov).*

7 *FDA Announces Food Safety (HACCP) Pilot Program*, FDA press release P95-3, U.S. Department of Health and human Services, FDA, Washington, D.C., May 8, 1995 (see http://www.cfsan.fda.gov).

8 *Managing Food Safety: A HACCP Principles Guide for Operators of Food Establishments at the Retail Level*, 1998, Food and Drug Administration, Center for Food Safety and Applied Nutrition, Washington, D.C. (See *http://www.cfsan.fda.gov).*

9 *General Advice on Handling Frying Fats* (SLV FS 1990:2), National Food Administration, Uppsala, Sweden, 1990.

10 Anon., 'DGF meeting offers frying oil recommendations', *INFORM,* 2000 **11**(6) 630–31 (see also *http://www.gdch.de/dgf/recomm.htm).*

11 *Official Methods and Recommended Practices of the American Oil Chemists' Society*, 5th Ed., 1998, American Oil Chemists' Society, Champaign, IL, USA, Official Method Cd 20-91.

12 *Official Methods and Recommended Practices of the American Oil Chemists' Society,* 5th Ed., 1998, American Oil Chemists' Society, Champaign, IL, USA, Official Method Cd 22-91.

13 STIER, R F, 'Quick tests for fats and oils', *Baking & Snack*, 1996 **18**(10) 62–66.

5

Health issues

B. Ruiz-Roso and G. Varela, Universidad Complutense de Madrid

5.1 Introduction

As a prominent component of the modern diet, fat receives very close attention because of its relationship to several chronic degenerative diseases. An association between dietary fat, excessive energy consumption, and obesity has been noted in some studies.[1,2] Excessive consuption of fat (especially saturated fat) has been linked to the development of cardiovascular disease.[3,4] Excessive intake of fat has also been associated with certain types of cancer, although the interpretation of the data is limited by the difficulty in distinguishing high-fat from high-energy diets.[5,6] As a result of such studies, the consumer's perception is that low-fat diets are automatically healthier and that vegetable oils are healthier than animal fats (although it is important to realize that extreme low-fat diets can cause health problems, leading to deficiences of fat-soluble vitamins and essential fatty acids).

Currently, most of these studies are derived from epidemiologic or experimental studies in which lipid intake is calculated using food-consumption tables or databases. In most of these tables the quoted lipid content is that of the raw food, whereas most foods are usually consumed only after being subjected to culinary processes such as heating or frying. However, it is known that in the course of these processes the lipid content of such foods may undergo important qualitative and quantitative changes. There is, in addition, often no indication of the type of fat used in the cooking of raw food. Failure to take account of these issues may bc an underlying cause of the conflicting results of studies trying to establish the relationship between lipid intake and health.[7,8] Any major change in lifestyle habits in a population can have a variety of consequences, many of them unforeseen. As a result, recommendations for change in dietary habits must have a solid scientific foundation, suggesting the need for continuing research in this area.[9]

5.2 Dietary lipids: structure and function

Lipids are compounds insoluble in water but soluble in organic solvents. The term 'fat' is used in discussing diets, whereas lipids is a more appropiate term to use in discussing metabolism. Dietary fat is classically defined as triglycerides (TG), phospholipids (PL) and sterols. The sources of lipids in food products are of both plant and animal origin, and can be considered as having similar nutritional value with a few exceptions. Food products of animal origin contain cholesterol (CH), whereas vegetable foods may supply nonabsorbable carbohydrates and phytosterols, which may interfere with CH absorption. Associated nutrients consist primarily of fat-soluble vitamins and related compounds (e.g., carotenoids and tocotrienols).

Dietary fat is a concentrated energy source relative to carbohydrate and protein; carbohydrate and protein each have 4 kcal/g, whereas dietary fat has 9 kcal/g. Fat is an efficient way of storing energy in the body, for example, it is hydrophobic and therefore requires less water for storage than does either protein or carbohydrate. Body fat provides insulation against temperature extremes and protects vital organs from physical trauma. In food, fat functions as a carrier of flavor components and helps to tenderize food. High-fat foods are associated with rich flavor and high overall palatability. Dietary fat is a carrier of fat-soluble vitamins A, D, K and E and facilitates their absorption. Vitamins A and D are found predominantly in butter and fish oils. Vegetable oils contain vitamin E. Cellular lipids are also important as structural components of cells. Phospholipids (PL), which form an interphase between water and other lipids, serve a vital role in cells and blood by binding water soluble compounds such as protein to a lipid-soluble substance. Furthermore, the PL of the outer cellular membrane can undergo lysis by various phospholipases and release polyunsaturated fatty acids which are the precursor fatty acids for the biosynthesis of eicosanoids. Some lipids are also precursors for steroid hormone synthesis.

Dietary lipids consist mainly of triglycerides (TG). TG are made up of three fatty acids (FA) esterified to a glycerol molecule. FA are classified according to a number of systems. They can be classified acccording to chain length: short (4 to 6 carbon atoms), medium (8 to 10 carbon atoms), long chain (12 to 18 carbon atoms), and very long-chain FA (20 carbons and longer). Each group with different chain length is metabolized differently. FA are also classified according to presence or absence of double bonds. Saturated fatty acids (SFA) contain no double bonds. Unsaturated fatty acids contain at least one double bond.

Unsaturated FA are further divided into monounsaturated (MUFA one double bond) and polyunsaturated (PUFA two or more double bond). Unsaturated FA can be classified according to the position of the first double bond counting from the methyl end of the carbon chain. If the first double bond starts at the 3 carbon, it is designated an omega-3 FA ($n-3$). If the first double bond starts at the 6 carbon, it is designated an omega-6 FA ($n-6$). Different physiological functions have been ascribed to each of these series.[10,11] The FA of naturally occurring lipids have even-number carbon atom chains (C) with a typical length of 16 and 18. Notable exceptions are milk fat and coconut oil with a high percentage of short-chain (4 to 6) and medium-chain (6 to 12) carbon atoms, and palm oil, a C14 saturated FA. On the other side of the range are fish oils with 4–35% C20:1, C22:1, and C24:1; and with 25–50% of very long PUFA C20: 5$n-3$, C22:6 $n-3$, and C24:6 $n-3$ all cis. Olive oil, one very important factor in the Mediterranean Diet (MED), has 63–83% of C18:1 $n-9$, and 3–14% of polyunsaturated C18:2 $n-6$.

Animals, including man, cannot synthesize certain FA, termed essential FA. They are defined as FA that the body cannot synthesize in amounts adequate for optimal health. Linoleic acid, arachidonic acid, and the $n-3$ PUFA, (-linoleic acid, are all considered essential to maintain health. They serve as dietary precursors of the formation of eicosanoids, and are thus of great significance in health and the modulation of disease conditions.[12,13]

5.3 Sources of dietary lipids

Genetic and climatic differences are responsible for a wide variation in the composition of vegetable oils. The composition of animal feed determines to a great extent animal fat composition, especially that of nonrumiants. Most unsaturated FA have the double bonds in the cis geometric configuration. Trans-FA are found in ruminant fats as a result of bacterial action in the rumen, and in shortenings and spreads produced during the hydrogenation of oils.[14] Table 5.1 provides information on FA composition of different types of lipids used in human foods, obtained from the US Department of Agriculture and may be accessed from the Home Page (HYPERLINK http://www.nal.usda.gov/fnic/foodcomp).[15] Edible fats and oils also contain sterol and phospholipids (PL) which are integral parts of all animal and plant biomembranes. Most PL are derivatives of glycerol, with FA on the sn1 and sn2 positions and phosphorylcholine, phosphorylethanolamine, phosphorylserine, or phosphorylinositol on the sn3 position. The PL from vegetable sources, although absorbed as well as those from animals, normally have a different FA composition. During the processing of vegetable oils, most of the PL and sterols (phytosterols) are removed for technological reasons as well as because of taste. However, virgin olive oil and other lipid-containing foods have varying amounts of these compounds. These phytosterols diminish absorption of CH. This effect is presumably the result of competition with CH for incorporation into micelles or for transport across the intestinal cell membrane. CH is not considered a nutrient because it can be synthesized in the body. Dietary CH, however, is critical for optimal body function. It is used as a substrate for sex hormones, bile acids, and

Table 5.1 Different types of fatty acids in human foods

Systematic name		Common name	Typical source
Saturated fatty acids			
4:0	Butanoic	Butyric	Butterfat
6:0	Hexanoic	Caproic	Butterfat
8:0	Octanoic	Caprylic	Coconut oil
10:0	Decanoic	Capric	Coconut
12:0	Dodecanoic	Lauric	Coconut
14:0	Tetradecanoic	Myristic	Butter, Coconut
16:0	Hexadecanoic	Palmitic	Most fats, oils
18:0	Octadecanoic	Stearic	Most fats, oils
20:0	Eicosanoic	Arachidic	Lard, peanut oil
22:0	Docosanoic	Behenic	Peanut oil
24:0	Tetracosanoic	Lignoceric	
Unsaturated fatty acids			
10:1 n-1	9-Decenoic	Caproleic	Butterfat
12:1 n-3	9-Dodecenoic	Lauroleic	Butterfat
14:1 n-5	9-Tetradecenoic	Myristoleic	Butterfat
16:1 n-7t	*trans*-Hexadecenoic	Palmitelaidic	Hydrogenated oil (HO)
16:1 n-7	9-Hexadecenoic	Palmitoleic	Fish oils
18:1 n-9	9-Octadecenoic	Oleic	Most fats, oils (olive)
18:1 n-9t	*trans*-Octadecenoic	Elaidic	Butterfat, beef, HO
18:1 n-7	11-Octadecenoic	Vaccenic	Butterfat, beef
18:2 n-6	9,12-Octadecadienoic	Linoleic	Most vegetabe oils
18:3 n-6	6,9,12-Octadecadienoic	Gamma-linoleic	Borage oil
18:3 n-3	9,12, 15-Octadecadienoic	Alpha-linoleic	Soybean, canola oils
20:1 n-11	9-Eicosaenoic	Gadoleic	Fish oils
20:1 n-9	11-Eicosaenoic	Gondonic	Rapeseed oil
20:3 n-6	8,11,14-Eicosaenoic	Dihomogamma -linoleic	
20:4 n-6	5,8,11,14-Eicosatetraenoic	Arachidonic	Meat, fish oil
20:5 n-3	5,8,11,14,17-Eicosapeutaenoic	EPA, timnodonic	Fish oils
22:1 n-9	13-Docosaenoic	Erucic	
22:4 n-6	7,10,13,16-Docosatetraenoic	Adrenic	Brain
22:5 n-3	7,10,13,16,19-Docosapeutaenoic		Brain
22:5 n-6	4,7,10,13,16-Docosapeutaenoic	DPA, clupanodonic	Fish oils, brain
22:6 n-3	7,10,13,16,19-Docosahexaenoic	DHA, cervonic	Fish oils, brain

vitamin D. It is also essential for the function of cellular membranes and the structure and function of lipoprotein particles.[11]

5.4 Digestion and absorption

TG and PL, the two major groups of ingested lipids, are poorly absorbed and require enzymatic conversion into more water-soluble and polar metabolites for uptake by the gut mucosa. Emulsification in the stomach is mediated by a combination of mechanical and physicochemical mechanisms. Chewing releases fat from other food components and gastric contractions expose the lipids to lingual lipase. Lingual lipase hydrolyzes approximately 30% of ingested TG after a meal, in combination with gastric lipase.[16] The two lipases have similar properties, including a pH optimum in the range of the physiologic postprandial gastric pH. Both lipases have a preference for cleaving the FA at the sn-3 ester linkage. Short- and medium-chain FA are the preferred substrates and these are partially absorbed from the stomach. The postprandial luminal content of the stomach therefore contains TG, diglycerides (DG), PL and FA. Emulsification is enhanced by dietary PL and hydrolyzed FA. In emulsified particles, TG and DG are located in the center of the small droplets with a monolayer of PL and FA on the outside. Intermittent delivery of gastric chyme in small quantities to the duodenum by gastric peristalsis, and intermittent relaxation of the pyloric musculature, facilitate further digestion and absorption in the small bowel.[11]

TG and DG will not permeate the absortive mucosal membrane of the small bowel. Only 2-monoglycerides (2-MG) and nonionized FA can pass through the membrane by diffusion as free monomers in the aqueous phase adjacent to the wall cell of entherocite.[17] This absorption can be accomplished only if the following conditions are met:

- There must be more complete hydrolysis to MG and FA than occurs in the stomach.
- The surface area of TG droplets is enlarged by detergents for faster enzymatic hydrolysis.
- Because MG and FA are poorly soluble in water, a solubilizing transport system is required.

The gastric chyme induces the release of cholecystokinin (CCK) and secretin from the duodenal mucosa into the circulation. CCK stimulates primarily the synthesis and release of exocrine pancreatic enzymes. To a lesser degree, the release of electrolytes also induces sustained gallbladder contraction and the synthesis and release of hepatic bile, containing bile salts, PL, and CH. Secretin is the physiologic stimulant for release of electrolytes (mostly $NaHCO_3$) and to a minor degree intestinal digestive enzymes.[11]

The mixed micelles consisting of bile salts (BS), PL, and CH present in bile have a strong affinity for the surface of the emulsified lipid droplets, thereby displacing lipase from its substrate. However, lipolysis is effective because procolipase is also released by the pancreas simultaneously with lipase in a ratio 1:1. In presence of TG or FA, colipase complexes firmly with lipase and also binds to the surface of lipid droplets. Thus, colipase gives lipase access to its substrate. Micellar aggregates, as present in bile, are highly efficient in absorbing the 2-MG and FA released by the pancreatic lipase from the surface of the TG droplets.[16] The rate at which pancreatic lipase hydrolyzes FA from positions 1 and 3 of TG depends on a number of physical and chemical characteristics of the FA; in general, the longer the chain length, the slower is the release. The degree of unsaturation itself seems to have minimal influence.

Short- and medium-chain FA are rapidily released. However, in the long term, the digestion of almost all dietary TG is complete.

The final step in the digestive phase of absorption is the uptake of free monomers of FA and 2-MG by passive diffusion from a water phase located between the micelle and cell membrane. This is a rate-limiting step. The process by which long-chain FA leave the mixed micelles and traverse the gut wall into enterocytes is still unclear. Dietary PL play only a minor role in the digestive process, because the average diet contains approximately only 2g. Pancreatic phospholipase A2 hydrolyzes the FA primarily at position 2, with release of a free FA and lysophospholipid (i.e. lysophosphattidylcholine) which diffuse into the mucosal cells. Dietary cholesterol is dissolved into mixed micelles in the gut, but biliary CH is solubilized in part into micelles and in part into PL vesicles. Vesicles then dissolve into the mixed micelles. Approximately one-half of the CH in the bowel lumen is of endogenous origin. Physiologic malabsorption of CH (50%) is related to its poor micellar solubility. Plant sterols diminish the absorption of CH in animals, presumably because of their competition with CH for incorporation into micelles and for transport across the intestinal mucosa. Although blood CH increases in response to a high-CH diet, most of the serum CH is of endogenous origin. Thus, the serum CH level is regulated mainly by endogenous synthesis. However, in the long run, a diet low in CH and saturated fat leads to significantly lower serum CH levels in a high percentage of the population.[11,18]

The maximum capability for fat absorption in adults is greater than that amount of fat present in an average meal. With an increasing load of fat in adults, absorption is completed somewhat more distally in the small bowel. Newborns, however, have no such reserve. For infants receiving mother's milk, fat excretion is similar to that in adults, but infants reared on cows' milk may have a certain degree of fat malabsorption for up to one year. In contrast to cows' milk, human milk also contains lipase resistant to gastric acid and pepsin. Elderly individuals also have a limited capacity for lipids absorption, but because their appetite also decreases, fat intake usually has decreased also. Maldigestion can occur during malnutrition or disease when the pancreas fails to secrete enough lipase, the liver fails to supply sufficient bile acids, or emulsification of food fats in the stomach is inefficient. Malabsorption can also occur, even when digestion is functioning normally, as the result of defects in the small intestine that affect the absorbing surfaces. It may also occur during severe bacterial infection of the gut or sensitization of the gut to dietary components, such as gluten in celiac disease, or to allergens. A major problem in severe fat malabsorption is essential FA deficiency.[11,18]

5.5 Transport and metabolism

The first step in mucosal transport is re-esterification. The second step is synthesis of transport particles, the so-called lipoprotein particles. Once the digestion products are inside the enterocyte, the cell has to take steps to protect itself from their highly disruptive detergent properties. This is accomplished by their attachment to a small molecular-mass, FA binding protein as soon as they enter the cell. Lipid digestion products are then reconverted into TG in the enterocyte by sequential esterification.[17] Most TG are resynthesized in the enterocyte by the monoacylglycerol pathway. The second pathway, accounting for 20% of enterocytic TG, is the α-glycerophosphate pathway.[11] However, the FA with chain lengths shorter than 14 C atoms are bound to albumin and preferentially transported directly to the liver by way of the portal vein. The medium-chain FA are not re-esterified and are metabolized rapidly. CH is esterified in the mucosa shortly before the chylomicron enters the lymph. Esterification is accomplished by incorporation of acyl-CoA into the CH molecule.[18]

The biological problem of how to transport apolar lipids in the predominantly aqueous enviroment of the blood has been solved by stabilizing the lipid particles with a coat of amphilic compounds; PL and proteins.[19] The protein moieties are known as apolipoproteins and have much more than a stabilizing role. They also confer specificity on the particles. This allows them to be recognized by specific receptors on the surfaces of cells in various body tissues and organs, thereby enabling them to be taken up from the blood as well as regulating their metabolism. The major lipoprotein classes are chylomicrons, very low-density lipoproteins (VLDL), low-density lipoproteins (LDL), and high-density lipoproteins (HDL).

Chylomicrons primarily shuttle dietary TG and other fat-soluble components from the gut to the peripheral tissue. These chylomicrons, which are secreted into the intestinal lymph, are large lipoproteins with lowest density (<0.95 g/ml). They are assembled in the enterocytes during fat absorption, all the components essentially having been synthetized or resynthetized *in situ*,[19] These consist of about 86% TG, 8.5% PL, 3% CH, and CH ester, and 2% protein.[11,20] The surface coat contains the structural apo-B-48 (the main apoprotein of chylomicrons), and the transferables apo C-II, apo C-III, apo E, apo A-I and apo A-IV in small amounts. In the peripheral circulation, the first port of call is the adipose tissue, where the chylomicrons, identified by their apo-B48 (a large hydrophobic protein), are depleted of some of their TG by the enzyme lipoprotein lipase (LPL). LPL is an enzyme synthesized in adipose tissue and skeletal muscle and anchored to the capillary endothelium of these tissues, which take up much of the FA released. Apo C-II acts as the activator for the interaction between the chylomicrons and the LPL. This process is repeated until most of the TG are hydrolised, leaving the chylomicron in smaller particles called remnant particles. Undegraded remnant particles are then picked up by the liver for further processing.[20,21] Normally, the concentration of chylomicrons in plasma rises quickly after a meal, reaches a peak two to three hours after a meal, and then subsides. Chylomicron remnants are removed from plasma in two ways. The preferred route is by interaction with receptors on the surface of liver cells for further degradation in the liver. The apo B-48 on the remnant surface binds to a site on a liver cell by a process involving apo E to effect remnant uptake. Alternatively, the remnants can be processed by interaction with another type of plasma lipoproteins, high-density lipoproteins (HDL). Genetic defects in any of the chylomicrons apoproteins lead to failure to clear chylomicrons from the blood and hyperlipemia occurs.[20,21]

Another type of TG-rich lipoprotein, very low-density lipoproteins (with density between 0.95–1.006 g/ml), carry lipid that has been synthesized in the liver, mainly from circulating nonesterified fatty acids. VLDL comprises TG (50%), PL (18%), CH (6%), CH ester (15%) and various lipoproteins.[11] The major protein component of newly synthesized VLDL particles is the structural apolipoprotein-B100, also containing the transferables apo E and apo C. CH and CH-ester may be derived from exogenous sources or may be produced *de novo* in the liver. 3-Hydroxy-3-methylglutaryl coenzyme A reductase (HMG-CoA red) is a rate-limiting enzyme in CH biosynthesis. FA of both exogenous and endogenous origin can be incorporated into TG and PL destined for incorporation into VLDL. Different FA appear to be specifically channelled into the synthesis of these lipids.[20] The incorporation of TG and CH into VLDL appears to be coordinately regulated. Thus, the metabolic rate of different FA in the liver may influence their effect on VLDL secretion.[22] The microsomal trygliceride transfer protein (MTP) also plays an important role in the incorporation of all of these lipids and apo-B100 into the VLDL particle. Once VLDL is secreted into the circulation it may undergo further modification with the transfer of apolipoproteins and lipids into and out of the particle. The transfer of CH-ester from the HDL fraction into VLDL plays an important role in the reverse CH transport pathway.[23] Much of the TG core of VLDL is then hydrolysed by LPL mediated by apo-C. On loss of TG and many of its surface components VLDL is released back into circulation as intermediate-density lipoprotein (IDL) which contains apo E and apo B100. IDL is removed either directly or indirectly. It is removed directly from the circulation via interaction with a Goldstein & Brown receptor (apo B100, E receptor) in the liver. Alternatively, it is further metabolized, probably by hepatic triglyceride lipase (HTGL) which makes loss remnant TG and apo E, to produce cholesterol-rich low-density lipoprotein (LDL).[20,22]

LDL (with a density between 1.019–1,063 g/ml) comprises approximately TG (8%), CH esters (35%), CH (7%), PL (25%), and apo B100 (25%).[11] LDL is removed from the circulation after interaction with the LDL (apo B, E) receptor in the liver. Lesser amounts are also taken up after interaction with receptors apo B, E in peripheral tissues. While virtually all tissues of the body make some contribution to LDL uptake, the liver is quantitatively the most important, accounting for more than 60–80% of LDL turnover. The LDL-receptor also plays a major role in determining the rate of LDL production as it binds not only

LDL itself but also IDL. Thus, when receptor expresion is high, more particles are removed from the circulation as IDL and less LDL is formed. Conversely, if the LDL-receptor is down regulated, in addition to reducing the rate of removal of LDL itself, fewer IDL particles are removed and hence more are converted to LDL. Transcription of the LDL-receptor gene is regulated by intracellular CH level.[24] Non-esterol mediated regulation of LDL-receptor transcription has also been demonstrated and may be tissue specific.[25] Plasma LDL cholesterol concentrations are potentially dependent on the rate of VLDL production, the rate of lipolytic degradation of VLDL and its remnants, transfer of CH directly to the LDL particle and the rate of receptor-dependent and independent removal of IDL and LDL itself, primarily by the liver.[22]

High-density lipoproteins (HDL), the most abundant lipoproteins in circulation, show much less dramatic fluctuations in concentration than chylomicrons. HDL exist in several forms and their role is to act as vehicles for lipid exchange. These lipoproteins have a density between 1.063 and 1.21 g/ml due to high content of apoproteins (approximately 48%). HDL tends to promote CH flux from peripheral tissue and is negatively correlated with coronary heart disease risk.[20] In addition, a greater percentage is distributed in the extravascular space compared to LDL. Originating from the liver and the intestinal tract as nascent discoid particles, HDL exchange surface lipids and apoproteins with chylomicrons and VLDL, particulary during lipolysis when these particles decrease in size by losing their TG by LPL. FA are transferred from phosphatidylcholine to HDL cholesterol to form CH- ester by the lecithin-cholesterol-acyltransferase (LCAT) system which moves CH-ester into the hydrophobic core. Thus more CH can be incorporated in the surface layer of the HDL particle to form CH ester-rich HDL. The change of composition of HDL facilitates uptake of CH from tissue membranes and transfer of CH and PL from other lipoprotein particles. The effectiveness of HDL as a CH acceptor is enhanced further by the action of a neutral exchange protein, which exchanges TG for CH-ester to form larger TG-rich HDL particles, a process termed neutral lipid exange.[20,21] The CH-enriched HDL becames susceptible to hydrolysis by circulating lipases and the particle has renewed capacity for CH uptake. The HDL particles function so effectively because they have a relatively large surface area, and because their apoproteins enhance contact with other particles' membranes and lipophilic enzymes. Approximately 25% of HDL is catabolized by the liver. Other organs with a high requirement for CH, such as the adrenals and ovaries, also have a high uptake of HDL.[11,20]

Fat from the diet is primarily used for energy. Some fat is synthesized from carbohydrate and protein as it is metabolized in the liver. If there is little lipid intake in the diet, a large proportion of fat is synthesised from carbohydrate for storage. The FA released from chylomicrons and VLDL in adipose tissue are activated by forming coenzime A derivatives and transferred to glycerol-3-phosphate to form TG. To be made available for energy use, FA are released from adipose tissue by action of a hormone-sensitive lipase. The lipase is activated by adrenaline, glucagon and adenocorticotrophic hormone. FA are released into the circulation where they bind to albumin. FA are oxidized in mitochondria.[14] The process of oxidation of FA for cellular energy known as β-oxidation is a controlled breakdown of FA to yield metabolic energy. During this process, adenosine triphosphate is generated.[21]

Animal tissues contain desaturases, enzymes that insert double bonds, normally beginning at position 9, into saturated FA. Also present are enzymes that catalyze the insertion of more double bonds to produce polyunsaturated FA, but this can occur only at positions between the first double bond and the carboxyl group. Further elongation and desaturation result in PUFA with up to twenty-two carbon atoms and six double bonds, further metabolized to biologically active compounds termed eicosanoids.[21] The rate-limiting step in the formation of eicosanoids seems to be the release of free arachidonic acid from the cell membrane PL mediated through activation of phospholipase A. Arachidonic acid, through the role in the eicosanoid synthesis, is an important mediator of a number of physiological phenomena. The key enzymes in controlling the extent of eicosanoids biosynthesis are ciclooxygenases, key enzymes in prostaglandins (PG) and thromboxanes (TX) biosynthesis, and lipoxygenases that form a series of products named leukotriens (LT). The eicosanoids possess the most potent and most diverse biological activities of any naturally occurring compounds. Eicosanoids are powerful autocrine and paracrine regulators of cell and tissue functions. These functions include: thrombocite aggregation, inflammatory reactions and leukocyte functions, vasoconstriction and vasodilation, blood pressure, bronchial constriction, etc.[14] From a pathophysiological viewpoint, an absolute or relative deficiency of PG relative to TX has been implicated in the aetiology of hypertension, thrombosis and atherogenesis.[26]

5.6 Health issues relating to fat and oil intake

Dietary fat intake and its effects on human health is a hotly debated issue in nutrition research and practice today. Since the beginning of the twentieth century, the amount of fat in the human diet has been increasing. How much and what type of fat should be consumed by individuals are subjects of controversy. Diet can influence the regulation of lipid metabolism in several ways.[21] One way is by increasing or restricting the availability of molecules that are substrates for the enzymes in a lipid metabolic pathway. Another is the supply of a coenzyme (the enzymes of FA biosynthesis, for example, require pantothenic acid and biotine as coenzymes). Deficiency of some vitamins, therefore, leads to defects in lipid biosynthesis, which can have profound effects on health. An important effect of diet in the regulation of lipid metabolism is to bring about changes in the concentrations of specific circulating hormones, particularly insulin, that induce or repress the biosynthesis of some important receptors and enzymes of lipid metabolism (i.e. LPL).[27,28] Currently, there is much research into the ability of different FA and other lipids to regulate the expression of various genes.[22] Humans in developed countries generally consume diets containing a high proportion of energy as fat. Under these conditions, the enzymes of fat biosynthesis are switched off or work at a low level, and the need for both structural and storage fats is satisfied from dietary intake. However, FA synthesis from carbohydrate, at least in adipose tissue, is an important process in humans, even when there is little fat in the diet.[21]

There is substantial variability in dietary lipids availability between and within major world regions. Two-thirds of the world's population, mostly in developing countries, have access to <60g lipids/d per person, whereas inhabitants of most developed countries have access to >130g/d per person.[29] FAO data on dietary lipids availability by source in the various world regions are shown in Table 5.2.[30] Among visible dietary lipids, vegetable oils have been gradually assuming greater importance, with North America showing the largest absolute increase in availability per capita over the past 30 years. The gross patterns evident in Table 5.2 hide considerable within-region variability by country, locality, and type of dietary lipids. With respect to per capita dietary lipids intake at a national level, countries can be considered to fall into one of four patterns.[29]

- Pattern 1: high total dietary lipids, high intake of mostly saturated animal lipids (e.g., most countries in Europe, North America, and Oceania)
- Pattern 2: high total dietary lipids, low intake of animal lipids (e.g., most Mediterranean countries)
- Pattern 3: moderate total dietary lipids, low to moderate animal lipids (e.g., countries of the Near East and South America)
- Pattern 4: low total dietary lipids, low animal lipids (e.g., most countries in Africa and the Far East).

Pattern 1 is more clearly associated with high incidences of cardiovascular disease and several forms of cancer. Pattern 2 is characterized by low incidences of these diseases but is still compatible with the dietary preferences of the

Table 5.2 Dietary lipids availability according to source in major world regions[30] (Data from the FAO)

Region	Total dietary lipids available g/d. person
Africa	43
Far East	45
Near East	72
South America	75
Former Soviet Union	107
Oceania	138
Europe	143
North America	151
World	68

Western world. Patterns 3 and 4 are distinguished by their low cardiovascular and cancer rates but may not be compatible with the long-established dietary preferences of populations in developed countries. Over time, intake of animal lipids tends to decrease in pattern 1 countries, whereas in most of the other populations, consumption of saturated animal lipids and polyunsaturated plant lipids, the latter frequently partly hydrogenated, tends to increase.[29] The American Heart Association reports[31] and the USDA Dietary Guidelines for Americans[32] have all recommended that total dietary fat intake make up 30% or less of total energy intake for individuals over the age of two years.

Some dietary saturated FA (SFA) raise serum CH concentrations. These include palmitic acid, myristic acid, and lauric acid. These three FA, however, make up about two-thirds of the SFA in the American diet. In contrast, another SFA, stearic acid, does not raise serum CH, and it is uncertain whether the medium-chain SFA do or not.[33] The current data suggest that SFA exert their effect on CH-raising by suppressing the expression of LDL-receptors in the liver.[34] For the general population it seems reasonable and practical to consider a reduction of the contribution of SFA to total energy intake to 7–8%.[33]

Results from human feeding studies and large-scale epidemiologic surveys suggest that dietary trans-FA enhance the risk of developing coronary heart disease.[35] Different authors have shown that trans-FA raises LDL-cholesterol concentrations about two-thirds as much as does palmitic acid,[33] and they may have a small HDL-lowering action as well. The adverse effect of trans-FA on the ratio of total CH to HDL cholesterol is higher than that of SFA.[36] In addition, intake of trans-FA increases plasma concentration of lipoprotein-(a), another risk factor of coronary artery disease.[37] Moreover, they may compete with essential FA for elongating and desaturating enzymes and thereby interfere in the formation of eicosanoids.[38] On the basis of these metabolic effects, Ascherio and Willett (1997) estimate conservatively that 30,000 premature deaths each year in the United States are attributable to consumption of trans-FA. Because there are no known nutritional benefits of trans-FA and clear adverse metabolic consequences exist, prudent public policy would dictate that their consumption be minimized.[36]

The $n-6$ and $n-3$ PUFA are essential nutrients. Intake of relatively small amounts of these FA prevent nutritional deficiences. The $n-6$ PUFA, predominantly linoleic acid, lowers total CH concentrations in all of the lipoproteins fractions (VLDL, LDL and HDL). Although linoleic acid lowers LDL-cholesterol concentrations slightly more than does oleic acid, high intakes of linoleic acid can promote chemical carcinogenesis and suppress the immune system, which is not true for oleic acid.[39] Limited epidemiologic data further suggests that high linoleic acid comsumption can increase the risk of human cancer. Finally, linoleic acid enriches membrane PL and predisposes them to free radical oxidation. This could lead to harmful effects. An example is the increased susceptibility of LDL to oxidation associated with high intakes of linoleic acid.[40,41] Therefore, consumption of PUFA probably should not exceed current intakes, 7–10% of total energy.[33] Below this ceiling there is little evidence that high dietary intake of PUFA implies health risks.[12] However, when $n-3$ FA are included in the diet, EPA (C:20; $5n-3$) inhibits conversion of linoleic acid to arachidonic acid and competes with arachidonic acid for the 2-acyl position in membrane PL, reducing plasma and cellular levels of arachidonic acid. In addition, EPA competes with arachidonic acid as the substrate of cyclooxygenase, inhibiting PG and TX. EPA inhibits the synthesis of TXA2, the PG that causes platelet aggregation and vasoconstriction.[13] Moreover, dietary $n-3$ PUFA act to prevent heart disease through a variety of actions.[12] They prevent arrhythmias, inhibit synthesis of cytoquines and mitogens, stimulate endothelial-derived nitric oxide, and inhibit atherosclerosis. Diets abundant in fatty fish also lower plasma VLDL and TG concentrations through depression of syntheis of TG in the liver and suppress postprandial lipemia. Effects of $n-3$ FA on LDL and HDL have been variable.[42] The intake of $n-3$ FA recommendations for primary prevention to coronary artery disease could be 2-3g/d of fish oil. This will occur if there is regular consumption of 200–300g fish and shellfish/wk.[43]

Converging evidence from scientific fields as varied as biochemistry, clinical, and epidemiology have led the scientific community to reexamine the role in health of monounsaturated (MUFA) lipids. Oleic acid appears to be the most important FA in the dietary prevention of atherosclerosis, both lowering CH concentrations and minimizing oxidative damage to LDL.[40] Some workers have identified carbohydrate instead of oleic acid as the neutral nutrient. However, carbohydrates seem to affect lipoprotein metabolism entirely differently from fats. They enhance VLDL concentrations by enriching VLDL particles with TG. They reduce LDL cholesterol concentration by reducing the CH content of LDL, not by reducing the number of circulating LDL particles as dietary oleic acid does. Furthermore, carbohydrates reduce HDL-cholesterol concentrations whereas oleic acid does not.[33,44] Higher intakes of oleic acid at the expense of carbohydrate are well tolerated in the Mediterranean diet (MeD) where rates of both coronary disease and cancer are relatively low. The same, however, is true for populations consuming low-fat, high carbohydrate diets. A reasonable compromise for the general public that considers both practicality and safety is to limit total fat to 30% of total energy. Such a diet will allow for an intake of oleic acid of 15–16%.[33]

5.7 The role of deep-frying in the fat intake

In the past, dietary guidelines in relation to fat intake have been dominated by two principles: a reduction in dietary lipids intake and an increase in the ratio of polyunsaturated to saturated FA. However, the Mediterranean diet (MeD) can be considered an example of how ideas about eating habits are changing. Until recently, the diet of Mediterranean people did not enjoy much prestige. At a time when height was considered to equate with optimal health, the average height of the Mediterranean people was believed to be indicative of poor diet. Some staple foods in that diet, such as olive oil, did not have a good reputation. Nor were some of the cooking methods properly understood, such as deep-frying, one of the characteristics of the MeD.[7]

The differences in the fat intake of the Mediterranean countries and more northern countries are not due merely to the composition of the fat consumed. It is also a question of how this fat is consumed. One good example is Spain,[45] where total fat intake is high at 54.9g/person d (per day). This is on a par with other developed countries. However, 95% of the intake is in the form of vegetable oils, mainly because of the widespread use of deep frying. Olive oil (32.9g/person d), figuring prominently among these oils, accounts for 60% of total intake, whereas aggregate consumption of corn, sunflower, and soybean oils amount to 19g/person d. In recent years, there has been a slight drop in the consumption of olive oil, which has been replaced by other oils. The use of butter (0.86g/person d) and margarine (1.99g/person d) is extraordinarly low. From a health point of view, the quality of Spanish fat intake is excellent.[7] The lipids profile lies within recommended limits, although it must be remembered that it is assessed in the context of a high-fat diet. The high intake of MUFA and the fairly moderate intake of SFA and PUFA play a fundamental part in the quality of this diet.[7] One particularly interesting feature of fat intake in the MeD is the high percentage of total fat provided by cooking (e.g., oils in the fying process). As is known, fat intake is made up of two basic components: the fat contained in food, and the fat used to prepare food. In MeD roughly 50% of total fat intake comes from cooking fat. This is an advantage in that it gives great scope for adjusting lipid intake, something which is not possible in other countries where cooking fat accounts for a much smaller proportion.[7] This prompts a question. How is this cooking fat, which is mainly derived from deep-frying, absorbed by the body?

First confined to the Mediterranean, deep frying (DF) has spread to all parts of the world. Fried food was traditionally said to be indigestible and there was even talk of toxicity.[46] Nowadays, the situation is different. Frying is now one of the most important unit operations in the developed countries, be it in the home, catering or the food-processing industries. This is due in part to studies which have confirmed that deep frying is less nutritionally damaging to foods than other methods, and to epidemiological findings on the MeD. However, another reason for its great popularity is that frying also generates very palatable foods and needs short preparation time.[47]

DF is a extremely complex process and depends on numerous factors, some of which are dependent on the process itself, and others on the food and the type of fat used.[48] It is mainly a dehydration process in which part of the water in the food is replaced by cooking fat. The high temperature of the oil (around 180ºC) allows rapid heat transfer and very short cooking time, but the temperature inside the food does not usually exceed 100ºC. If frying is done correctly (especially as regards frying times and temperatures, food surface/volume ratio and food weight/cooking fat volume ratio), when the hot fat starts to penetrate the food a crunchy crust forms on the outside. This prevents large amounts of fat from passing through to the inside.[47] 'Crunchiness' is a quality attribute in specific foods.[49] It is attracting more and more attention because of the role it plays in palatability and acceptance (one only has to think of the current success of some snacks and breakfast cereals whose crunchiness is one of the most important characteristics of their acceptance). In summary, some of the key characteristices of DF are as follows:[7,48]

1. The actual length of time that the hot cooking fat acts on the food is quite short. The entire frying process lasts only a few minutes. Most of that time corresponds to the phase when the water is evaporating. During this phase the temperature inside the food is around 100ºC. The hot cooking fat really acts on the food for an extremely short period of time, and that action is limited to the outside of the food.[50] Another point to remember is that the process takes place virtually in the absence of oxygen. The end result of these two factors is that DF affects the thermolabile components of the food less aggressively than do other culinary processes. As an example, vegetables (potatoes and peppers) retain much more vitamin C when they are fried than when they are cooked by other methods such as sautéeing or stewing.[51]
2. Fat is one of the strongest influences on the palatability of fried foods. The exchange of water for cooking fat and the crusty surface formed in the frying process create a 'crunchy' quality that can be very important for the consumer.[52]
3. If frying is done properly, the amount of fat consumed in fried foods is no greater than when other cooking methods are used, because of the distinctive kinetics of penetration during frying. The use of batter and breading when frying fish and meat produces fried foods with an increased energy densitity.[53]
4. Important quantitative and qualitative changes occur in the fat composition of the food during the frying process. As a result, the real lipid intake when consuming fried food differs greatly from lipid intake when eating raw food.[48,54]

This last point must be kept in mind when considering the influence of frying on the real lipid profile of a diet. Frying modifies the lipid profile of the food, depending on whether lean or fatty foods are fried. In the case of lean food, the cooking oil penetrates the food, enriching it with fat. The fat composition of the fried food will be virtually that of the cooking fat.[8,55] Plant foods which initially have high water and low fat content, absorb more frying fat than animal foods. In addition, unlike the intercellular spaces of animal tissues which are filled with fluid, those of plant tissues are filled with air and this results in a greater capacity of plant foods to retain absorbed fat.[56] Since much of the fat taken up by the fried food is located on the surface and in the crust, size and geometry are important features to take into account when considering fat uptake. Greenfield *et al.*[57] found that decreases in French fries size significantly increased fat content of fries, with a linear relationship between size decrease and increase in fat content.

Matters become much more complex in the case of fatty foods such as meat or fish. From the quantitative point of view, the quantity of fat that passes from the food to the cooking fat and vice versa is usually almost the same. No great difference exists between the total amount of fat in fried or raw foods, but qualitative changes seem to depend, to a large extent, on the differences between the concentration gradients of the different FA in the cooking fat and food.[7,8] When sardines are fried in olive oil or sunflower oil, the concentration of SFA (lower in these oils) decreases in the fish due to the exchange between food and cooking fat. Moreover, if the sardines are fried in olive oil, there is a notable increase in their MUFA concentration as olive oil has a higher content of these FA than the fish. Conversely, if the fish is fried in sunflower oil, its concentration of total PUFA increases, especially the w-6 fatty acids, because sunflower oil is particularly rich in this fraction. Both oils cause a decrease in the *n*–3 PUFA fraction of the fish. However, this drop is smaller when using olive oil and greater when using sunflower oil.[58,59] In contrast, in pan-fried breaded trout fillets, there was a marked increase in fat content due to the absorption of frying oil by the breading. However, changes in the FA composition of the trout fillets *per se* were negligible, while the FA composition of the breading resembled that of the frying oil.[56] Consequently, from a nutritional point of view, great care should be taken regarding the FA composition of the frying fat. Olive oil rich in MUFA will give a food rich in these FA, whereas foods fried in animal fats will be enriched in SFA and CH.[8]

5.8 The impact of repeated frying

In DF the oil is usually reused several times to fry new portions of the same or different foods. This question of repeated frying (RF) is interesting and ties in closely with the 'useful life' of the different cooking fats. It is not easy to determine this 'useful life' because it depends on many factors, especially the fat composition of the food and of the type of oil used. Olive oil and hydrogenated fats are acknowledged to be much more stable than PUFA-rich oils.[59,60]

An important factor in RF is the number of times that the cooking fat can be reused without affecting the quality of fat intake. At some point in time, the oil used in RF can no longer be used and has to be discarded. A very large amount of oil is discarded in this way depending, among other things, on the composition of the cooking fat and of the food.[61] Discarded oil does not have the same composition as the raw oil because it is enriched by the FA from the lipids in the food. As an example, if the discarded olive oil (cooking fat) has been used to fry meat, it contains such compounds as SFA. On the other hand, the FA composition of the food (e.g., meat, eggs) is improved considerably because MUFA that penetrates from the olive oil is very beneficial.[8] Losses of FA from the frying oils also occur. Tyagi and Vasishtha (1996) found changes in FA profile during frying in soybean oil, mainly among the PUFA (after 70h, 52% of linoleic acid was lost). This decrease could be attributed to the destruction of double bonds by oxidation, scission and polymerisation.

5.9 Measuring the impact of frying on fat intake

As has been noted, the information available on the relationship between fat intake and degenerative diseases comes from epidemiological or experimental studies in which fat intake is calculated according to the food consumed, using food composition tables or databases. In the majority of these tables, the fat composition cited is that of the raw food, but food is usually eaten after undergoing various types of processes. In these processes, major quantitative and qualitative changes occur.[62] Clearly, these changes in the quantity and fatty acid composition of FI must affect the way in which it is possibly linked with various diseases. Failure to take these changes into account could be the cause of the difficulties, and in many cases the contradictory results, encountered when studying this relationship between FI and health.

It is also necessary to take into account discarded frying oil in estimating real quantitative and qualitative fat intake when food consumption is studied on the basis of food records.[48,61] This has been done for various population groups in Madrid (Fig. 5.1). It was found that discarded frying oil (DFO) in the 96 households studied accounted for 19.3% of recorded oil consumption. This value of DFO has been used in studies of the nutritional status of the Spanish population in cooperation with the National Statistics Institute.[45,63] According to food records, average oil consumption per person per day in the households

SAMPLE CHARACTERISTICS

Households: 96
Two districts of Madrid (different socioeconometric levels)

1. Salamanca
2. San Blas

Institutions: 13

- 4 kitchens of public hospitals (540 mean meals/day)
- 5 public schools (225 mean meals/day)
- 4 restaurants
  - 2 conventional (100 and 200 mean meals/day)
  - 2 fast-food (250 and 800 mean meals/day)

Fig. 5.1 Study of the quantity and quality of raw and discarded culinary oils used by the population of Madrid.

Table 5.3 Mean content of polar compounds in discarded frying oils from homes and catering services (g per 100g)

	Schools	Hospitals	Restaurants	Households
Total polar content (g/100g)	9.6±2.8	10.5±3.0	19.9±12.6	6.6±2.6 [ab]
Triacylglycerol polymers (%) [c]	5.9±2.4	5.6±1.7	15.8±12.2	3.2±1.4 [ab]
Triacylglycerol dimers (%)	30.7±4.9	31.4±3.9	31.8±6.3	16.9±6.6 [abd]
Oxidized Tri-acylglycerols (%)	38.8±6.9	41.4±6.0	28.2±6.3 [ab]	27.2±4.3 [ab]
Diacylglycerols (%)	17.9±3.4	15.7±4.5	19.5±9.2	40.8±7.2 [abd]
Fatty acids plus polar unsaponifiable (%)	6.6±1.7	5.8±1.2	4.5±1.9	12.6±5.9 [abd]

Mean ± standard deviation
$p<0.05$ Student's t test
[a] Significantly different to schools
[b] Significantly different to hospitals
[c] Percent of total polar
[d] Significantly different to restaurants

studied was 62.9g. After correction to deduct the oil discarded and not therefore consumed, real average oil intake amounts to 53.0g per person per day. This correction means a change in calorie intake from 2212 to 2120 kcal/person/day and in total dietary fat intake from 121 to 110 g/person/day. When the fatty acid composition of DFO is taken into account, not only the quantity of fat intake differs with respect to the data in the household food records but the quality of intake is also different. However, in catering services the percentages of oil discarded were much higher than in households. In all the institutions studied (restaurants, hospitals and schools) the percentages are much higher, with averages of approximately 26% for frying pans and 42.4% for industrial deep-fryers. In most institutions, however, food is fried in deep-fryers. Frying pans are used only for special dishes.

The FA composition of the raw and discarded oils showed no significant differences. Table 5.3 shows the content of degradation compounds in frying oil discarded by households and institutions. The total polar content and polymeric materials in the discarded oil from homes and catering services is below the maximum limit of 24% and 12% respectively suggested for human consumption of fried foodstuffs in the recommendations of the Deutsche Gesellschaft für Fettwissenschaft.[64] However, the results are different when food such as fish, which has a high content of unsaturated fats, is fried. Dobarganes and Marquez-Ruiz[65] found oils with an oxidative deterioration of more than 24% when they analyzed samples from fried-fish shops. These results indicate that all the oils discarded after domestic frying still would be usable and that, with exceptions, those discarded after frying in catering services also could be reused. These comments naturally lead on to the need to broaden studies of this kind. Such

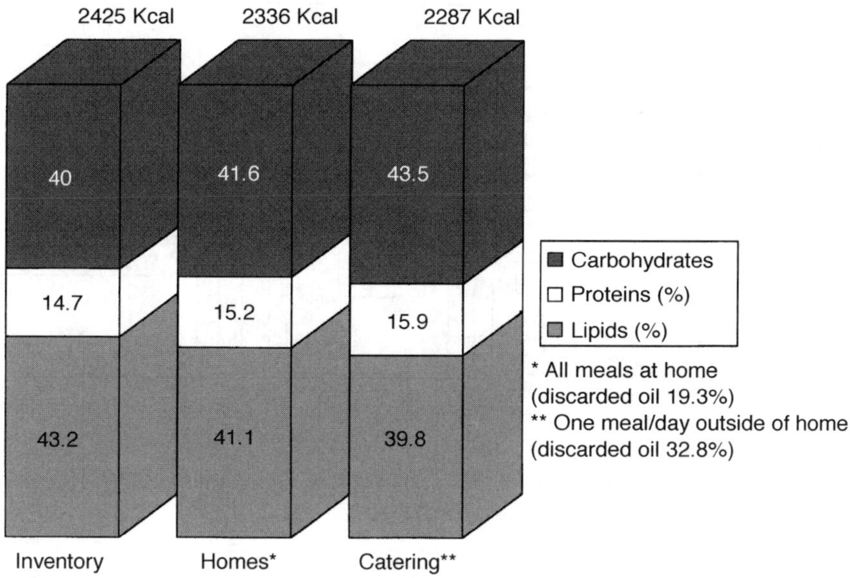

Fig. 5.2 Differences in the total calorie intake and the mean calorie profile, depending on whether or not data on discarded frying oil are considered.

studies should not include only the factors mentioned.[66] As an example of the practical repercussions of these results, Fig. 5.2 shows how the calculation of total energy intake and the average calorie profile of the Madrid population are affected by whether or not discarded oil is taken into account. According to data from Varela *et al.*,[45] consumption is theoretically 2425 kcal/person d. When the oil discarded in households is taken into consideration, consumption is lowered at 2336 kcal/person d. If one meal a day is eaten out, the value is even lower, at 2287 kcal/person d. When DFO is taken into account, total fat intake is, therefore, lower. Dietary protein and carbohydrate intake are also modified, although not to such a marked extent.

A recent study monitored the real fat intake of a group of 28 students, between 20 and 25 years, according to the precise weighed method, and performed chemical analysis of duplicate portions and of any food left on their plates. At their lunch during six consecutive days the students ate fried food and food prepared by other methods provided by a catering business in Madrid.[48] Figure 5.3 shows that fried food of the meals included in the different menus had (in dry matter) less fat than the same products when stewed. Because all of them were prepared with olive oil, the FA composition of the dishes showed only small variations across the type of cooking method used. In looking at the real intakes of fat and other macronutrients of the participants in the study (Table 5.4) it was found that fat intake and the percentage of total calories provided by fat were significantly lower when the second dish on the menu was fried as opposed to being cooked by other ways. These differences were very marked in

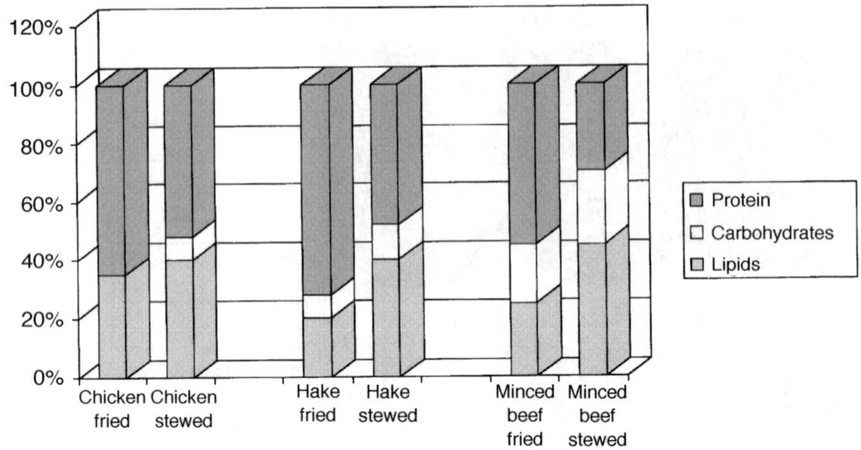

Fig. 5.3 Protein, carbohydrates, and fat content of fried and stewed foods (g/100g of dry weight).

the case of the two lean foods studied, chicken and hake. When they were fried, fat intake was almost half of what it was when they were eaten stewed.

5.10 Conclusions

The intake of dietary fats and oils, their types and amounts, have important effects on health. Over the past 30 years, animal lipids intake has tended to decrease in most developed countries and consumption of saturated animal lipids and PUFA plants lipids (the latter frequently partly hydrogenated) has tended to increase in most of the other populations. Long-term clinical trials indicate more positive effects from supplementation with MUFA-rich fats and also with $n-3$ PUFA than to diets low in total and saturated fats. The SFA raise LDL cholesterol concentrations by suppressing the expression of LDL receptors. The intake of SFA should be reduced to 7–8% of total energy intake. Low-fat, high carbohydrate diets lower plasma LDL but also lower HDL concentrations and raise plasma VLDL. In contrast, diets low in SFA but high in MUFA improve the ratio of HDL to LDL in plasma and thus reduce risk of coronary disease. A reasonable compromise for the population that considers both practicality and safety is to limit total fat intake to 30% of total energy, allowing for an intake of MUFA of 15–16%. The $n-6$ PUFA lowers total CH concentrations, enriches membrane phospholipids in linolenic acid and predisposes them to free radical oxidation. However, $n-3$ PUFA from fish oils inhibits the synthesis of thromboxane A2, VLDL productions, and apo-B synthesis. This will occur if there is regular consumption of 30–40g fish/d. Thus the intake of PUFA should not exceed 10% of total energy. Because there are no known nutritional benefits of trans FA and clear adverse metabolic

Table 5.4 Daily mean intake of macronutrients and energy by participants in the study of two cooking methods[48]

	Monday Fried hake	Tuesday Stewed mince beef	Wednesday Chicken fried	Thursday Stewed hake	Friday Fried mince beef	Saturday Stewed chicken
Meal (g)[a]	690.5±94.4	675.1±96.1	714.3±136.7	670.4±96.9	624.9±69.7	711±125.7
Rest (g)	85.0±45.1	138.7±58.8	125.2±53.1	112.1±46.4	120.7±69.1	192.8±63.0
Intake (g)	605.5±104.9	536.4±117.3	589.1±160.4	558.3±125.7	504.2±114.4	519.1±124.8
Intake (dry matter, g)	185.5±42.5	146.9±37.1	151.9±38.7	166.6±42.6	144.8±33.7	133.1±39.3
Fat (g)	19.8±4.6	46.1±12.4	21.4±9.5	32.9±8.6[b]	37.5±11.0	36.9±15.5[b]
Carbohydrates (g)	107.7±26.6	68.3±21.0	71.2±17.7	92.2±24.3	59.9±16.7	55.4±16.1
Proteins (g)	48.6±11.4	25.4±5.7[b]	47.5±15.1	31.0±9.0[b]	38.6±9.7	39.0±97.9
Energy (kcal)	776±179	772±184	650±186	765±193	761±170	696±229

[a] With reference to participant's daily lunch
[b] Significant difference with the fried food ($p<0.05$)

consequences exist, their consumption should be minimized and information on the trans FA content of foods should be available to consumers.

Deep-frying, until a few years ago, had little popularity outside the Mediterranean area. It is now one of the most important unit operations in both the catering and food processing industries. The major opportunities that frying with olive oil offers are the possibilities of lowering fat intake and improving the quality of fat actually consumed, which are important in preventing cardiovascular and other diseases. It has been experimentally confirmed that an average 19.3% of frying oil is discarded in households and that this figure is even higher in the catering services. Consequently, in a Mediterranean country, where a good proportion of total fat intake is from frying oil, total fat intake is significantly overestimated if the quantity of oil discarded after frying is not taken into account when studying dietary intake. In households and catering, frying oils are generally discarded before their fatty acid composition changes and the average values for their degraded compounds are well below the maximum 25% limit. These conclusions naturally confirm the need to carry on with this kind of study, working with the diversity of eating habits in both the Mediterranean area and in countries with other patterns of dietary lipids intake.

5.11 References

1. LISSNER L, HEITMANN BL. Dietary fat and obesity: evidence from epidemiology. *Eur J Clin Nutr* 1995 49 79–90.
2. ROLLS BJ. Carbohydrate, fats, and satiety. *Am J Clin Nutr* 1995 61 960S–67S.
3. OLIVER MF. It is more important to increase the intake of unsaturated fats than to decrease the intake of saturated fats: evidence from clinical trials relating to ischemic heart disease. *Am J Clin Nutr* 1997 66 980S–86S.
4. ERNST ND, SEMPOS CT, BRIEFEL RR, and CLARK MB. Consistency between US dietary fat intake and serum total cholesterol concentrations: the National Health and Nutrition Examination Surveys. *Am J Clin Nutr* 1997 66 965S–72S.
5. WYNDER EL, COHEN LA, WINTERS BL. The challenges of assessing fat intake in cancer research investigations. *J Amer Diet Assoc* 1997 97 S5–8.
6. MOREIRAS O, MUÑIZ J, CARBAJAL A, CAMPO M, VARELA G, GRANDE F. Estudio nacional de nutrición y alimentación 1991. Tomo II. Relación entre el consumo de alimentos, energía y nutrientes y la mortalidad por enfermedades degenerativas seleccionadas. Instituto Nacional de Estadística 1996. Madrid.
7. VARELA G. and RUIZ-ROSO B. Some effects of deep frying on dietary fat intake. *Nutr Rev* 1992 50 256–262.
8. VARELA G and RUIZ-ROSO B. Influence of the frying process in the real fat intake. *Grasas y Aceites* 1998 49 366–69.

9. LICHTENSTEIN AH, KENNEDY E, BARRIER P, DANFORD D, ERNST ND, GRUNDY SM, LEVEILLE GA, VAN HORN L, WILLIAMS CL, BOOTH SL. Dietary fat consumption and health. *Nutr Rev* 1998 56 S3–S28.
10. LAIRON D. Dietary fatty acids and arteriosclerosis. *Biomed Pharmacother* 1997 51 333–6.
11. LINSCHEER WG, VERGROESEN AJ. Lipids. In: *Modern Nutrition in Health and Disease*. Eighth edition 1994. Lea & Febiger. Philadelphia. 47–88.
12. ERITSLAND J. Safety considerations of polyunsaturated fatty acids. *Am J Clin Nutr* 2000 71 197S–201S.
13. JAMES MJ, GIGSON RA, CLELAND LG. Dietary polyunsaturated fatty acids and inflamatory mediator production. *Am J Clin Nutr* 2000 71 343S–348S.
14. DUPONT JL. Fats and Oils. Nutritional value. In: *Encyclopedia of Human Nutrition*. SADLER MJ, STRAIN JJ and CABELLERO B. Acad Press. San Diego, CA. 1999, 719–29.
15. US DEPARTMENT OF AGRICULTURE, AGRICULTURAL RESEARCH SERVICE USDA. Nutrient Database for Standard Reference, Release 11-1. 1997. Home page http://www.nal.usda.gov/fnic/foodcomp.
16. DESNUELLE P. The lipases. In: *The Enzimes*. Boyer PD. Academic Press. New York 1972 575–616.
17. LEHNER R, and KUKSIS A. Biosynthesis of triacylglycerols. *Prog Lipid Res* 1996 35 169–201.
18. GURR MI, and HARWOOD JL. *Lipid Biochemistry: An introduction*. 4th ed. Chapman & Hill. London.1991.
19. FIELD FJ, and MATHUR SN. Intestinal lipoprotein synthesis and secretion. *Prog Lipid Res* 1995 34 185–198.
20. BLAZQUEZ E. Estructura y clasificación de las lipoproteínas. In: *Metabolismo Lipídico*. Fundación Jiménez Díaz. Madrid. 1994 1–13.
21. GURR MI. The role of lipids in human nutrition. In: *Handbook of Olive Oil*. Harwood J, & Aparicio eds. Aspen Publishers Inc. Gaithersburg (Maryland). 2000 521–63.
22. SALTER AM, and WHITE DA. Effects of dietary fat on cholesterol metabolism: regulation of plasma LDL concentrations. *Nutr Res Rev* 1996 9 241–57.
23. FIELDING CJ, and FIELDING PE. Molecular physiology of reverse cholesterol transport. *J Lipid Res* 1995 36 211–228.
24. SANCHEZ HB, YIEH L and OSBORNE TF. Cooperation by sterol regulatory element-binding protein and Sp1 in sterol regulation of LDL receptor gene. *J Biol Chem* 1995 270 1161–69.
25. LIU JW, SHOYAB M, and GROVE RI. Induction of Egr-1 by oncostatin M precedes up-regulation of low density lipoprotein receptors in Hep G2 cells. *Cell Growth and Differentiation* 1993 4 611–16.
26. KATAYAMA S, LEE JB. Prostaglandins and leukotrienes. In: *Encyclopedia of Human Nutrition*. SADLER MJ, STRAIN JJ and CABELLERO B. Acad Press. San Diego, CA. 1999 1639–46.
27. KERN P.A. Potential role of TNF-alpha and lipoprotein lipase as candidate

genes for obesity. *J Nutr* 127 1917–22 (1997).

28. KERN P.A. SAGHIZADEH M. ONG J.M. BOSCH R.J. DEEM R. SIMSOLO R.B. The expression of tumor necrosis factor in human adipose tissue. Regulation by obesity, weight loss, and relationship to lipoprotein lipase. *J Clin Invest* 95 2111–19 (1995).
29. TRICHOPOLOU A, LAGIOU P. Worldwide patterns of dietary lipids intake and health implications. *Am J Clin Nutr* 1997 66 961S–4S.
30. US DEPARTMENT OF HEALTH AND HUMAN SERVICES. *The Surgeon Generaĺs report on nutrition and health*. Washington, DC: US Government Printing Office, 1988. (PHS publication 89-50210).
31. KRAUSS RM, DECKELBAUM RJ, ERNST N, FISCHER E, HOWARD BV, KNOPP RH, *et al.* Dietary guidelines for healthy American adults: a statement for health professionals from the Nutrition Committee, American Heart Association. Circulation 1996; 94: 1795–800.
32. USDA. Nutrition and your health. Dietary guidelines for Americans. U.S. Dept Health and Human Services: 4th ed, Home and Garden Bulletin N° 232;1995.
33. GRUNDY SM. What is the desirable ratio of saturated, polyunsaturated, and monounsaturated fatty acids in the diet?. *Am J Clin Nutr* 1997 66 988S–90S.
34. DIETSCHY JM, TURLEY SD, SPADY DK. Role of liver in the maintenance of cholesterol and low density lipoprotein homeostasis in different animal species, including humans. *J Lipid Res* 1993 34 1637–59.
35. NELSON GJ. Dietary fat, trans fatty acids, and risk of coronary heart disease. *Nutr Rev* 1998 56 250–252.
36. ASCHERIO A and WILLETT W. Health effects of trans fatty acids. *Am J Clin Nutr* 1997 66 1006S–10S.
37. MENSINK RP, ZOCK PL, KATAN MG, HORNSTRA G. Effect of dietary cis and trans fatty acids on serum lipoprotein(a) levels in humans. *J Lipid Res* 1992 33 1493–501.
38. PEDERSEN JI, JOHANSSON L, THELLE DS. Trans-fatty acids and health. *Tidsskr Nor Laegeforen* 1998 118 3474-80.
39. ROSE DP. Dietary fatty acids and cancer. *Am J Clin Nutr* 1997 66 998S–1003S.
40. BERRY EM, EISENBERG S. HARATZ D. *et al.* Effect of diets rich in monounsaturated fatty acids on plasma lipoproteins-The Jerusalem Nutrition Study: High MUFAs vs high PUFAs. *Am J Clin Nutr* 1991 53 899–907.
41. ABBEY M, BELLING GB, NOAKES M, HIRATA F, NESTEL PJ. Oxidation of low-density lipoproteins: intraindividual variability and the effect of dietary linoleate supplementation. *Am J Clin Nutr* 1993 57 391–8.
42. VON SCHACKY C. n-3 Fatty acids and the prevention of coronary atherosclerosis. *Am J Clin Nutr* 71 224S–7S.
43. CONNER SL, CONNOR WE. Are fish oils beneficial in the prevention and treatment of coronaryartery disease? *Am J Clin Nutr* 1997 66 1020-S-31S.

44. KATAN MB. High-oil compared with low-fat, high-carbohydrate diets in the prevention of ischemic heart disease. *Am J Clin Nutr* 1997 66 974S-9S.
45. VARELA G. MOREIRAS O. CARBAJAL A. CAMPO M. Encuesta de Alimentación y Nutrición en España 1991. Instituto Nacional de Estadística (INE). 1995. Madrid.
46. CUSTOT F. Toxicité des graisses chauffées. Le probleme des huiles de friture. *Ann Nutr Aliment* 1959 13 417–48.
47. MOREIRAS-VARELA O, RUIZ-ROSO B, VARELA G. Effects of frying on the nutritive value of food. In: VARELA G, BENDER AE, MORTON ID, eds. *Frying of food. Principles, changes, new approaches.* Chichester, UK: Ellis Horwood Ltd, 1988: 93–102.
48. VARELA G and RUIZ-ROSO B. Some nutritional aspects of olive oil. In: *Handbook of Olive Oil. Analysis and Properties.* Harwood J. and Aparicio R. Aspen Publishers, Inc. Gaithersburg MA. 2000: 565–80.
49. PROSISE WE. Reduced oil french fried potato products and process for preparing U.S. Patent. *Food Abstract* 1991 29.
50. GUILLAUMIN R. Kinetics of fat penetration in food. In: VARELA G. BENDER A.E. MORTON I.D. eds. *Frying of Food. Principles, Changes New Approaches.* Chichester UK. Ellis Horwood Ltd. London. 1988: 112–28.
51. MOREIRAS-VARELA O, RUIZ-ROSO B, BELMONTE S, PÉREZ M. Influencia de dos procesos culinarios utilizando aceite de oliva y margarina, sobre la bioutilización de la proteína y el contenido en vitamina C de algunos alimentos. *Rev Agroquim Technol Aliment* 1990 3 387–96.
52. VARELA G, BENDER AE, MORTON ID, eds. *Frying of food. Principles, changes, new approaches.* Chichester, UK: Ellis Horwood Ltd, 1988.
53. CUESTA C, SANCHEZ MUNIZ FJ, VARELA G. Nutritive value of frying fats. In: *Frying of Food. Principles, changes, new approaches.* VARELA G, BENDER AE, MORTON ID, eds. Ellis Horwood Ltd, Chichester, UK: 1988: 112–28.
54. VARELA, G. RUIZ-ROSO, B. and MOREIRAS, O. Changes in the quantity and quality of the fat content of some foods after processing: Influence on the intake. Report No. 32 National Food Administration, Sweden 1994: 42–44.
55. VARELA G, PÉREZ M, RUIZ-ROSO B. Changes in the quantitative and qualitative composition of fat from fish, due to seasonality and industrial and culinary processing. *Bibl Nutr Dieta* 1990 46 104–9.
56. FILLION L, and HENRY CJK. Nutrient losses and gains during frying: a review. *Internat J Food Sci Nutr* 1998 49 157–68.
57. GREENFIELD H, MAKINSON J, WILLS RBH. Lipid in french fries: a retail and laboratory study. *J Food Technol* 1984 19 239–45.
58. PÉREZ M. El proceso de enlatado de sardinas. Repercusiones nutricionales y sensoriales de la fritura y de las diferentes fases de elaboración y maduración. Doctoral Thesis. Universidad Complutense. Madrid, 1990.
59. RUIZ-ROSO B, CUESTA I, PÉREZ M, BORREGO E, PÉREZ-OLLEROS L, VARELA G. Lipid composition and palatability of canned sardines in olive oil. *J Sci Food Agric* 1998 77 244–50.

60. VARELA G, MOREIRAS-VARELA O, RUIZ-ROSO B, CONDE R. Influence of repeated fryings on the digestive utilization of various fats. *J Sci Food Agric* 1986 37 487–90.
61. RUIZ-ROSO B, CUESTA I, VARELA-MOREIRAS G, CARMENA R, PÉREZ-OLLEROS L. Comparative study of two techniques for assessing fat intake in a Mediterranean diet. The influence of culinary oils changes. *Int J Food Sci Nutr* 2000 51 in press.
62. RUIZ-ROSO B. Vitamins. *Grasa y Aceites* 1998 49 347-51.
63. VARELA G, MOREIRAS-VARELA O, REQUEJO A. *Estudios sobre nutrición.* Madrid: Publicaciones del Instituto Nacional de Estadística, 1985.
64. DEUTSCHE GESELLSCHAFT FÜR FETTWISSENSCHAFT. Recommendations for Frying Oils. 3rd International Symposium on Deep-Fat Frying. March 2000. Hagen/Westphalia. Germany.
65. DOBARGANES, M.C. and MARQUEZ-RUIZ, G. Calidad de las grasas de fritura en el sector de restauración de Andalucía. *Grasas y aceites* 1995 46 115–120.
66. VARELA G and RUIZ-ROSO B. Frying process in the relation fat/degenerative diseases. *Grasas y Aceites* 1998 49 359–65.

Part II

Frying oils

6

The composition of frying oils

S. P. Kochhar, Good-Fry International NV, Rotterdam

6.1 Introduction

The oil or fat used for frying becomes part of the food being fried. The quality of the frying oil is, therefore, of great importance with regard to quality of the fried food. There is complicated interaction between the frying oil and the food, quite apart from the cooking process. Frying oil quality influences oil absorption and the types of by-products and residues absorbed by food. The type of food being fried affects the frying life of the oil. During the deep-fat frying process, oil is continuously or repeatedly subjected to high temperatures, 160–190ºC, in the presence of air and moisture. Under these conditions a variety of oil degradation reactions occur such as autoxidation, thermal polymerisation, thermal oxidation and hydrolysis. Food, when fried, can introduce various components in the oil, such as carbohydrates, phosphates, sulphur compounds, trace metals, etc. Many of these compounds contribute to colour formation and other changes, which may be deleterious, in the frying medium by reacting with the oil or its breakdown products.

The design aspects of cooker/fryer can also have a profound effect on the working life of a frying oil and consequently on the food quality and shelf life of fried product on storage. The design features should eliminate known factors of heat degradation/oxidation in the oil while maintaining output of fried products of consistent quality. For instance, copper or brass valve fittings must not be employed as copper is a pro-oxidant catalyst. Continuous removal of debris and maintenance of uniform oil temperature lead to a lower rate of free fatty acids development, better taste, colour and appearance of the product. Oils and fats contain a range of minor components, for example, tocopherols, certain sterols, phospholipids, etc. which, when in appropriate concentrations, are beneficial to oil stability during frying. Generally, less stable liquid oils are partially hydrogenated to enhance their oxidative stability for the catering industry and fast-food sectors. However, considerable amounts of *trans* and positional isomer fatty acids are formed during hydrogenation, which are nutritionally undesirable.

This chapter discusses various types of frying oils and fats for both catering and industrial operations, new frying oils on the horizon, influence of a variety of minor components on the properties and stability of frying oils.

6.2 Types of frying oils and fats

There are many types of oils and fats available for frying. Over the last three decades, in many countries, the once traditional use of groundnut/peanut oil for frying purposes has become rare or special. These days in the UK and other European countries, alternative vegetable oils namely refined rapeseed oil, partially hydrogenated rapeseed oil, palm oil/rapeseed oil or soybean oil blends, and palm olein or 'super' olein are used. Due to regional choice and for special product applications, animal fat – beef tallow or lard – is still employed as a cooking medium. In exceptional cases, sunflower seed oil and/or partly hydrogenated sunflower seed oil is used as a frying medium. Generally, palm oil or slightly hydrogenated palm oil is used for producing pre-fried French fries, frozen chips and other convenience foods. Increasingly, palm olein because of its good performance due to natural high oxidative stability is becoming the oil of choice for the major snack food manufacturers in many EU countries, whilst the more unsaturated oil products or special long-life oils containing blends of hydrogenated oils are usually used by caterers and fast-food service industry.

Important characteristics of industrial frying oils are high oxidative stability, high smoke point, low foaming, low melting point, bland flavour and nutritional value. Characteristics of some selected frying oils and fats, which are used by the catering industry and snack manufacturers, are given in Table 6.1. It can be seen that currently the *trans* fatty acids content of partially hydrogenated rapeseed oil could comprise up to 22%. Generally, in North America, the frying industry and fast-food restaurants employ frying fats and shortenings based on cottonseed oil, partially hydrogenated soybean oil and/or rapeseed oil. Several new frying oils having good stability are emerging on the horizon (Table 6.2), for example, Nu-Sun, mid-oleic sunflower oil grown principally in the USA (Gupta 1998), and high-oleic sunflower seed oil grown mainly in Spain, Italy and Southern France. High-oleic sunflower seed oil (HOSO) contains 80–82% oleic acid, C18:1 and 9–11% linoleic acid C18:2. High-oleic safflower oil containing 82% C18:1 and 7% C18:2 is also grown to a limited extent. As expected, these oleic rich oils are relatively stable at frying temperature because they have fatty acid composition similar to that of olive oil. Extremely high oleic-rich HOSO, C18:1 ~ 91% and very low in C18:2 ~ 3% is also cultivated in Europe and the USA but the oil is expensive and might have limited applications in frying. This is because when oil is heated to 180ºC, mild initial oxidation of

Table 6.1 Characteristics of some selected frying oils and fats

Fatty acid	RSO	Hyd. RSO	Palm oil (% wt)	Palm olein	Tallow
C14:0	–	–	1.1	1.1	3.2
C16:0	4.5	4.5	44.0	39.8	24.3
C18:0	1.5	5.5	4.5	4.4	18.6
C18:1	59.0	72.0	39.2	42.5	42.6
C18:2	21.0	14.0	10.1	11.2	2.6
C18:3	11.0	1.0	0.4	0.6	0.7
Others	3.0	3.0	0.7	0.4	8.0*
IV	116	90	53	58	47
Slip point	–	13°C	38°C	22°C	43–45°C
IP (hrs) at 120°C	3–4	8–9	15–16	12–14	2–6

RSO = refined and deodorised rapeseed oil; Hyd. RSO = partially hydrogenated rapeseed oil, which contained 22% trans fatty acids
IV = iodine value; IP = induction period which indicates oxidative stability of the oil
* C16:1 3.7%, C17:0 1.5, C17:1 0.8%

Table 6.2 Typical fatty acid composition (%wt) of new frying oils and of normal sunflower seed oil

Fatty acid	Sunflower seed oils			Good-Fry® Oil*
	Normal	Nu-Sun	HOSO	
C16:0	7.0	8.8	4.3	4.5
C18:0	4.5	2.3	4.2	3.7
C18:1	18.7	64.5	81.2	78.7
C18:2	67.0	22.1	8.2	10.8
C18:3	0.8	0.4	0.1	0.1
Others	2.0	1.9	2.0	2.2
IV	134	95	83	86
IP (hrs) at 110°C	4.5	12	18	19

Nu-Sun = mid-oleic sunflower seed oil; HOSO = high-oleic sunflower seed oil
* Comprises mainly HOSO and a small portion of 'dedicated' produced sesame seed oil and rice bran oil (Silkeberg and Kochhar 2000)

linoleic acid, C18:2 is essentially required to produce low concentrations of oxidation products which contribute significantly to deep-fat fried flavour (Perkins 1996; Hamilton and Perkins 1997; Kochhar 1999). However, although high-oleic sunflower seed oil does have an attractive fatty acid composition, it does not have the excellence in frying that might be expected of it. As explained in Chapter 7 this is because it has only the natural antioxidant α-tocopherol, and almost zero concentrations of the more powerful antioxidants γ- and δ-tocopherols. In comparison Good-Fry® edible oil is a different approach and was, initially, created for two purposes, firstly, to be as healthy as 'virgin' olive oil, and secondly, for use in industrial frying – to be more stable or at least as stable, as partly hydrogenated liquid oils and/or palm oil. However, this 'healthy' stable oil for the 21st century can be used for many other food applications. The Good-Fry® edible oil (Table 6.2) comprises mainly high-oleic sunflower seed oil, to which a small proportion of 'dedicated' refined sesame oil and specially produced rice bran oil has been blended (Silkeberg and Kochhar 2000). This Good-Fry® edible oil has been approved as 'dietetic' oil by the *Federal Institute for the Health Protection of the Consumers in Berlin*, Germany, with implications for the rest of the EU. Good-Fry® edible oil is now commercially available in Europe and used by the catering industry and health institutes in several EU countries.

Normally, as mentioned above, less stable liquid oils are hydrogenated to enhance their oxidative stability for catering industry and fast-food sectors. However, considerable amounts of *trans* and positional isomer fatty acids are formed during hydrogenation, which are nutritionally undesirable. Recent analytical data on the fats extracted from French fries bought from three major UK fast food sectors are presented in Tables 6.3 and 6.4. It can be seen that the frying fats can contain 16–37% saturated and 15–23% *trans* fatty acids. From a nutritional viewpoint, the fries sample A containing 60.3% saturated plus *trans* fatty acids and 9.1% free fatty acids (as oleic) is probably the worst. Moreover, the high levels of acidity determined in samples A and B indicated that the frying fats used to produce these fries had degraded badly. Surprisingly there are no regulatory controls on the sale of fries containing degraded fat in the UK, in

Table 6.3 Analysis of French fries bought from three major UK fast food services

	A	B	C
Fat (%)	13.5	15.0	13.8
Free fatty acids (as % oleic)	9.1	8.1	3.1
Peroxide value (PV) (mEq O_2 /kg)	9.4	11.4	5.2
Anisidine value (AnV)	20.8	33.5	19.9
TOTOX value (= 2 × PV + AnV)	39.6	56.3	30.3

Table 6.4 Fatty acid composition (%wt) of fats from French fries from three major UK fast food services

Fatty acid	A	B	C
Saturates	37.3	15.7	27.0
Mono-unsaturated			
cis	33.5	57.5	55.0
trans	22.0	17.7	13.7
Ploy-unsaturated			
cis, cis	5.3	5.3	1.5
cis, trans	1.0	2.6	1.2
Others	0.9	1.2	1.6
Iodine value (calculated)	58.6	78.3	63.7

contrast to regulations concerning abused frying fats in many EU countries. It is worth pointing out here that recently the 3rd International Symposium on Deep-Fat Frying, arranged by the German Society of Fat Research, made eight recommendations about the optimal frying operation (Gertz and Kochhar 2000). The definition of a long-life frying oil claim is linked to two recommendations; (1) the principal quality index for deep-fat frying should be sensory parameters of the food bring fried and (2) the analysis of suspect frying fats and oils should utilise two tests to confirm abuse. The recommended analyses should be: total polar materials $<24\%$ and polymeric triglycerides $<12\%$.

6.3 Minor components and frying oil stability

As mentioned briefly in chapter 7, all oils and fats contain a variety of minor components for example, hydrocarbons, sterols, tocopherols, colour compounds, trace metals, etc. Some of these components e.g. tocopherols (particularly γ-tocopherol), phospholipids (at less than 100 mg/kg), carotenoids (at low levels), squalene, and certain sterols such as Δ^5-avenasterol are beneficial to oil stability during frying (Boskou and Morton 1975, 1976; Gordon and Magos 1983, 1984; Gordon 1989). It has been shown (Yuki *et al.* 1978) that phospholipids, particularly phosphatidylcholine, inhibit the degradation of tocopherols and their dimers. A range of other components such as sesamolin (antioxidant precursor), sesaminol and its isomers, sesamol and its dimer (present in sesame seed oil) and oryzanol (a group of ferulic acid esters of sterols present in rice bran oil) have been shown to possess strong stabilising/beneficial effects during frying operations (Appelqvist 1997; Kochhar 1998). The beneficial effects of adding dimethyl-polysiloxane, DMPS, (E900) as an anti-foaming agent to frying oils (Freeman *et al.* 1973), and the usage of MirOil Life Powder as a processing aid to enhance the frying oil's properties are also reported in the literature (Rossell

1998). The improvement in frying stability of oils, in low-turnover restaurants, by the addition of DMPS (1–2 mg/kg) is generally considered to be due to slowing down the convection currents of the frying oil and thus suppression of thermal deterioration of the oil. The importance of convection currents in the oxidation of oils was studied by Rock and Roth (1964), who showed that the rate of oxidation was strongly dependent on the extent of convection currents. Moreover, during the quiet period, some protective effect of DMPS may also come from the inhibition of oil oxidation by the formation of an inert surface layer. In the opinion of the author, such protective effects of DMPS in frying oils are minimised or little during a continuous industrial frying operation. It is claimed that in a continuous large-scale frying operation the usage of MirOil Life Powder has very little or no beneficial effect (Rossell 1998).

6.3.1 Beneficial minor components

Tocopherols and tocotrienols

There are several factors that affect the efficacy of natural antioxidative components at room temperature and during the frying operations. These include the fatty acid composition of the oil, the temperature of frying, the amount and type of natural antioxidant, the presence of synergists, chelators, sequesterants and pro-oxidant trace metals, light, and product manufacturing conditions. The mode of action of tocopherols and tocotrienols as antioxidants at room and moderate temperatures is well known, which can be explained by their role in free radical chain-breaking by donating hydrogen from their phenolic group to radicals to stabilise them (Kochhar 1988, 1993). The resulting tocopheryl semiquinone radical-molecule (see Fig. 6.1, radical (a) or (b)), which does not have antioxidant properties, still shows possibilities of further different reactions at higher temperature. For instance, in the case of γ-tocopherol/tocotrienol a reaction between the two such radicals can give rise to two tocopherol dimers (Fig. 6.1), namely γ-tocopherol biphenyl-dimer (b+b) and γ-tocopherol ether-dimer (a+b), both of which possess antioxidative properties. α-Tocopherol degrades much faster in oils at frying temperature and produces four oxidation products only one of these, α-tocopherol ethane-dimer, shows antioxidant activity. It can thus be concluded that γ-tocopherol is superior to α-tocopherol because it oxidises to more stable compounds which are still effective as antioxidants. That is, γ-tocopherol improves frying oil stability more than α-tocopherol and thus can provide better protection to fried food in storage.

Sesamolin, sesamol and sesaminol isomers

A range of potent components such as sesamolin (an antioxidant precursor), sesamol, sesaminol and its isomers, etc., largely retained in 'dedicated' refined sesame seed oil, are shown to possess strong stabilising effects during frying operations. Figure 6.2 illustrates the mechanistic pathway for the liberation of sesamol and sesaminol isomers from sesamolin during frying operations. It has been shown (Fukuda *et al.* 1986) that sesaminol and related isomers are formed

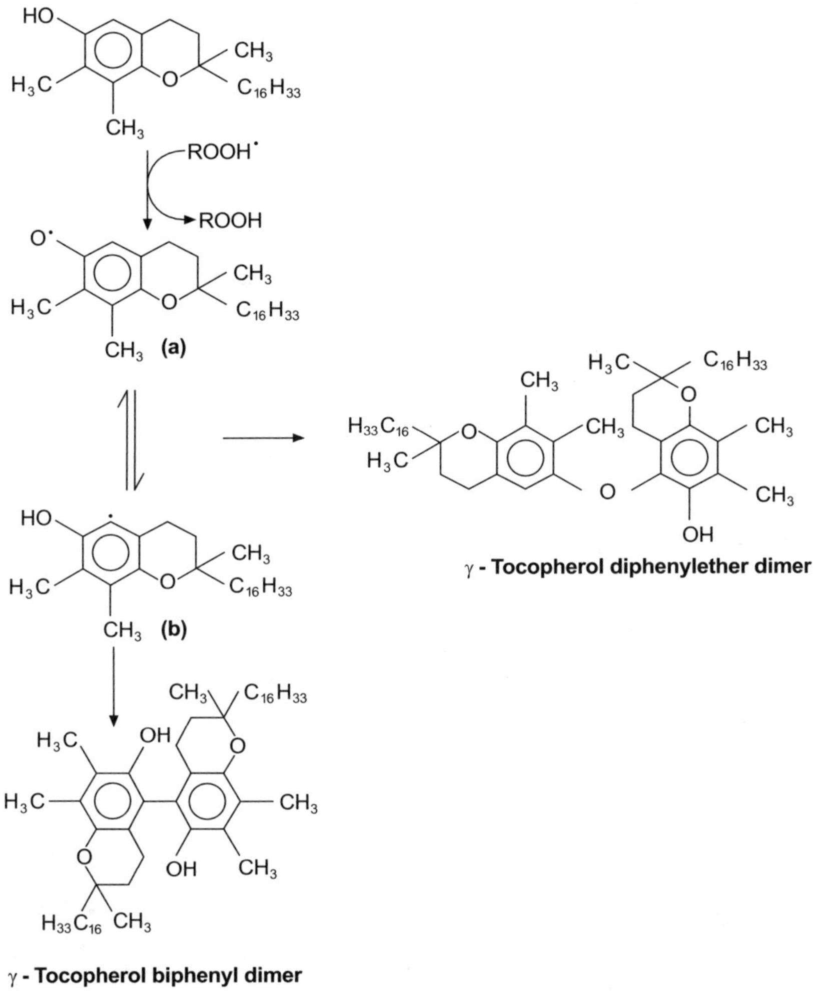

Figure 6.1 γ-Tocopherol acting as a chain-breaking antioxidant and formation of two antioxidative dimers.

from sesamolin by intermolecular transformation under anhydrous conditions in the presence of an acid. During food frying, in the presence of moisture, sesamolin is decomposed by protonlysis to sesamol and an oxonium ion is formed. When moisture is taken out by heat, these components are combined at C2 via a carbon-carbon bond to form sesaminol and related isomers through intermolecular transformation. Both sesamol and sesaminol are also shown to have significant synergistic effects with tocopherols during thermal oxidation of the oil (Fukuda *et al.* 1994). Moreover it has been noticed that sesaminol and sesamol are preferentially used as radical scavengers thus depressing the degradation of tocopherols in the oil being used for frying. Sesamin has little antioxidant activity

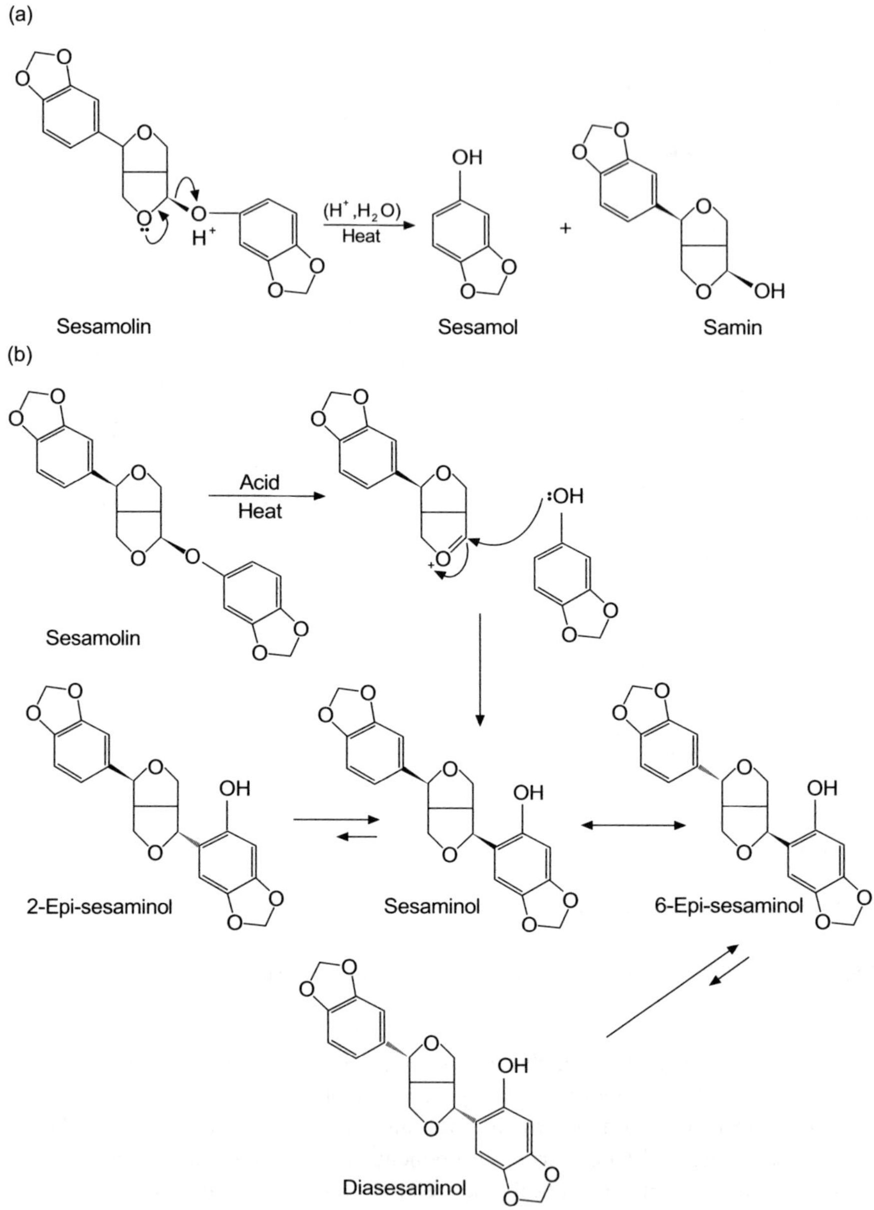

Figure 6.2 Mechanisms for the liberation of sesamol (a) and sesaminol isomers (b) from sesamolin during frying.

in laboratory experiments, but sesamin lignans are reported to be responsible for many unique health and physiological properties of sesame oil.

Oryzanol
Oryzanol is a group of compounds, which occur naturally in rice bran oil, 1.5 to 2.1%, which are shown to have antioxidant activity (Diack and Saska 1994). These five compounds are actually sterol esters of ferulic acid, namely cycloartenyl ferulate (I), 24-methylene cycloartanyl ferulate (II), cycloartanyl ferulate (III), β-sitosteryl ferulate (IV) and campesteryl ferulate (V). The triterpene alcohol esters of ferulic acid (I) and (II) make 60 to 70% of oryzanol. The sterol ferulates have been found to have substantial synergistic effects with tocopherols. Therefore, the beneficial effects of oryzanol during the frying operations are possibly due to a reduction in the degradation of tocopherols at high temperature. The antioxidant activity of oryzanol compounds is ascribed to the ferulic acid moiety, which contains a phenolic group (Marinova and Yanishlieva 1994). The protective role of sterol ferulates (oryzanol) can also be explained by the formation of five resonance-stabilised structures (see Fig. 6.3), which are very effective at interrupting the free-radical chain reactions by reacting with alkyl/lipid radicals. It is worth mentioning here that oryzanol components have been reported to have potential health benefits, e.g. significantly lower concentrations of plasma and liver cholesterol, and a decrease fatty streaks in arteries (McCaskill and Zhang 1999).

Other phenolic compounds
There are various phenolic antioxidants that have been studied for the protection of frying oils against thermal oxidation. These include the synthetic antioxidants

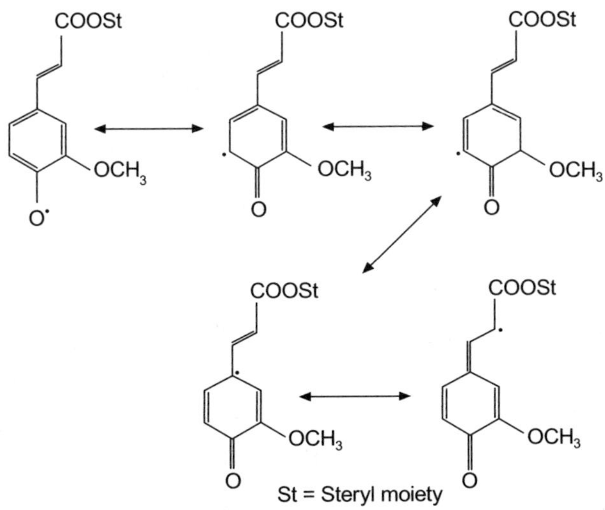

Figure 6.3 Antioxidant activity of sterol ferulate by the formation of five resonance stabilised radicals.

butylated hydroxyanidsole (BHA, E 320), butylated hydroxytoluene (BHT, E 321), tertiary butyl hydroquinone (TBHQ) and propyl gallate (PG, E 310). TBHQ is not permitted in EU countries and Japan, but is permitted in the USA and some other countries. These synthetic antioxidants retard lipid oxidation at room and moderately high temperatures but they are volatilised and/or decomposed at the temperature of frying. Therefore, these volatile antioxidants may not prevent deterioration of polyunsaturated oils during frying operations (Hawrysh *et al.* 1990; Tyagi and Vasishtha 1996). Nevertheless these antioxidants are sometimes included in commercial frying fats and shortenings. This is so because there is some evidence that these antioxidant are partially retained during frying and thus extend shelf life of the fried foods, e.g. potato crisps, noodles, banana chips (Rasit and Augustin 1982; Augustin and Berry 1983; Asap and Augustin 1986). Moreover, the small addition of TBHQ to palm olein may cause minimal changes in total tocopherols in fried potato chips (crisps in UK terminology) on storage of the finished chips (du Plessis and Meredith 1999). Many herbs and spices, namely rosemary, sage, mace, oregano

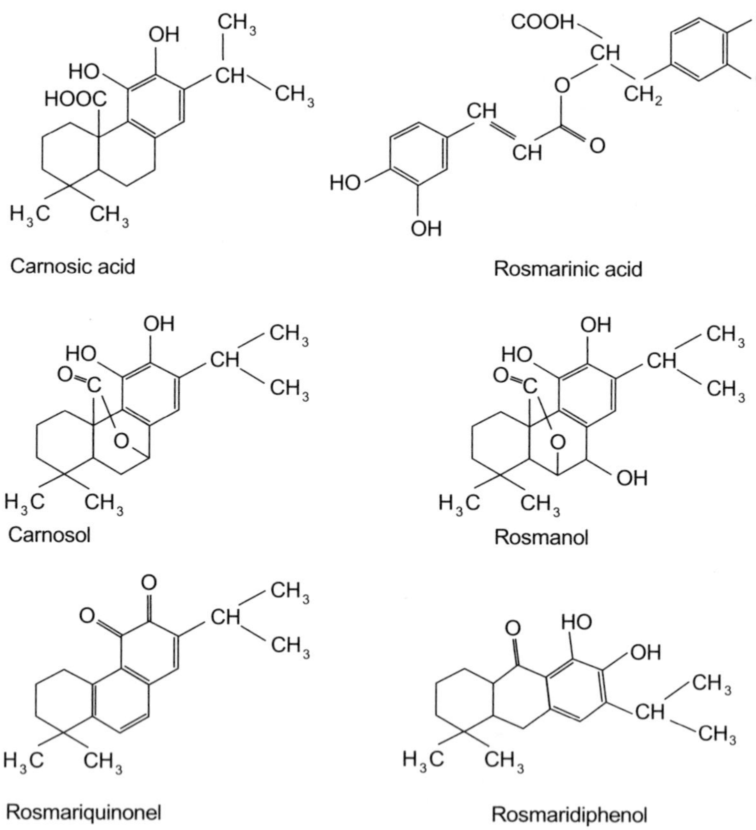

Figure 6.4 Structures of rosemary antioxidant components.

and clove have been shown to possess antioxidant properties (Kochhar 1988). Several types of rosemary extracts are now commercially available, and Fig. 6.4 illustrates the structures of rosemary antioxidant compounds. The potency of the main component, carnosic acid, has been found to be more than twice that of any other diterpene compound identified in various rosemary products (Richheimer *et al.* 1996). In order to meet the ever-growing demand of consumers the uses of natural antioxidants are being evaluated by many researchers in heated oils. The effectiveness of two commercially natural antioxidants, oleoresin rosemary and sage extract has been evaluated, in comparison with synthetic antioxidants BHA and BHT, during deep fat frying of potato chips in palm olein (Che Man and Tan 1999). The order of activity found for antioxidants in palm olein during deep fat frying of crisps and in a storage study of fried snacks was rosemary > BHA> sage extract > BHT. Moreover, the addition of rosemary extract at a level of 0.1% to refined rapeseed oil has also been reported to have caused a marked reduction in the rate of tocopherol degradation during deep fat frying of potato chips in the oil at 162°C (Gordon and Kourimska 1995).

Sterols containing ethylidene group

Unlike most other sterols, Δ^5-avenasterol (and related sterols such as Δ^7-avenasterol and citrostadienol) acts as an antioxidant at elevated/frying temperatures. This antioxidant activity has been ascribed to the formation of an allylic-free radical at C29 followed by isomerisation to a relatively stable (see Fig. 6.5a) tertiary free radical at C24 (Gordon and Magos 1983). These ethylidene side-chain sterols protect the oil against polymerisation at high temperature such as that in frying. It should be pointed out clearly that these ethylidene group-containing sterols show little or no antioxidant activity at room temperature, 100°C or at 120°C. As a consequence, if one attempts to measure oxidative stability of the oil containing such sterols by traditional methods, for example, Rancimat or oxidative stability index (OSI) the results of their antioxidative efficacy will be practically zero/negligible. This is probably due to

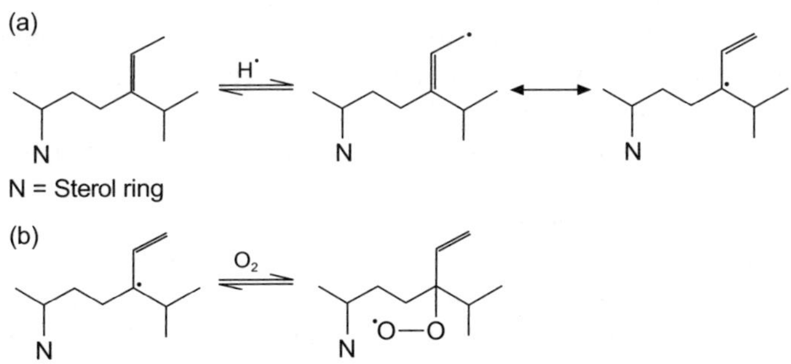

Figure 6.5 Mechanisms for antioxidant activity of sterols containing ethylidene side-chain group, (a) formation of stable tertiary free radical at C24 and (b) formation of less-stable peroxy radical.

possible formation of a less hindered peroxy radical, which is ineffective at interrupting free-radical chain reactions (Fig. 6.5b). The formation of this peroxy radical from the stable hindered radical derived from Δ^5-avenasterol is favoured at high partial pressures of oxygen, while during frying of food, oxygen deficiency caused by steam volatilisation in the frying oil favours the formation of a stable tertiary radical at C24 (Fig.6.5a). Another beneficial effect of ethylidene group-containing sterols is that they retard the loss of tocopherols in heated oils and thus enhance the frying life of the oil and subsequently prolong the shelf life of fried snacks in storage.

Squalene

Squalene, a highly unsaturated hydrocarbon $C_{30}H_{50}$, has been reported to possess moderate antioxidant activities (Govind Rao and Achaya 1968; Boskou and Katsikas 1979; Malecka 1991). Squalene is present in considerable quantities in shark liver oil and also occurs in small amounts in olive oil (10–1200 mg/100g), rice bran oil 100–330 mg/100g), wheat germ oil and yeast. The information about the effect of squalene on oil stability at frying temperatures is rather scant. For example, the addition of 0.5% squalene retarded the degradation of unsaturated fatty acids in safflower oil heated at 180ºC (Sims *et al.* 1972). The effect of squalene on model lipid systems and rapeseed oil during heating at 170ºC was also marked, but not as strong as sterols with an ethylidene side chain, in limiting the extent of polymerisation (Malecka 1994).

Recently, Abdalla (1999) has studied the effect of unsaponifiable matter (UM) extracted from olive oil distillate on the stability of sunflower oil during frying and on the quality and storage of potato chips (crisps). The UM extract contained squalene 77.4%, total sterols 13.1% and total tocopherols 4.9%. After 80 fryings at 180ºC, the results of three different levels of UM showed that the frying operation caused losses from 64 to 79% in squalene content, from 42 to 69% in total tocopherols, and from 53 to 69% in Δ^5-avenasterol. These losses were considered to be due to different antioxidant mechanisms of these compounds. The exact antioxidative mechanistic pathway of squalene at frying temperature is not clearly understood. It is, possibly, suggested that tocopherols and squalene act together in combination as chain-braking antioxidants. They scavenge the chain-carrying peroxy radicals, interrupt the chain propagation, but are themselves modified/oxidised during this reaction process. At first squalene is preferentially consumed, so its loss is greater than the tocopherol loss.

Carotenoids and phospholipids

Carotenoids are a group of oil soluble pigments that contribute to its yellow/deep orange colour. Crude palm oil from South America may contain up to 4,900 mg/kg of carotenoids, and Malaysian palm oil contains 500–1000 mg/kg carotenoids. The carotenoids profile of palm oil is similar to that of carrots which is rich in α-carotene (30–40%) and β-carotene (50–60%). The role of β-carotene as pro-vitamin A is well known. β-Carotene is very sensitive to oxidative decomposition when exposed to air, and its antioxidant action is limited to low oxygen partial pressure. β-Carotene and related oxygen-free carotenoids act as singlet oxygen quenchers and free-radical trapping agents (Jorgensen and Skibsted 1993; Handelman 1996). This type of antioxidant action of β-carotene needs further research to clarify whether or not it can be used in prolonging shelf life of fried snack products in bags particularly at very low level. However, it should be pointed out, as discussed later, that at high concentration (>0.01%) and at frying temperatures β-carotene shows no protective effect but displays pro-oxidant behaviour. At room temperature, β-carotene also inhibits lipid peroxidation by forming a transit free radical complex with peroxy or hydroxy radical. The addition of another peroxy radical to this complex results in the formation of stable product. This antioxidant action of β-carotene has been ascribed to the presence of its conjugated double bond system.

Phospholipids such as cephalin possess moderate antioxidant activity, which is generally related to their ability to form chelating effects with trace metals and thus inhibit their catalytic effect on the formation of peroxides and their decomposition to free radicals (Dziedzic and Hudson 1984; Pokorny *et al.* 1992; Kourimska *et al.* 1993). The effect of phospholipids on the efficacy of tocopherols in free-radical termination has also been reported by Hildebrand *et al.* (1984). Therefore, in some applications, phospholipids removed during processing of oils could be added back to the oil at a low level after final processing in order to increase oil stability (Gopalakrishna and Prabhakar 1985). There is rather scant information about the use of lecithins to improve stability of frying oils. Some researchers had studied the effects of adding 0.1% lecithin to oils used for frying (Chu 1991; Kourimska *et al.* 1995). Although this level showed adverse effects on oil colour and foaming, significant antioxidant effects were demonstrated by various parameters during the frying operations. The mechanism of this protective effect of phospholipids ascribed to synergism with the tocopherols and other phenolic antioxidants naturally present in the oils. It is generally accepted that at low level addition (< 0.01%), lecithins do not exhibit adverse effect on the oil colour and foaming, but can in fact improve frying oil stability noticeably.

Ascorbyl palmitate and metal sequestrants

Ascorbyl palmitate is a weak antioxidant when used alone, but it shows increased activity in mixtures with natural tocopherols and/or sequestrants. Chelating agents or sequestrants such as citric acid, ascorbic acid, phosphoric acid, amino acids, etc., form complexes with trace metals such as copper and iron, which catalyse the decomposition of hydroperoxides to initiate free radical chains. For example, the addition of 0.01% citric acid decreased the loss of tocopherols in soybean oil and significantly increased its oxidative stability. Moreover, this addition of citric acid was sufficient to eliminate the pro-oxidant effect of iron added at the level of 0.3 mg/kg. The chelators are sometimes referred to as synergists as they greatly enhance the action of phenolic antioxidants. Ascorbyl palmitate (AP) has been reported to have the ability to inhibit thermal degradation of frying oils and fats (Gwo *et al.* 1985). AP was claimed to be a better antioxidant than rosemary extract, BHA, BHT and δ-tocopherol in rapeseed oil during a heating and deep fat frying operation (Gordon and Kourimska 1995). In contrast, other researchers found no significant effect on oil deterioration during frying of prawn crackers (Augustin *et al.* 1987), and after addition of AP at 0.02% level to palm oil during frying (Ibrahim *et al.* 1991).

6.3.2 Detrimental minor components

Many chemical reactions take place in oil during the frying operation due to the elevated temperature involved. These may be grouped into oxidation, polymerisation and hydrolysis (Blumenthal 1991; Chang *et al.* 1978; Fritsch 1981; White 1991). The mechanisms of these various reactions during the frying operation have been discussed in detail by Hamilton and Perkins (1997). Oxidation proceeds rapidly at frying temperatures; the higher the temperature, the more rapid the oxidation. The rate of oxidation of the frying oil is also affected by factors other than temperature, such as oil turnover, surface exposure to air, presence of pro-oxidant trace metals (especially copper or iron), quality and type of fat, type of food being fried, and the presence of heat stable/or novel antioxidants. The reaction of water, introduced from the food being fried, with frying oil results in development of free fatty acids (FFAs) and partial glycerides. Such high acidity oils will also dissolve more oxygen, thus promoting oxidation.

Hot oil is always a potential hazard and must be treated with care. For instance, if excessive foaming occurs for any reason the heating source must be switched off to reduce the risk of fire. Excessive foaming may be caused by one or more of a combination of several factors such as oil breakdown, rapid heating up, contact with copper or brass, frying of foods with excessive moisture, and presence of soap or detergent residue after cleaning the fryer. Table 6.5 lists minor detrimental components that may occur in the frying oil or may migrate from the food product into the oil during the frying process.

Table 6.5 Effect of several detrimental components on frying oil stability

Component	Effect	Safe Limit
Alkaline material (as sodium oleate)	oxidation/ deterioration	< 20 mg/kg
Free fatty acids	smoke/pro-oxidant	< 0.4 %
Lecithin	foaming/darkening	< 100 mg/kg
Partial glycerides	foaming	$< 6\%$
Proteinaceous residues	burnt flavours	
Trace metals	pro-oxidant	Cu < 0.01 mg/kg Fe < 0.1 mg/kg
Pigments	pro-oxidant	

Free fatty acids

In both catering and industrial frying operations, smoke haze can be a problem. Volatile break-down products, such as the free fatty acids (FFAs) formed in the oil at higher temperatures, give rise to smoke, which if not controlled, can enhance the fire hazard due to their low flash point. The smoke, flash, and fire points of oils and fats are indirect measures of the safety of the frying oil when it is heated in contact with air. The smoke point is defined as the temperature at which the sample starts to smoke under the specified conditions of the test. The flash point is the temperature at which volatile products are evolved in sufficient quantity to allow instantaneous ignition. The fire point is the temperature at which the production of the volatile products is sufficient to support continuous combustion. The smoke point is of most benefit in assessing the quality of frying oil, since it is mainly FFAs produced in the frying operation which contribute to the smoke haze. Table 6.6 presents the relationship between smoke, flash and fire points and FFAs, which indicates that the free fatty acid content of an oil has a substantial effect on its smoke point (Weiss 1983). Quantitatively, the smoke point of used frying oil or fat decreases in proportion to the amount of FFAs present (Bracco *et al.* 1981). The smoke point of the frying fat also affects oil absorption by the fried food. A significant negative correlation between oil absorption by doughnuts and smoke point of frying oil has been reported (Orthoefer *et al.* 1996).

Development of FFA arises from hydrolysis – reaction of water from the frying food with the frying oil, and partly as an end product of oxidation. The rate of free fatty acid development, partly, depends on:

1. Amount of water, from the frying food, added to the fryer.
2. Temperature of the frying oil. The higher the frying temperature, the more rapid the rate of FFAs formation.
3. Rate of turnover of the oil. The higher the turn-over rate, the slower the production of free fatty acids.

Table 6.6 Effect of free fatty acid content on smoke, flash, and fire points of frying oils and fats

Free fatty acid (%)	Smoke point (°C)	Flash point (°C)	Fire point (°C)
0.04	218	327	366
0.06	210		
0.08	205		
0.1	200	313	363
0.2	190		
0.4	177		
0.6	171		
0.8	165		
1.0	160	307	360

Source: Weiss (1983).

4. Accumulation of debris/burnt food particles in the fryer. These deposits tend to accelerate the development of FFAs.
5. Amount of FFA and partial glycerides, particularly di-glycerides, in oils. Increased amounts of FFA and partial glycerides increase the solubility of water in the oil, thus promoting additional hydrolysis.

The presence of free fatty acids in the frying oil has been shown to catalyse further hydrolysis of triglycerides. They are also shown to promote oxidation, perhaps by increasing the solubility of oxygen, so their development in the frying operation should be controlled. It has been suggested that the free carboxyl group of the fatty acid molecule is responsible for the pro-oxidant activity of FFAs (Mistry and Min 1979). The catalytic effect of the carboxyl groups is probably due to the formation of free radicals via homolytic decomposition of hydroperoxides (Miyashita and Takagi 1986). As explained in chapter 8 the FFA content of the oil does not correlate well with fried food quality.

Specific end-points for frying oil depend on the type of oil in the fryer and on the food being fried. Normally, in a batch frying operation, 0.5–0.8% free fatty acid is encountered. If the frying rate is insufficient to remove water from food at a sufficient rate, the FFA content increases rapidly. High FFA content frying oil will smoke progressively at a lower temperature. The key elements of good frying oil are bland flavour, pale colour, good oxidative stability and good thermal stability during the frying operation. While providing adequate resistance to oxidation/degradation during continuous use in a fryer, the frying oils or shortenings should have a minimum smoke point of 220°C. Therefore, if unrefined butter oil, olive oil or animal fat is employed in a frying process, its smoke point will be quite low even when the fat or oil is fresh (e.g. smoke point of virgin olive oil is 169°C). The usage of such fats and of lauric-rich oils will lead to unsatisfactory frying at elevated temperatures of 180–200°C.

Lecithins and partial glycerides

These components can cause excessive foaming if they are present at moderate or high concentrations. The leaching of lecithin from egg yolk batters and/or doughnut mixes into the oil will cause therefore foaming. Moreover, phospholipids (lechithins) may also cause darkening of the oil at higher frying temperatures. The safe level of lechithins in a frying medium is generally considered to be less than 10mg/100g oil. In contrast, the beneficial effects of low concentrations of phospholipids are probably due to their synergistic action with tocopherols thus enhancing the antioxidant properties (Dziedzic and Hudson 1984). The chelating action of phospholipids with transition trace metals is also suggested, thus inactivating their metal catalytic pro-oxidant effect.

Emulsifiers, e.g. mono- and di-glycerides (E 471) are often used in the production of foods and if they escape into the oil during frying operations they will cause excessive foaming. Good-quality frying oil should have less than 0.4% monoglycerides content. Excessive foaming in the frying operation that does not disperse is an indication that the oil has reached the end-point and should be discarded, from a safe operation/fire hazard viewpoint.

Trace metals
Trace metals have a great detrimental influence on both oxidative and flavour stability of oils, particularly at high temperatures during frying. Subsequently, upon storage of the fried product, trace metals such as copper and iron catalyse the decomposition of hydroperoxides into secondary oxidation products which are actually responsible for the development of off-flavours/odours or rancid/unpleasant tastes and aroma in the fried products. Table 6.7 shows that copper is the most effective pro-oxidant trace metal, which is about ten times more active than iron in causing a detrimental effect. Therefore, copper and its alloys such as brass and bronze, *must not* be used in the frying equipment, including brass valve fittings which come in contact with the oil. Furthermore, if the thermocouples, frying baskets, etc. are manufactured from plated copper or brass material, these must be inspected regularly for normal wear and tear in order to avoid any premature breakdown of the fat due to the strong detrimental effect of copper. Some food products may introduce substantial quantities of trace metals into the oil. For example, potato 'whiteners' may improve the appearance of fried chips, but they can degrade the frying oil, possibly by introducing some trace metals promoting oxidation.

Proteinaceous residues/debris
Debris of the fried food is normally responsible for proteinaceous residues, which if not removed by effective filtration or skimming, would give rise to burnt flavours/unpleasant after taste. Moreover, the modern bread-crumb coatings are more friable and can easily break down to give increased amounts of debris in the oil. Furthermore, excessive quantities of bread-crumb prepared from bread-flour fortified with iron for nutritional purposes, escaping into the frying oil are detrimental. Obviously, the iron in the bread-crumbs will contaminate the oil and thus catalyse oxidative deterioration. Proteinaceous

Table 6.7 Pro-oxidant catalytic/detrimental effect of certain trace metals*

	(mg/kg)
Copper	0.05
Iron	0.60
Manganese	0.60
Chromium	1.20
Nickel	2.20
Vanadium	3.00
Zinc	19.60
Aluminium	50.00

* To reduce the keeping time of lard by 50% at 98°C.
Source: Sonntag (1979).

residues may be present in the starting frying medium when unrefined oils such as cold-pressed groundnut oil or beef dripping are used for any frying process. Therefore care should be taken and such unrefined oils should be filtered free from these harmful components prior to their usage. There is another important point which should be mentioned, namely that the oil must not continue to flow through a filter bed of hot debris, as this will negate the advantages of filtration.

Alkaline material

Alkaline material has been shown as a possible cause of deterioration in frying oils and fats (Blumenthal *et al.* 1985). This is probably through migration of alkali metal ions from the food into the frying oil, where they combine with free fatty acids to form soaps. Usually, in catering/restaurant frying operations the concentration of alkaline material (AM) can increase during the use of frying oil, as given in Table 6.8 (Blumenthal and Stockler 1986). The problem of formation of detrimental AM should be minimised by salting the food after frying. But if salted foods were fried, the sodium chloride used would not be of analytical grade purity, and this would introduce minute quantities of copper and iron salts into the oil. Such pro-oxidative trace metals will undoubtedly degrade the oil, thus affecting the discard oil-point and quality of the fried food.

Colour compounds

Oils and fats contain varying amount of colour components such as carotenoids and chlorophyll. For instance, virgin olive oil may contain 2–16 mg/kg chlorophylls and derivatives, and crude palm oil may have 600–1000 mg/kg of carotenoids. Chlorophylls and related pigments, like pheophytins, are very active promoters of photosensitised oxidation, and must therefore be completely removed by a bleaching process to prevent the damaging effects of photo-oxidation. However, these photosensitisers may show slight antioxidant effects on the oils in the dark – probably by donating hydrogen to break the free-radical chain reaction (Endo *et al.* 1985; Gutierrez-Rosales *et al.* 1992). At frying temperature, these chlorophyll components are likely to have pro-oxidant effects. Carotenoids and especially β-carotene, which impart a yellow colour to oils, are effective inhibitors of photosensitised oxidation by quenching singlet oxygen. At low levels (less than 5 mg/kg), under photo-oxidising conditions, β-

Table 6.8 Concentration of alkaline material (AM) in oil during its use in a restaurant

Days of oil use at 180°C	AM (as sodium oleate, mg/kg)
0 (fresh)	7.5
1	15
2	18
3	37
4 (discard)	43

Source: Blumenthal and Stockler (1986).

carotene was shown to have a protective effect (Clements *et al.* 1973; Levin and Mokady 1994). At higher concentrations 0.01–0.08%, the pro-oxidant effect of β-carotene on oils heated at 100–120°C were noticed by various workers (Meara and Weir 1976; Kim *et al.* 1995; Wagner and Elmadfa 1998). From their study on the effects of heating and frying operations on the retention of β-carotene in crude palm oil, Manorama and Rukimi (1992) found that the fourth frying operation resulted a total loss of β-carotene in the oil. These findings suggest that 'red' palm olein with its rich content of β-carotene should not be used for repeated deep frying operations.

6.4 Combined effects of natural products on stabilisation of frying oils

Sometimes tocopherols, the naturally occurring antioxidants, do not on their own provide the antioxidant effect required in a particular oil application or food product. In such a case, the antioxidant activity of tocopherols may be enhanced considerably by the addition of other natural and/or permitted synthetic substances, commonly called synergists. For example, the addition of rosemary and sage extracts to palm olein effectively retarded oil deterioration during intermittent frying of potato chips (Che Man and Jaswir 2000). The synergistic effect of rosemary, sage, and citric acid on unsaturated fatty acid retention of palm olein during deep-fat frying has also been studied (Jaswir *et al.* 2000). Based on the ratio of linoleic acid (C18:2)/palmitic acid (C16:0), a combination of 0.076% rosemary extract, 0.066% sage extract and 0.037% citric acid was found to provide the optimal retention of C18:2. Some speciality oils, containing a group of natural antioxidant components, e.g. olive, rice bran and sesame seed oils possess excellent frying oxidative stability and performance. Individually, each oil contributes a characteristic/pleasant flavour to the fried food. Roasted sesame oil is very popular in Chinese, Korean and Japanese cooking because of its flavour. The strong resistance to oxidation of roasted sesame oil, which contribute to its excellent frying stability, has been attributed to the production of sesamol from sesamolin during roasting/frying operation, combined with the presence of γ-tocopherol and some oil soluble Maillard browning products, especially those formed at temperature above 190°C during roasting of the seed.

The branded group of potent components 'Good-Fry' Constituents (GFC) comprise a blend of 'dedicated' refined sesame seed oil (SSO) and specially produced rice bran oil (RBO) in which a range of antioxidant precursors and compounds are substantially retained (Silkeberg and Kochhar 2000). Both of these speciality oils contain unique antioxidant components. For example, sesame seed and its oil contain two major sesame lignans, sesamin and sesamolin, while rice bran oil contains oryzanol – a group of ferulic acid esters of sterols. In addition, both rice bran oil and sesame oil comprise large concentrations of Δ^5-avenasterol and related ethylidene group containing sterols, which are present in the ranges 295–355 mg/100g in RBO and 84–265 mg/100g

Table 6.9 Stability of various oils by actual frying of French fries

Oil type	Endpoint (hours)	Criteria
Groundnut oil	20	greasy fries
HOSO	30–35	dark/greasy
Long-life[a]	40	greasy fries
Palm olein	40	greasy fries
Good-Fry® oil	65	foaming, fries still OK
Total polar material = 21.1–23.4 %		

a = partly hydrogenated rapeseed oil.

in SSO respectively. Moreover SSO oil contains 40–70 mg/100g tocopherols, mainly γ-tocopherol (> 97%), and RBO contains 70–140 mg/100g tocopherols, out of which γ-tocopherol and tocotrienols are 50–80%. The combined actions of these potent antioxidative components, largely retained in GFC, in stabilising frying oils are discussed below.

The results of the extensive frying study carried out by frying more than 10,000 portions of French fries in several oils are presented in Table 6.9. The frying performance of groundnut oil, high-oleic sunflower oil (HOSO), long-life frying fat (partially hydrogenated rapeseed oil), and palm olein were first compared using commercial pre-fried French fries. Each portion of French fries (100g) was fried at 175°C for 3 minutes in individual oil. Stainless steel electrical fryers of capacity 3 litres were used in frying operations, which continued for 5 hours per day. The sensory quality of the French fries produced was used as the end-point indicator. As expected, palm olein being a stable frying medium was satisfactory for up to 40 hours. After this time the French fries were judged to be greasy, a danger signal of olein degradation. Also, foaming and darkening was observed. Surprisingly, the groundnut oil gave only 20 hours of frying stability under these conditions. This could have been due to below-average quality of the groundnut oil used and/or commercial pre-fried potatoes. It is worth mentioning that all the frying oils, including HOSO, were solid in the morning after three days (15 hours) of frying, due to exchange of partly hydrogenated palm oil between the pre-fried potatoes and the frying oils. It was, however, noticed that at the discard point, the amount of total polar materials (TPM) determined in all of these oils had not exceeded 24–25% TPM, the regulatory limits for used-restaurant frying oils and fats in many EU countries.

The results of frying performance of Good-Fry® oil are also included in Table 6.9, where Good-Fry® oil was used both for pre-frying French fries and frying. It was found that the frying performance of Good-Fry® oil was very satisfactory up to 65 hours compared with 35 hours for high-oleic sunflower oil or 40 hours for palm olein. It can, therefore, be concluded that to achieve the maximum stabilisation potential of Good-Fry® constituents, it is important to use Good-Fry® oil in both frying and pre-frying. In other words, it should be emphasised that the use of French fries pre-fried in oil fortified with GFC-containing natural

Table 6.10 Effect of storage on potato chips produced industrially in palm olein at 180±2°C

	Storage time (weeks) at ambient temperature					
	0	4	8	12	16	20
FFAs (as % oleic)	0.07	0.15	0.22	0.29	0.18	0.18
PV (mEq O_2/kg)	0.15	2.05	2.0	3.0	3.1	3.6
AnV	–	6.9	16.9	23.4	18.2	19.3
TOTOX value (= 2 × PV + AnV)	–	21.0	20.9	29.4	24.4	26.5

Sensory: Slightly stale odour and taste; some panellists reported bitter/burnt after taste

Packaging material: metallised/polyester-coated

potent antioxidative components, as well as oil replenishment (topping-up), extends the frying performance of the stabilised/fortified oil substantially by supplying fresh antioxidant precursors and anti-polymerisation components during the entire frying operation.

Analytical and sensory evaluation data of potato chips (crisps in UK terminology) fried in palm olein (control) and palm olein fortified with 2% GFC are presented in Tables 6.10 and Table 6.11 respectively. Potato chips (crisps) were fried on an industrial scale, using a 2.5 tonne oil capacity fryer, at 180°C and packed in 50g bags that were metallised-polyester-coated. The data show clearly that the quality of chips (crisps) fried in the stabilised palm olein was improved significantly with storage. For example, after 8 weeks storage at ambient temperature the oil from the control sample gave a TOTOX value of 20.9 and some taste-panel members reported a flavour deterioration in the chips (crisps), namely staleness and a burnt taste, while the chips (crisps) fried in

Table 6.11 Effect of storage on potato chips produced industrially in palm olein plus 2% GFC* at 180±2°C

	Storage time (weeks) at ambient temperature					
	0	4	8	12	16	20
FFAs (as % oleic)	0.07	0.06	0.13	0.19	0.10	0.15
PV (mEq O_2/kg)	0.15	3.5	3.0	3.0	3.0	3.6
AnV	–	8.7	8.6	9.2	10.1	12.1
TOTOX value (= 2 × PV + AnV)	–	14.6	14.6	15.2	16.1	18.9

Sensory: Fresh flavour, taste and texture good

Packaging material: metallised/polyester-coated

* Blend of 'dedicated' refined sesame seed oil and rice bran oil – manufactured according to Silkeberg and Kochhar (2000).

Table 6.12 Effect of storage on potato chips produced on an industrial scale

	Storage time (weeks) at ambient temperature					
	Fried in					
	Palm Olein			Canola oil + % 6 GFC*		
	Initial	Wk8	Wk12	Initial	Wk8	Wk12
FFAs (as % oleic)	0.08	0.13	0.21	0.09	0.13	0.18
PV (mEq O_2/kg)	1.0	4.5	5.0	0.1	3.9	4.2
AnV	–	6.4	6.9	–	6.1	7.9
TOTOX value (= 2 × PV + AnV)	–	15.8	16.9	–	13.8	16.2

Sensory: No formation of any rancid/paint-like, off-flavour was noticed
Packaging material: transparent polypropylene

*Blend of 'dedicated' refined sesame seed oil and rice bran oil – manufactured according to Silkeberg and Kochhar (2000).

fortified palm olein were still fresh in taste, had a crispy texture after 16 weeks, and the oil was had a TOTOX value of 16.1. In a separate industrial scale experiment, analytical and sensory data obtained from chips (crisps) fried in palm olein and from those fried in canola oil (i.e. unhydrogenated rapeseed oil) stabilised with 6% GFC are given in Table 6.12. As it is well known that unhydrogenated rapeseed oil, being unstable, cannot be used for chips (crisps) manufacture this oil was not chosen for the control, but a comparison was conducted with palm olein. It can be seen from Table 6.12 that normal canola oil/rapeseed oil, when fortified with Good-Fry® constituents can be used safely for frying chips (crisps). Both the taste and smell of chips (crisps) fried in stabilised canola oil were found to be comparable with those fried in palm olein. Up to a storage period of 17 weeks at ambient temperature, no formation of rancid, paint-like off-flavour was noticed by trained taste-panel members. It is worth reporting here that both analytical and sensory data collected from the storage study on chips (crisps) fried in soybean oil (iodine value = 130; C18:3 = 7%) with and without GFC showed that the shelf life of chips (crisps) was increased substantially (from 8 to 20 weeks) by the addition of 5% Good-Fry® constituents.

6.5 Future trends

The type of oil selected and the length of time that the oil has been used for frying affect the desired flavour of fried foods and subsequently the shelf life of snack products. Frying oils degrade with continuous use. Therefore, as discussed above, a suitable frying oil must be very low in detrimental minor components e.g. free fatty acids, trace metals, etc. and must have a high oxidative stability (resistance to breakdown) during continuous use. The ideal frying oil would be low in saturated fatty acids, ultra-high (75–85%) in oleic acid, low in linoleic, and very low in linolenic acid (< 0.2%). The working life of this ideal oil, virtually free from detrimental components, can be further enhanced by the balanced presence of potent, natural, non-volatile antioxidant components, which are effective at frying temperatures. Antioxidants with these properties include Δ^5-avenasterol, γ-tocopherol, oryzanol, rosemary antioxidants, mixed tocopherols, sesamolin and related compounds. Such stable oils rich in oleic acids, containing zero *trans* fatty acids, should emerge in the new century.

In recent years, changes in frying fats are taking place due to consumer demand for healthier snack products and convenience pre-fried foods. The latest developments in healthy stable frying oils have been reported (Haumann 1996; Appelqvist 1997; Gupta 1998; Kochhar 2000). Several snack and convenience-food manufacturers are now attempting to make fried products with a healthier profile, using for example Good-Fry® oil or Good-Fry® Sunolive oil which are commercially available in European countries. Certainly, nutritious and high-quality, niche fried products can be prepared by frying in these healthy oils with their Mediterranean image of a healthy diet. Moreover, for low-fat-conscious consumers, these healthier stable oils can also be used for the production of low-fat snacks of different shapes and for many new convenience foods for healthier eating as part of a balanced diet, of different types and fancy shapes and which is gaining popularity among consumers.

6.6 Conclusions

New frying oils are emerging with better fatty acid profiles and better combinations of antioxidants. Some of these are achieved by plant breeding and others by careful, calculated, blending of selected oils. One of these is Good-Fry® Sunolive oil. The Good-Fry® dietetic oil is a blend of mainly high-oleic sunflower oil and small proportion of 'dedicated' refined sesame oil and specially produced rice bran oil.

The trace metal copper is the most damaging pro-oxidant metal catalyst for oils and fats, about ten times more so than iron. Copper and its alloys/bronze or brass valve fittings must not be used in frying equipment. If frying baskets/thermocouples, etc., in a batch fryer are manufactured from plated copper or brass material, these must be inspected regularly in order to avoid premature degradation of the oil. The formation of deleterious alkaline material, a possible cause of oil deterioration, should be minimised, e.g. by salting the food after frying. Debris/sediments arising from modern bread-crumb coatings and/or proteinaceous material which may be present in an unrefined frying medium must be removed by effective filtration. The potential benefits of a good filtration system are better-quality fried food, prolonged oil life, better heat transfer, reduced fryer cleaning and fewer oil discards.

Antioxidants present naturally in a frying oil play a very important role in stabilising the complex frying system, and can enhance frying oil performance and extend the shelf life of the fried food. The natural antioxidant components, namely antioxidant precursor (sesamolin), Δ^5-avenasterol and related sterols, squalene, oryzanol, γ-tocopherol and tocotrienol, largely retained in 'dedicated' refined sesame oil and specially produced rice bran oil, can improve the frying stability of soft oils tremendously. Such fortified soft oils, e.g. rapeseed oil, normal sunflower oil, soybean oil and blends, etc. can be used as good frying media with nutritional advantages over partially hydrogenated oils and shortenings containing *trans* fatty acids. The stabilisation of frying oils with Good-Fry® constituents (blend of specially refined sesame oil and rice bran oil), GFC, results in an improvement not only of the shelf life of the fried products but also in their flavour quality. It should be pointed out that to get the optimum antioxidant components of GFC during a French fries operation, the use of fries pre-fried in oil fortified with GFC is recommended. This is so because each portion of pre-fried fries adds fresh GFC components, each time, into the frying medium, thus prolonging the frying oil stability with little or no oils discarding.

Nutritious and high-quality, niche, fried products can be prepared by frying in speciality frying oils like Good-Fry® oil or Good-Fry® Sunolive oil. These give the Mediterranean image of a healthy diet. It is forecasted that the demand for 'healthier' snack products and convenience foods with good taste, texture and appearance, and the use of natural antioxidants in stabilising new oleic-rich and soft frying oils will grow tremendously.

6.7 References

ABDALLA A E M (1999), 'Antioxidative effect of olive oil distillate on frying oil and quality of potato chips', *Fett/Lipids*, **101**, 57–63.

APPELQVIST L-A (1997), 'Healthy frying oils', in S P KOCHHAR, *New Developments of Industrial Frying*, Bridgwater, PJ Barnes & Associates, 91–117.

ASAP T and AUGUSTIN M A (1986), 'Effect of TBHQ on quality characteristics of RBD olein during frying', *J Am Oil Chem Soc*, **63**, 1169–1172.

AUGUSTIN M A and BERRY S K (1983), 'Efficacy of the antioxidants BHA and BHT in palm olein during frying', *J Am Oil Chem Soc*, **60**, 1520–1523.

AUGUSTIN M A, CHUA C J J and HENG L K (1987), 'Effects of silicone and ascorbyl palmitate on the quality of palm olein used for frying prawn crackers', *J Sci Food Agric*, **40**(1), 87–93.

BLUMENTHAL M M (1991), 'A new look at the chemistry and physics of deep-fat Frying', *Food Technol*, **2**, 68–71.

BLUMENTHAL M M and STOCKLER J R (1986), 'Isolation and detection of alkaline contaminant materials (ACM) in used frying oils', *J Am Oil Chem Soc*, **63**(5), 687–688.

BLUMENTHAL M M, STOCKLER J R and SUMMERS P J (1985), 'Alkaline contaminant material in used frying oils: A new quick test', *J Am Oil Chem Soc*, **62**(9), 1373–1374.

BOSKOU D and MORTON I D (1975), 'Changes in the sterol composition of olive oil on heating', *J Sci Food Agric*, **26**, 1149–1153.

BOSKOU D and MORTON I D (1976), 'Effect of plant sterols on the rate of deterioration of heated oils', *J Sci Food Agric*, **27**, 928–932.

BOSKOU D and KATSIKAS H (1979), 'Effect of olive oil hydrocarbons and triterpene alcohols on the stability of heated cotton seed oil', *Acta Alimentaria*, **8**, 317–320.

BRACCO U, DIEFFENBACHER A and KOLAROVIC L (1981), 'Frying performance of palm liquid fractions', *J Am Oil Chem Soc*, **58**, 6–12.

CHANG S S, PETERSON R J and HO C-T (1978), 'Chemical reactions in the deep-fat frying of foods', *J Am Oil Chem Soc*, **55**, 718–727.

CHE MAN Y B and TAN C P (1999), 'Effects of natural and synthetic antioxidants on changes in refined, bleached, and deodorised palm olein during deep-fat frying of potato chips', *J Am Oil Chem Soc*, **76** (3), 331–339.

CHE MAN Y B and JASWIR I (2000), 'Effects of rosemary and sage extracts on frying performance of refined, bleached, and deodorised (RBD) palm olein during deep-fat frying', *Food Chem*, **69**, 301–307.

CHU Y-H (1991), 'A comparative study of analytical methods for evaluation of soybean oil quality', *J Am Oil Chem Soc*, **68**, 379–383.

CLEMENTS A H, VAN DEN ENGH R H, FROST D T and HOOGENHOUT K (1973), 'Participation of singlet oxygen in photosensitised oxidation of 1,4-dienoic systems and photo-oxidation of soybean oil', *J Am Oil Chem Soc*, **50**(8), 325–330.

DIACK M and SASKA M (1994), 'Separation of vitamin E and γ-oryzanols from rice bran by normal-phase chromatography', *J Am Oil Chem Soc*, **71**,1211–1217.

DZIEDZIC S Z and HUDSON B J F (1984), 'Phosphatidyl ethanolamine as a synergist for primary antioxidants in edible oils', *J Am Oil Chem Soc*, **61**(6),1042–1045.

ENDO Y, USUKI R and KANEDA T (1985), 'Antioxidant effects of chlorophyll and pheophytin on the autoxidation of oils in the dark II The mechanism of antioxidative action of chlorophyll', *J Am Oil Chem Soc*, **62**,1387–1390.

FREEMAN I P, PADLEY F B and SHEPPARD W L (1973), 'Use of silicones in frying oils', *J Am Oil Chem Soc*, **50**(4), 101–103.

FRITSCH C W (1981), 'Measurements of frying fat deterioration: A brief review', *J Am Oil Chem Soc*, **58**, 272–274.

FUKUDA Y, ISOBE M, NAGATA M, OSAWA T and NAMIKI M (1986), 'Acidic transformation of sesamolin, the sesame oil constituent, into an antioxidant bisepoxylignan, Sesaminol', *Heterocycles*, **24**, 923–926.

FUKUDA Y, OSAWA T, KAWAKISHI S and NAMIKI M (1994), 'Chemistry of lignan antioxidants in sesame seed and oil', in HO C-T, OSAWA T, HUANG M–T and ROSEN R T, *Food Photochemical for Cancer Prevention II,* Washington, DC, ACS Symposium Series no. 547, pp. 264–274.

GERTZ CH and KOCHHAR S P (2000), 'Review: 3rd Symposium on deep-fat frying – optimal operation', *Eur J Lipid Sci Technol*, **102**(4), 306–308.

GOPALAKRISHNA A G and PRABHAKAR J V (1985), 'Antioxidant efficacy of water washed gums as a function of relative humidity in peanut oil', *J Am Oil Chem Soc*, **62**(11), 1581–1583.

GORDON M H (1989), 'Plant sterols as natural antipolymerisation agents', in Proceedings of International Symposium, *New Aspects of Dietary Lipids – Benefits, Hazards, and Uses*, Sept. 17–20, SIK, Göteborg, Sweden.

GORDON M H and KOURIMSKA L (1995), 'Effect of antioxidants on losses of tocopherols during deep-fat frying', *Food Chem*, **52** 175–177.

GORDON M H and MAGOS P (1983), 'The effect of sterols on the oxidation of edible oils', *Food Chem*, **10**, 141–147.

GORDON M H and MAGOS P (1984), 'Products from the autoxidation of Δ^5-avenasterol', *Food Chem*, **14**, 295–301.

GOVIND RAO M K and ACHAYA K T (1968), 'Antioxidant activity of squalene', *J Am Oil Chem Soc*, **45**, p296.

GUPTA M K (1998), 'NuSun – The future generation of oils', *INFORM*, **9**(12), 1150–1154.

GUTIERREZ-ROSALES F, GARRIDO-FEMANDEZ J, GALLARDO-GUERRERO L, GANDUL-ROJAS B and MINGUEZ-MOSQUERA M I (1992), 'Action of chlorophylls on the stability of virgin olive oil', *J Am Oil Chem Soc*, **69**, 866–871.

GWO Y Y, FLICK G J, DUPUY H P, ORY R L and BARAN W L (1985), 'Effect of ascorbyl palmitate on the quality of frying fats for deep fat frying operations', *J Am Oil Chem Soc*, **62**(12), 1666–1671.

HAMILTON R J and PERKINS E G (1997), 'Chemistry of deep fat frying', in KOCHHAR S P, *New Developments in Industrial Frying*, Bridgwater, PJ Barnes & Associates, 9–33.

HANDELMAN G J (1996), 'Carotenoids as scavengers of active oxygen species', in Cadenas E and Packer L, *Handbook of Antioxidants*, New York, Marcel Dekker, Inc., 259–314.

HAUMANN B F (1996), 'Frying fats', *INFORM*, **7**(4), 320–334.

HAWRYSH Z J, SHAND P J, LIN C, TOKARSKA B and HARDIN R T (1990), 'Efficacy of tertiary butylhydroquinone on the storage and heat stability of liquid canola shortening', *J Am Oil Chem Soc*, **67** (9), 585–590.

HILDEBRAND D H, TERAO J and KITO M (1984), ' Phospholipids plus tocopherols increase soybean oil stability', *J Am Oil Chem Soc*, **61**(3), 552–555.

IBRAHIM K, AUGUSTIN M A and ONG A S H (1991), 'Effects of ascorbyl palmitate and silicone on frying performance of palm olein', *Pertanika*, **14**(1), 53–57.

JASWIR I, CHE MAN Y B and KITTS D D (2000), 'Synergistic effects of rosemary, sage, and citric acid on fatty acid retention of palm olein during deep fat frying', *J Am Oil Chem Soc*, **77**(5), 527–533.

JORGENSON K and SKIBSTEED L H (1993), 'Carotenoid scavenging of radicals', *Z Lebensm Unters Forsch*, **196**, 423–429.

KIM M, RHEE S H and CHIEG H S (1995), 'Effect of tocopherols and carotene on the oxidation of purified pine nut oil in the model system', *J Korean Soc Food Nutr*, **24**(1), 60–66.

KOCHHAR S P (1988), 'Natural antioxidants – A literature survey', Leatherhead

Food RA, Sci and Tech Surv No. 165.

KOCHHAR S P (1993), 'Deterioration of edible oils, fats and foodstuffs', in SCOTT G, *Atmospheric Oxidation and Antioxidants*, Vol. II, London, Elsevier, 71–139.

KOCHHAR S P (1998), 'Stabilisation of frying oils with novel natural components', paper presented at 89th AOCS Annual Meeting, May 10–13, Chicago, USA.

KOCHHAR S P (1999), 'Advances in chemistry and technology in deep-fat frying – European trends in 21st century', in Proceedings of the 1999 PORIM International Palm Oil Congress, *Emerging Technologies and Opportunities in the next Millennium, Chemistry and Technology,* 141–158.

KOCHHAR S P (2000), 'Stable and healthful frying oil for the 21st century', *INFORM*, **11** (6), 642–647.

KOURIMSKA L, POKORNY J and REBLOVA J (1995), 'Phospholipids as inhibitors of oxidation during food storage and frying', *Prehrambenotechnol Biotechnol Rev*, **32**, 91–94.

KOURIMSKA L, REBLOVA J and POKORNY J (1993), 'Stabilisation of dietetic oils containing phospholipids against oxidative rancidity', in LEVE G and PALTAUF R, *Phospholipids: Characterisation, Metabolism and Novel Biological applications*, Champaign, Illinois, AOCS Press, 372–377.

LEVIN G and MOKADY S (1994), 'Antioxidant activity of 9-cis compared to all-trans β-carotene in vitro', *Free Radic Biol Med*, **17**, 77–82.

MALECKA M (1991), 'The effect of squalene on the heat stability of rapeseed oil and model systems', *Die Nahrung*, **35**, 541–542.

MALECKA M (1994), 'The effect of squalene on the thermostability of rapeseed oil', *Die Nahrung*, **38**, 135–140.

MANORAMA R and RUKIMI C (1992), 'Crude palm oil as a source of beta–carotene', *Nutrition Research*, **12** (1) 223–232.

MARINOVA E M and YANISHLIEVA N V (1994), 'Effect of lipid unsaturation on the antioxidant activity of some phenolic acids', *J Am Oil Chem Soc*, **71**, 427–434.

MCCASKILL D R and ZHANG F (1999), 'Use of rice bran oil in foods', *Food Technol*, **53** (2), 50–52.

MEARA M L and WEIR G S (1976), 'The effect of beta-carotene on the stability of palm oil', *Riv Ital delle Sost Grasse*, **53**(7), 178–180.

MISTRY B S and MIN D B (1979), 'Effects of fatty acids on the oxidative stability of soybean oil', *J Food Sci*, **52**(3), 831–832.

MIYASHITA K and TAKAGI T (1986), 'Study on the oxidative rate and pro–oxidant activity of free fatty acids', *J Am Oil Chem Soc*, **63**(10), 1380–1384.

ORTHOEFER F T, GURKIN S and LIU K (1996), 'Dynamics of Frying' in Perkins E G and Erickson M D, *Deep Frying: Chemistry, Nutrition, and Practical Applications*, Champaign, Illinois, AOCS Press, p233.

PERKINS E G (1996), 'Volatile odour and flavour components formed in deep frying', in PERKINS E G and ERICKSON M D, *Deep Frying: Chemistry, Nutrition, and Practical Applications*, Champaign, Illinois, AOCS Press, 43–48.

DU PLESSIS L M and MEREDITH A J (1999), 'Palm olein quality parameter changes during industrial production of potato chips', *J Am Oil Chem Soc*, **76** (6), 731–738.

POKORNY J, REBLOVA Z, RANNY M, KANOVA J, PANEK J and DAVIDEK J (1992), 'Natural lechithins and phosphorylated acylglycerols as inhibitors of autoxidation of fats and oils', *Die Nahrung*, **36**, 461–465.

RASIT R and AUGUSTIN M A (1982), 'Effect of tertiary-butylhydroquinone on the stability of fried banana chips', *Pertanika*, **5**,119–122.

RICHHEIMER S L, BERNART M W, KING G A, KENT M C and BAILEY D T (1996), 'Antioxidant activity of lipid-soluble phenolic diterpenes from rosemary', *J Am Oil Chem Soc*, **73**(4), 507–514.

ROCK S P and ROTH H (1964) 'Factors affecting the rate of deterioration in the frying quality of fats II', *J Am Oil Chem Soc*, **41**, 531–533.

ROSSELL J B (1998), 'Industrial frying process', *Grasas y Aceites*, **49**(3–4), 282–295.

SILKEBERG A and KOCHHAR S P (2000), 'Refining of edible oil retaining maximum antioxidative potency', US Patent number: 6,033,706 March 7, 2000; EPA no. 98102528.1-2109, 13.02.98 and other worldwide patent applications pending.

SIMS R J, FIORITI J A and KANUK M J (1972), 'Sterols additives as polymerisation inhibitors', *J Am Oil Chem Soc*, **49**, 298–301.

SONNTAG N O V (1979), in SWERN D, *Bailey's Industrial Oil and Fat Products*, New York, John Wiley and Sons, Vol.1, pp152.

TYAGI V K and VASISHTHA A K (1996), 'Changes in characteristics and composition of oils during deep-fat frying', *J Am Oil Chem Soc*, **73**(4), 499–506.

WAGNER K H and ELMADFIA I (1998), 'Optimisation of the stability of plant oils by adding antioxidants', paper presented at the 52nd International Congress and Expo of the German Society of Fat Science, 13–15 Sept, Magdeburg, Germany.

WEISS T J (1983), *Food oils and their uses*, Westport, The AVI Publishing Co, pp16.

WHITE P J (1991), 'Methods for measuring changes in deep-fat frying oils', *Food Technol*, February, 75–80.

YUKI E, MORIMOTO K, ISHIKAWA Y and NOGUCHI H (1978), 'Inhibition effect of lecithin on thermal oxidation of tocopherols', *J Japan Oil Chem Soc*, (Yukagaku) **27**(7), 425–430.

7

Factors affecting the quality of frying oils and fats

J. B. Rossell, Leatherhead Food Research Association

7.1 Introduction

This chapter relates primarily to industrial frying and to the quality of industrially fried food rather than to domestic frying. It is maintained that the largest single influence on industrial frying is the quality of the frying oil, and the chapter therefore surveys the various aspects of industrial frying that can affect oil, and thus food, quality. In this chapter, the term 'frying oil' is preferred to the alternative descriptions 'frying fat' or 'shortening', as the frying media are used well above their melting points.

It is perhaps important to draw a distinction between industrial frying of foods for retail sale and restaurant frying. In the latter case the food is eaten hot immediately after frying and the chef can taste the food knowing that the customer will get food with the same taste as he approves in the kitchen. In industrial frying for retail sale, however, the food is cooled after frying, perhaps frozen and then kept in storage in warehouses, retail outlets and the customers' kitchen for a period that may be three months or more before it is reheated and then eaten. Any instability in the oil will have time to lead to the development of off flavour during this extended period between initial frying and final consumption, putting far more emphasis on the need for high quality in the oils used for industrial frying.

The quality of oil, fat or shortening used for frying is thus of paramount importance with regard to the quality of the fried food. Table 7.1 illustrates this by listing the amount of oil absorbed in different fried foods, where it can be seen that battered food, such as fish or chicken, absorbs about 15% frying oil, while breaded fish or chicken absorbs up to 20% frying oil.

Table 7.1 Oil absorption in fried foods

	% Absorption[a]
Frozen chips	5
Fresh chips	10
Battered food (fish/chicken)	15
Low-fat crisps	20
Breaded food (fish/chicken)	15–20
Traditional potato crisps	35–40
Doughnuts[b]	15–20

Notes
[a] As a percentage of the finished food.
[b] Doughnuts also contain about 10% fat used in preparation of the dough.

The amount of oil absorbed by doughnuts varies from 15–20% of their final weight. This is, of course, in addition to the shortening used in preparation of the dough, giving a final oil/fat content of up to 30%. Standard or traditional potato crisps absorb the highest quantity of oil, and up to 35 or 40% of the final food may be frying oil. Recently, low-fat crisps have been introduced, but these still contain about 20% fat. It should therefore be remembered that the fat used for frying becomes part of the food we eat, which further emphasises its importance with regard to the quality of the final food.

The most important aspect of industrial frying is therefore the frying oil, and, in surveying factors that affect frying oil quality, this chapter reviews (a) oil properties and composition; (b) transport, packaging and storage of oil; (c) the nature of the food fried and its interaction with the frying oil; (d) the frying equipment and the process of frying; and (e) the evaluation of the quality of the frying oil during use. Each of these factors is important in its own way, and it is of no advantage to concentrate on one or two, or even three, of these aspects without appreciating that there may be additional influences on the quality of the frying oil and thus the fried food.

7.2 Properties and composition of oils and the relationship between oil composition and its suitability as a frying oil

7.2.1 Fatty acid compositions and properties of unmodified oils

The compositions of some unmodified oils are indicated in Table 7.2. Some of these oils are frequently used in frying applications while others are not. The lauric-acid-rich oils, palm kernel (PKO) and coconut (CNO) for instance, are generally unsatisfactory as industrial frying oils since they contain large proportions of lauric and other fatty acids with fewer than 14 carbon atoms. These acids are quite volatile. If palm kernel or coconut oil is used in a frying application, the moisture in the fried food causes hydrolysis of the glycerides and liberation of the fatty acids. In the case of PKO and CNO these contain

Table 7.2 Fatty acid compositions of unmodified oils[7]

Fatty acid	Arachis oil*	Coconut oil	Cottonseed oil	Grapeseed oil	Lard[8]	Maize oil	Palm oil	Palm kernel oil
C6:0	ND	0.0–0.6	ND	ND		a	ND	0.0–0.8
C8:0	ND	4.6–9.4	ND	ND		a	ND	2.4–6.2
C10:0	ND	5.5–7.8	ND	ND		a	ND	2.6–5.0
C12:0	0.0–0.1	45.1–50.3	0.0–0.2.	0.0–0.5		0.0–0.3(a)	0.0–0.4	41.0–55.0
C14:0	0.0–0.1	16.8–20.6	0.6–1.0	0.0–0.3	1.5	0.0–0.3	0.5–2.0	14.0–18.0
C16:0	8.3–14.0	7.7–10.2	21.4–26.4	5.5–11	24.0	8.6–16.5	41.0–47.5	6.5–10.0
C18:0	1.9–4.4	2.3–3.5	2.1–3.3	3.0–6.0	14.0	1.0–3.3	3.5–6.0	1.3–3.0
C18:1	36.4–67.1	5.4–8.1	14.7–21.7	12–28	43.0	20.0–42.2	36.0–44.0	12.0–19.0
C18:2	14.0–43.0	1.0–2.1	46.7–58.2	58–78	9.5	39.4–62.5	6.5–12.0	1.0–3.5
C18:3	0.0–0.1	0.0–0.2	0.0–0.4	0.0–1.0	1.0	0.5–1.5	0.0–0.5(b)	
C20:0	1.1–1.7	0.0–0.2	0.2–0.5	0.0–1.0	0.5	0.3–0.6	0.0–1.09	
C20:1	0.7–1.7	0.0–0.2	0.0–0.1	ND	1.0	0.2–0.4	b	
C20:2	ND	ND	0.0–0.1	ND	0.3–0.5	0.0–0.1	b	
C22:0	2.1–4.4	ND	0.0–0.6	0.0–0.3		0.0–0.5	b	
C22:1	0.0–0.3	ND	0.0–0.3	ND		0.0–0.1	b	
C24:0	1.1–2.2	ND	0.0–0.1	0.0–0.1		0.0–0.4	b	
C24:1	0.0–0.3	ND	ND	ND		ND	b	

Table 7.2 Continued

Fatty acid	Olive oil[9]	Rapeseed oil (low erucic acid LEAR)	Rice bran oil[10,11]	Safflower-seed oil	Sesame-seed oil[12,13]	Soya-bean oil	Sunflower-seed oil	Beef tallow[8]
C6:0	ND	ND		ND	ND	ND	ND	
C8:0	ND	ND		ND	ND	ND	ND	
C10:0	ND	ND		ND	ND	ND	ND	
C12:0	ND	ND	ND–Tr	ND	ND	0.0–0.1	0.0–0.1	
C14:0	0.05% max	0.0–0.2	0.2–0.5	0.0–0.2	0.0–0.1	0.0–0.2	0.0–0.2	2.5
C16:0	7.5–20.0	3.3–6.0	14.8–17.0	5.3–8.0	7.9–10.2	8.0–13.3	5.6–7.6	24.5
C18:0	0.5–5.0	1.1–2.5	1.8–2.0	1.9–2.9	4.8–6.1	2.4–5.4	2.7–6.5	18.5
C18:1	55.0–83.0	52.0–66.9	40.5–43.9	8.4–21.3	35.9–42.3	17.7–26.1	14.0–39.4	40.0
C18:2	3.5–21.0	16.1–24.8	35.0–37.6	67.8–83.2	41.5–47.9	49.8–57.1	48.3–74.0	5.0
C18:3	1.0 max	6.4–14.1	1.1–1.7	0.0–0.1	0.3–0.4	5.5–9.5	0.0–0.2	0.5
C20:0	0.6 max	0.2–0.8	0.6–0.8	0.2–0.4	0.3–0.6	0.1–0.6	0.2–0.4	0.5
C20:1	0.4 max	0.1–3.4	0.5–0.6	0.1–0.3	0.0–0.3	0.0–0.3	0.0–0.2	0.5
C20:2								0–0.1(d)
C22:0	0.2 max(c)	0.0–0.5	0.2–0.4	0.2–0.8	0.0–0.3	0.3–0.7	0.5–1.3	
C22:1		0.0–2.0	Tr–0.1	0.0–1.8	ND	0.0–0.3	0.0–0.2	
C24:0	0.2 max	0.0–0.2	Tr–1.2	0.0–0.2	0.0–0.3	0.0–0.4	0.2–0.3	
C24:1		0.0–0.4		0.0–0.2		ND	ND	

* Also known as groundnut and peanut oils (a) range of acids C6:0 to C12:0 totals 0.0–0.5%. (b) Range of acids C18:3 to C24:1 totals 0.0–0.1%.
ND = Not detected; Tr = trace(<0.05), (c) raised to 0.3% max for olive pomace oils. (d) one sample had 0.1% C20:2, a trace being found in one other sample in Food RA work.

appreciable quantities of the short-chain fatty acids. These volatilise at frying temperatures, causing excessive smoke development. Fatty acids are, of course, liberated in the same way if palm oil or some other non-lauric oil is used for frying, but in this case the liberated fatty acids are of higher molecular weight, reflecting the constituents of the parent oil, and are therefore not so volatile. There is therefore less hazard of smoke formation. Coconut oil, however, is sometimes used as a frying medium where there is a local preference, e.g. in the Philippines, or if the particular flavour attributes of coconut oil are desirable, and a smoke problem can be tolerated; thus it is often used in the Pacific Islands to fry banana slices or nuts, which are then used in western foods such as muesli. These operations do not normally, however, correspond to industrial frying as carried out in Europe. Lauric oils also are more expensive in Europe than alternative frying media and in general, therefore, PKO and CNO are not normally used in European frying applications.

Palm oil (PO) has no such problem of short-chain acids giving a smoke problem, but it contains a large quantity of saturated acids. The sample illustrated in Table 7.2 has over 45% saturated acid, which some people may see as a disadvantage from a nutritional point of view. Palm oil is also a solid fat in temperate climates, and this too can cause problems if the oil-storage facility in a frying plant is located out of doors. Bulk fats held in storage tanks seldom solidify during normal food factory storage, but solidification may take place in pipes or valves leading from an outside storage facility to the indoor frying installation. In locations where these aspects are not a problem, palm oil is an excellent frying oil. It has a low iodine value and a low level of polyunsaturated fatty acids. This is important, as the polyunsaturated fatty acids are, in general, responsible for oxidation and off-flavour development.

Olive oil is a premium frying oil. It has a low level of saturated fatty acids, so is not criticised on nutritional grounds; in fact it is said to be nutritionally advantageous and is extolled as part of the Mediterranean diet. Its good frying performance is linked to its relatively low melting point, which means that it is easily stored in cold climates. It also drains from the fried food readily, there being no danger of solidification of the frying oil leading to excess fat on the fried food. It has low concentrations of linoleic (C18:2) and linolenic (C18:3) acids, giving the oil a low iodine value (IV) of about 80, which renders it resistant to oxidation. Its long shelf life and high resistance to oxidation are also, in part, related to the advantageous combination of various phenolic antioxidants present in olive oil. Olive oil is available in several grades, extra virgin olive oil being the best and usually the most expensive. This high-grade olive oil is seldom used for industrial frying, although some high-class restaurateurs may justify the expense. For industrial frying, the cheaper refined olive oil grades are usually satisfactory.

Soya-bean oil and rapeseed oil are similar to one another in that they have high iodine values and low levels of saturated fatty acids, and are fully fluid even at low ambient temperatures. The main difficulty with soya-bean and rapeseed oils is the 5–14% or so of linolenic (C18:3) acid present, as this makes these two oils prone to oxidation and off-flavour development. In this respect, it must be remembered that, with industrial frying, it is not only oxidation during the frying process that must be considered but in many cases also oxidation of the oil during subsequent storage of the fried food. Crisps (chips in USA parlance) are a particular problem in this respect as they contain a high level of absorbed oil, are often stored in transparent bags and expose the oil on a large surface area to the surrounding atmosphere. However, rapeseed oil and soya-bean oil are cheap and are widely used in many frying operations, especially in the fast-food area, where storage of the fried product is not necessary.

Rice bran oil is frequently used in Pacific area countries. It is essentially a by-product of the rice polishing industry as the bran that is removed during polishing has a moderate oil content. Since a large quantity of rice is grown worldwide the potential production of rice bran oil could equal that of soya-bean oil.[1] However the bran deteriorates quickly as a result of the action of enzymes liberated when the bran in bruised during its removal from the rice grains. The liberated enzymes cause rapid hydrolysis of the oil and increase in FFA; in fact it is reported that the FFA can increase to 30% within one week of storage.[2] Rice bran oil must, therefore, be recovered very quickly after production of the bran if the oil is to be of any use in the food industry. However, good quality rice bran oil has several beneficial features. It contains sterol esters called γ-Oryzanol which are claimed to have several health benefits[3,4] and as explained later the oil also contains higher amounts of the sterol δ-5 avenasterol than other oils, a sterol that is reported to have an antioxidant action at frying temperatures.[4,5] Rice bran oil also has a good content of natural antioxidants, and as explained later it contains more tocotrienol than any other common vegetable oil except perhaps palm.

Groundnut oil was at one time a premium frying oil. It is sufficiently fluid at ambient temperature to cause few problems of pipe blockage due to solidification, although it might solidify in pipes or drums exposed to very cold winter weather. Its main attraction is its low iodine value, and low level of polyunsaturated fatty acids. Its most unsaturated component is linoleic acid (C18:2), which ranges from 15–43% depending on the origin of the oil. The oil should contain less than 0.1% linolenic acid. However, some manufactured products may contain trace contamination with other oils, and slightly higher levels of linolenic acid, up to 0.3%, might at times be experienced. In terms of frying performances, the low iodine value and the near zero level of linolenic acid make it an admirable oil. Unfortunately, groundnut oil has declined in popularity, owing in part to its cost but also to production problems related to the aflatoxin problem with groundnuts, and to public worries about the peanut allergy to which some people are sensitive.[6]

Sesame seed oil is a specialist commodity, much used in oriental cooking. Sesame seed oil has an attractive fatty acid composition as illustrated in Table 7.2. It has approximately equal amounts of both oleic (C18:1) and linoleic (C18:2) acids and a very low linolenic acid content. Its fatty acid composition is therefore closely related to that of groundnut oil. However, sesame seed oil naturally contains several powerful antioxidants.[4,12,13] In addition to a total of about 800 mg/kg of tocopherols, mainly γ-tocopherol,[4,12,13] the compounds sesamolin, sesamin and sesamol all give antioxidant properties. The properties of sesame seed oil have been discussed by Kochhar[4,13] and Abou-Gharbia *et al.*[14]

Cottonseed oil has been a popular oil for frying in the USA. Its main problem is that it has a moderately high amount of saturated fatty acid, namely 21–26% palmitic together with up to 4% of longer chain length saturated acids. This leads to the oil developing an unsightly deposit of solid fat when stored at refrigerator temperatures – an aspect criticised by consumers. It was discovered that cottonseed oil that had been allowed to stand out of doors during the winter in unlagged tanks did not suffer this problem, and 'winterised' cottonseed oil became preferred. Nowadays, oils that have been industrially processed to remove high-melting triglycerides are still called 'winterised'.

Grapeseed oil is a preferred oil in France, even though it has a moderately high content of linoleic acid (C18:2). This is seen as a dietary advantage by some who consider a diet rich in linoleic acid to alleviate the risk of coronary heart disease (CHD). The oil is therefore related in fatty acid composition to sunflower and safflower seed oils. These oils are more easily oxidised than (say) groundnut or palm and are therefore not such good frying media. They are therefore sometimes hydrogenated to improve their frying performance, but this removes their main dietary advantage and the hydrogenated oils are then no better than hydrogenated soya or rapeseed oils, although they are much more expensive.

Maize oil has for a long time been considered a superior frying oil for domestic purposes. This may be due in part to its high sterol content[7] which could impart a lower viscosity and thus better drainage from the fried food. However it is an expensive oil and seldom used for industrial frying.

Animal tallows are also used for frying, beef tallow being popular in Yorkshire and eastern England, while lard is popular in Lancashire. Although animal fats have high levels of saturated fatty acid, as shown in Table 7.2, and are solid at room temperature; fresh fats have pleasant distinctive flavours, which are imparted to the fried food. There is a low tendency to oxidation of the fresh fat as the level of polyunsaturated fatty acids is quite low, beef tallow being slightly better than lard in this respect as shown in Table 7.2. Unlike vegetable oils, however, animal fats contain no natural tocopherol antioxidants. Some fast-food chains have used beef tallow in their frying operation, usually blended with a small amount of liquid vegetable oil. Tallow used for frying is allowed to contain mutton as well as beef tallow. Mutton tallow generally contains more stearic acid than beef tallow, and it often has a higher *trans* fatty acid content. These two effects may be due to the different butchering techniques used with the two types of carcass and tend to give mutton tallow a higher melting point. Frying tallow may therefore have properties slightly different from those expected for pure beef tallow. The mutton tallow of most blends is usually kept below about 10%.

Fish oils are seldom used for frying. They have extremely high iodine values, reflecting the high levels of polyunsaturated fatty acids with 20 or more carbon atoms. These long-chain polyunsaturated acids render fish oils extremely prone to oxidation which in turn leads to off-flavours of the food and foaming of the oil. In general, fish oils are unsuited for frying operations. Producers of fried fatty fish therefore ensure that their normal frying oil does not become accidentally contaminated with fish oil.

7.2.2 Properties of oils modified to increase their suitability for industrial frying

Some oils are modified to improve their industrial frying performance. This modification usually involves processes such as hydrogenation and fractionation, but the modification of one oil (sunflowerseed oil) by changing the fatty acid composition of the plant seeds by plant breeding has also been included.

Sunflower seed oil has a poor frying performance due partly to its high content of polyunsaturated fatty acid (linoleic) and also to its somewhat low content of natural antioxidants. This is mainly alpha-tocopherol present at a concentration of about 700 ppm.[7] Soya-bean oil in comparison has a tocopherol content, comprising mainly gamma-tocopherol, at a concentration of about 1500 ppm,[7] gamma tocopherol being a better antioxidant than alpha.[17] In order to improve the frying performance of sunflowerseed oil it may be hydrogenated, but this increases the *trans* fatty acid content, an aspect frowned upon by those who link *trans* fatty acids with coronary heat disease. An alternative has been to modify the sunflower plant so that it produces seeds with a high oleic acid content.[19] High oleic sunflowerseed oil is commonly traded as 'Trisun Oil' and has experienced a rapid growth in sales since its introduction in 1988. While this involves modification of the genetic makeup of the plant it is not necessary, under EU regulations, to label the plant or oil as a 'genetically modified organism (GMO)' since the modification was carried out by traditional techniques, which included plant breeding, and did not involve gene transfer. The fatty acid composition of high oleic sunflowerseed oil is shown in Table 7.3. The tocopherol content of high oleic sunflowerseed oil is unchanged and still comprises mainly alpha-tocopherol.[19] The fivefold reduction in linoleic acid content gives high oleic sunflowerseed oil a 3.5 fold increase in its resistance to oxidation. High oleic sunflowerseed oil therefore has a better frying performance than conventional sunflowerseed oil,[17] and can be still further improved by blending with rice-bran and sesame seed oils[4,13] as discussed later.

Palm oil is an excellent frying oil but, as mentioned above, it has a high amount of saturated fatty acids and a high solid fat content at room temperature. It may be fractionated to give a liquid fraction, called palm olein, which has a reduced amount of both saturated fatty acids and solid fat content. Its compostion[15] is shown in Table 7.3.

Palm olein is still of low iodine value (55–65 units), and has an even lower content of polyunsaturated fatty acids than groundnut or sesame seed oils. Its

Table 7.3 Typical industrial frying oils

	Hydrogenated rapeseed oil	Olein from hyd. soya	Olein from hyd.blend cotton/soya oils (e.g. Durkex 500)	Hydrogenated sunflowerseed oil	High-oleic sunflowerseed oil	Blend of sesame and rice bran oils (Good-fry®)	Palm olein	Palm super olein
IV	86	93.0	75–81	Ca 110	78.4–88.6	86	56–60	60.1–67.5
Slip point	25°C max	25°C max	21°C max	25°C max		20°C max	19.2–23.6	12.9–16.6
Fatty acid composition (% by weight)								
C12:0	–	–	1	–			0.2–0.4	0.2–0.4
C14:0	–	–	1	–	Tr		0.9–1.2	0.9–1.1
C16:0	5	14.2	9	7.0	2.6–4.1	4.5	36.8–43.2	30.1–37.1
C18:0	2.5	8.0	5	6	2.9–6.2	3.7	3.7–4.8	3.2–4.3
C18:1	81	45.6	78	76	77.4–90.7	78.6	39.8–44.6	43.2–49.2
C18:2	9	29.3	5	10	2.1–12.4	10.8	10.4–12.9	10.7–15.0
C18:3	max 2	1.7	Tr	0	Tr–0.1	0.1	0.1–0.6	0.2–0.6
C20:0	0.5	–	–	–	0.2–0.5	0.4	0.3–0.5	0–0.04
trans acids	ca 20	22	40–50	ca20	less than 4	less than 0.5	less than 4	less than 4
Reference		Hastart[20]	Durkee[21]		Griffith, Farmer & Rossell[19]	Kochhar[4,13,14]	Siew Wai Lin[15]	Siew Wai Lin[15]

Tr = trace

most unsaturated component is linoleic acid, which occurs at concentrations of around 10%. However, palm olein still has a relatively high level of saturated fatty acids in comparison with (say) soya or groundnut oils, and may thus still be frowned on from a dietary point of view. However, it is more fluid than palm oil and should seldom cause problems of pipe blockage or lack of pourability due to solidification unless, as with groundnut oil, oil pipes or drums are exposed to cold wintry weather. In batch installations, palm olein may partly solidify in drum storage, and it is usually recommended that unopened drums should be warmed before use, e.g. by being placed alongside the frying kettle.

In an effort to overcome this partial solidification, and to reduce the saturated fatty acid content still further, super palm oleins have been introduced.[15] These are produced by double fractionation. They have higher iodine values and a lower tendency to crystallise during storage than standard palm oleins. Siew Wai Lin has reviewed the properties of both standard and super palm oleins.[15]

One disadvantage with palm olein is that the oil tends to darken during frying, causing kitchen staff to presume that the oil is deteriorating in quality. In fact, this has been shown by the Palm Oil Research Institute of Malaysia (PORIM) to be due to the presence of low levels of polar compounds such as parahydroxybenzoic acid and vanillin.[16] Darkening during frying can also result from the oxidation of tocopherols to tocoquinones, which are red in colour.[17] The darkening of palm products during frying is therefore in no way a reflection of any inferior performance. The uses of palm oil and olein in frying have been reviewed by Berger.[16,18]

Rapeseed and soya-bean oils are illustrated in Table 7.2, and it has already been mentioned that a problem with these two oils is the high level of linolenic acid. Some of the newer varieties of rapeseed oil recently introduced into the UK do in fact have even higher linolenic acid contents.[7,22,23] Rapeseed oil and soya-bean oil can, therefore, be slightly hydrogenated during manufacture of oils for industrial frying, in order to reduce the linolenic acid content. This procedure, illustrated in Table 7.4 for soya-bean oil, can also be applied to rapeseed and sunflowerseed oils. The fatty acid compositions of the oils that may result are shown in Table 7.4. It should be remembered however that partial hydrogenation can be taken to any stage depending on the wishes of the plant manager, and the values shown in Table 7.3 are only examples of what may be produced.

These hydrogenated oils have reduced levels of polyunsaturated fatty acids. In the case of hydrogenated rapeseed and hydrogenated soya-bean oils, the level of linolenic acid is reduced to below 3%, the corresponding iodine values being around 100 units, similar to those of some varieties of unhydrogenated groundnut oil. Sunflowerseed oil has no linolenic acid, but the linoleic acid content can be reduced to about 40%.

Hydrogenated sunflowerseed oil may be attractive where a high polyunsaturated to saturated (P/S) ratio is needed for dietary labelling purposes, together with a good shelf life. Unlike rapeseed and soya-bean oil, at the time of writing, sunflowerseed oil has not yet been adversely linked with the GMO problem and consumers wanting a GMO free diet may therefore prefer hydrogenated sunflowerseed oil to hydrogenated frying media derived from rape or soya-bean oils. However, hydrogenated sunflowerseed oil will be more expensive than alternatives based on other vegetable oils, and attention should be paid to the *trans* acid content of the hydrogenated oil, an aspect currently causing concern in nutritional debates.[24] In fact, the use of hydrogenated sunflowerseed oil may be seen as a not very attractive option, since the main dietary attraction of sunflowerseed oil is its high linoleic acid content which hydrogenation destroys.

These hydrogenated oils may, of course, have higher melting points, and in some cases a somewhat harder frying fat is an advantage. In doughnut production for instance, such a fat will solidify quickly, giving better adhesion to the sugar dusting. Palm oil is therefore also attractive here. In cases where visual clarity and/or cold weather fluidity are important, the hydrogenated oils can be fractionally crystallised and filtered to remove high-melting components, as in the production of palm olein. This is illustrated in the case of soya-bean oil in Tables 7.4 and 7.5.[20] Soya-bean oil with an initial iodine value (IV) of 131 can be selectively hydrogenated at either of two temperatures, viz. 170°C or 130°C, to IVs of 106.6 and 104.0, respectively. In both cases, the residual linolenic acid content will be close to the target level of 2%. However, there is an appreciable *trans* isomer content in both oils, and a danger of solid-phase formation during storage. There is also a criticism of increased concentrations of saturated and *trans* fatty acids which give rise to dietary and nutritional worries amongst those consumers who consider themselves to be at risk of contracting heart disease. The oils are therefore slowly cooled to approximately 5°C and then held at this temperature overnight to allow complete crystallisation. Several methods may be employed to separate solid crystals and liquid oil, a rotating drum filter being the cheapest to install, but other types of flat-band filter and membrane/diaphragm units are also used. A more sophisticated and more expensive procedure is to mix the oil with aqueous detergent solution, which preferentially

Table 7.4 Production of partially hydrogenated soya-bean oil[20]

	Feedstock oil	Hydrogenation at 170°C	Hydrogenation at 130°C
IV	103.9	106.6	104.0
		Fatty acid composition (%)	
C16:0	11.0	11.1	10.9
C18:0	4.0	4.4	5.3
C18:1	23.9	45.4	46.7
C18:2	52.6	36.1	34.3
C18:3	7.6	2.2	1.9
C20:0	0.4	0.5	0.5
Others	0.5	0.3	0.4
Trans isomers	0.5	21.9	19.7

Table 7.5 Properties of hydrogenated winterised soya-bean oil[20]

Fraction	170°C hydrogenation Solid	170°C hydrogenation Liquid	130°C hydrogenation Solid	130°C hydrogenation Liquid
Yield	24.3	75.7	23.1	76.9
IV	96.3	108.9	93.0	105.3
		Fatty acid composition (%)		
C16:0	13.9	10.3	14.2	10.3
C18:0	6.1	3.7	8.0	4.4
C18:1	45.6	45.7	45.6	46.9
C18:2	31.4	37.3	29.3	35.8
C18:3	1.9	2.3	1.7	1.9
Others	1.1	0.7	1.2	0.7
Trans isomers	24.4	21.6	22.0	19.5

wets the solid phase, enabling centrifugal separation. This is the so-called Lipofrac system. Winterised oils of the type illustrated in Table 7.5 may be obtained. These have good fluidity and resistance to oxidation and off-flavour development.

The production of frying oils from soya and rape is reviewed by Hastert[20] and in Leatherhead Food RA Symposium Proceedings No 31.[25]

In the 1950s the Glidden company developed a cocoa butter substitute (CBS) by hydrogenation under *trans* promoting conditions of a blend of soya and cottonseed oils to an iodine value of 50–60 units and solvent fractionation of the hydrogenated oil to give a CBS as a middle melting fraction.[26] This was called Kaomel. Its success depended in no small part on its very high content of *trans* fatty acids. It was found that the liquid fraction by-product, i.e. the olein, had an outstanding resistance to oxidation. It was sold as a high-grade frying oil under the trade name Durkex 500, in recognition of the fact that Glidden had in the meantime merged with the Durkee Company. The number 500 was attached to the brand name to draw attention to the fact that the oil was claimed to have an AOM 'life' or Induction Period[27] of 500 hours. Its composition is shown in Table 7.3, where it will be seen that its *trans* fatty acid content is 40–50%. Despite its very good resistance to oxidation it has never sold widely owing perhaps to its high price. Alternatively, it may be that it was always of low availability owing to the restricted capacity of the solvent fractionation plant.

A relatively new innovation is the blended oil 'Good-fry®'. This is a blend of carefully selected batches of specially prepared rice bran and sesame seed oils. Rice bran oil, as explained earlier, has a higher than usual content of both gamma-tocotrienol and delta-5-avenasterol, both effective antioxidants in the frying process. Sesame seed oil has a high content of tocopherols together with several related natural antioxidant compounds such as sesamin, sesamol, etc., which are also effective antioxidants in the frying process. It is claimed[4,13,14] that the blend is an exceptional frying oil. Like Durkex 500 it is quite expensive, owing perhaps to the need to specially prepare and select batches of the raw materials. It is sometimes blended with high-oleic sunflowerseed oil to provide a more economic blend. Unlike several of the other manufactured frying oils discussed in this chapter Good-fry® has a zero *trans* fatty acid content, it is also free from synthetic antioxidants and GMOs. Table 7.3 includes the compositions of some oils manufactured for industrial frying, as well as those of oils used for batch frying.

7.3 Oil authenticity

While the main criterion of frying oil quality is of course performance, the price and authenticity of the oil are also very important. Buyers will judge the price they pay against the known cost of the relevant raw materials and the degree of processing involved. With many frying oils, the raw materials will be self-evident and the buyer expects to get what he pays for. In some cases he may get an impure oil. While it may be held that he is not losing if the oil has the appropriate performance, it is not proper for him to order one oil and get a different oil, even if the substituted oil has the same performance. If a cheaper oil or blend does have the same performance, the buyer should be able to order this and gain the benefit, unless of course the delivered oil is a patented blend sold under a brand name and with a declared specification. Several authenticity issues and the analytical tests that can be applied to clarify these are relevant to the frying oil industry, as explained below.

7.3.1 Current authenticity issues

While olive oil is one of the most expensive oils and therefore a prime target for adulteration or misrepresentation, other less expensive oils and fats are also at times fraudulently adulterated. This usually involves adding a cheaper oil to one that is more expensive. There is seldom any problem of food safety, but there is one of misrepresentation and false labelling if the resulting blend is offered or traded as a pure oil, or as a different blend. For instance, there was for several years a worry that groundnut and sunflowerseed oils from the continent of South America might have been contaminated or adulterated with soya-bean oil, which was cheaper.[7,11,12,19,22,23] This worry was based on the observation that the traded oils often contained low concentrations of linolenic acid although it was held by many experts that pure sunflowerseed and groundnut oils should be free of this fatty acid. Unfortunately, the Codex Alimentarius Fats and Oils Committee had issued specifications[28] for these two oils that permitted up to 1% of linolenic acid. These specifications have now been revised to show much lower levels.[29] At the time, however, there was a temptation for traders to adulterate these premium oils with up to 10% of cheaper soya-bean oil, which contains up to 10% linolenic acid, and still claim that the adulterated oils were within the 1% range set by the Codex Alimentarius Fats and Oils Committee.[28]

A major problem in the early 1980s was the so-called Singapore Cocktail. This arose indirectly as a result of the policy of the Malaysian Government to encourage a local palm oil processing industry. At the time, Malaysia produced almost all the world's palm oil. It was exported to many countries in the world and formed a major part of the diet in countries such as India, Pakistan, China, the Gulf States and large parts of the Western world. In earlier times, crude palm oil was imported by the countries concerned, refined and used to produce margarines, dough fats and above all frying oils. However, the Malaysians wanted to carry out the downstream refining and processing themselves. Tax incentives were therefore provided to encourage Malaysian industry to carry out the refining and processing in Malaysia and for them to export processed oil. As explained earlier, one form of processing is to fractionate the semi-solid palm oil to give a liquid palm olein that will stay fluid at ambient temperatures, and give a more flexible frying medium. The solid by-product fraction is called palm stearin.

At the time in question, the olein was refined and deodorised by the Malaysian industry to give a fully processed, finished, palm olein for use in a variety of applications, such as frying. The palm olein was exported to a number of countries, including Singapore, and attracted the beneficial tax allowance. The stearin, on the other hand, was not processed further and some of it was also exported, in the crude form, to Singapore. The stearin has only a limited number of applications and as a result attracts a lower price that the original palm oil, or palm olein. Fortunately for the Singapore dealers, Western buyers still wanted unprocessed, whole, palm oil. The opportunity therefore arose to blend the palm stearin and olein back together again to form a 'reconstituted' palm oil. Owing to the tax regime in force in Malaysia, this was still profitable. Western dealers called this blend the 'Singapore Cocktail'.

Unfortunately, there was no easy way to ensure that the two components corresponded to one another, and it was quite likely that in most cases the stearin came from a different batch, perhaps even a different factory, from that of the olein. The properties of the reconstituted palm oil therefore varied considerably and led to increasing difficulties in the receiving food factories. There were also problems over the state of oxidation of the oil as the stearin fraction was much more likely to become oxidised and lower the quality of the otherwise high-quality olein when the two were combined. This problem with the stearin led to the cocktail being much more severely oxidised than unprocessed whole palm oil. Unfortunately, as both the fractions had originated from Malaysian palm oil and were usually blended back in something approaching the right ratio, it was almost impossible to prove by conventional analytical methods that the manipulation had taken place. Palm olein itself was also, at times, adulterated with small amounts of the cheaper palm stearin but still sold as the more expensive palm olein to those not astute enough to know the difference.

A related case of adulteration occurred in 1983, when cottonseed oil was diluted with palm olein.[30] One story of the incident was that a Singapore dealer secured an order from the Egyptian Government for the supply of 10,000 tonnes of cottonseed frying oil in two lots. He intended to satisfy the contract by purchase of cottonseed oil from Australia. However, there was a shortage of Australian cottonseed oil at the time and he was able to procure only about 4,500 tonnes. He therefore added 500 tonnes of palm olein to make the quantity up to the requisite 5,000 tonnes for the first lot. Apparently, this deception went unnoticed, and when the time came for the second delivery he reversed the ratio. This time he was detected and ended up in gaol. If his fraud had been successful it would have netted him an extra US$14m. in illegal profit. The incident was widely publicised by the Malaysian Palm Oil Processing Industry at the time[30,31] as, although the Malaysians had supplied the palm olein they had done so as part of an honest transaction and had not been part of the deception.

Maize oil is a premium frying oil but is sometimes imported into the United Kingdom at prices below those of the crude oil on the international market. UK dealers have suspected that the oil may be impure but have been unable to prove any fraud. The problem with maize oil has been that its fatty acid composition overlaps that of other vegetable oils and it is possible to concoct a blend of oils with a fatty acid composition close to that of maize oil itself. Adulteration of maize oil with this blend is therefore difficult to detect by consideration of the fatty acid composition, the normal first line of approach. The same is true of the tocopherol composition, a useful line of attack in some cases.

Maize oil has a much higher level of sterols than other oils and as a consequence the maize oil sterols predominate in any blends of maize oil with other vegetable oils. An analysis of the sterols in any such blend therefore differs by only a trivial amount from that of the original maize oil, the main effect of blending being to dilute the maize sterols to a lower concentration, the actual fall in concentration depending on the level of adulteration. Such a fall in concentration could equally well be caused by some form of processing, such as deodorisation. The final concentrations of sterols are often still within the natural variation of the pure oil bearing in mind the possible effects of processing. The ratios of the concentrations of the individual sterols, useful in resolving purity issues with most other oils, change hardly at all in these blends of maize oil with minor amounts of other vegetable oils. For these reasons, it is difficult to detect adulteration of maize oil by conventional analytical techniques.

Rape and soya-bean oils have somewhat similar fatty acid compositions as shown in Table 7.2. Although they are among the cheaper of the vegetable oils, they are traded at slightly different prices, depending on the season of the year, soya-bean oil usually being the dearer. Dilution of one oil by the other may alleviate a temporary shortage of one of the oils, or provide a small commercial advantage, but such dilutions have been difficult to identify when the levels of addition of foreign oil are below 20%.

The last example of this type of fraud relates to the animal fats beef tallow and lard. These too have similar properties and are both very variable, depending to a large extent on the diets of the animals. For many purposes, the fats are interchangeable in food production or manufacture, especially frying, but certain religions ban food containing one or other of the fats. The Jewish and Muslim communities, for instance, do not permit foods containing pig fat or lard, while in India Hindus do not permit food to contain beef tallow. In the vegetable oil cases discussed above, it has been held that a 3% contamination of one vegetable fat with another may be tolerated bearing in mind the capabilities of Good Manufacturing Practice in modern bulk-handling oil and fat facilities.[32] However, when the distinction of animal fats is made on religious grounds, there is no room for tolerance of contamination of one of these fats with the other. Unfortunately, however, there is no reliable, scientific way of identifying either beef or pig fats contaminated with very small amounts of the other fat. The situation is further confused when it is accepted that other animal fats, e.g. chicken, mutton or buffalo fat, may be permitted.

A second type of fraud, which can be a problem with olive oil, is when the frying oil is all of the stated type but the grade of oil is not properly declared. Thus oils may be prepared from selected oilseeds by cold-pressing. Oils prepared by mechanical means alone, without the application of extra heat and in the absence of further processing are described as 'cold-pressed'. These oils normally have a fine flavour, dependent on the quality of the seeds used as raw material. They are produced in relatively low yield and are therefore more expensive. They sell at a corresponding premium. Oils obtained by hot-pressing and/or solvent extraction are obtained in higher yield and are therefore cheaper, but they will not have the fine flavour of a cold-pressed oil. Furthermore, oils can be solvent-extracted from inferior, cheaper, seeds and then refined and deodorised to give a bland, manufactured oil, which is quite wholesome but does not have the fine flavour and is not cold pressed. These oils, solvent-extracted in high yields, from bulk quality, cheaper, seeds will be even cheaper than those produced, for example, by hot pressing from high-grade seeds. There is a corresponding temptation to dilute a cold-pressed oil with a hot-pressed or solvent-extracted and subsequently refined oil.

An equally difficult fraud to detect can arise when an oil initially has too high an acidity. High-acidity oils can arise through processing of poor-quality oilseeds or by poor storage conditions of the oil. For instance, storage of oil in the presence of water enables hydrolysis, and liberation of the constituent fatty acids in the triglycerides that constitute 95% or so of edible oils. This leads to the presence in the oil of free fatty acids (FFA), which are not welcome in oils used in the frying industry. Crude oils are therefore refined to remove the free fatty acids, which leads to a loss of oil. However, the amount of free fatty acid initially in an oil can be used as an indicator of the quality of the original oilseed raw material and/or the conditions used to store the oil. A high acidity implies that something may be amiss, and crude oils are often purchased on contracts that specify a maximum FFA content. If, as a result of some mishap, an oil has an FFA above the contractual maximum, there is a temptation to refine part of the oil to remove the FFA and then to blend back unprocessed oil to give a supposed crude oil that is now within the contractual limit. This is clearly a fraud as the buyer is misled about the original quality of the crude oil. The part-processed oil may be difficult to refine and/or may give a refined oil of substandard quality. Furthermore, it is not correct to call the oil crude if it is in fact a blend of crude and refined oils.

In some cases, a particular origin of oil may be preferred; for example, West African groundnut oil is preferred to that of other origins, mainly because it has a lower iodine value which can confer a longer fry-life. Olive oils from specific locations are said to have a better, or more reliable, flavour, or a longer shelf life. None of these aspects are easily assessed when placing a contract. It is clearly a fraud to pass off a product of one origin as coming from another more favoured origin, but tests to distinguish the different origins are rare.

In the cases described the blended oils are wholesome and often fit for food use but it is a fraud to trade them as if they were pure oils. In some applications, however, the blended oil may be unsuitable for the intended use but the food manufacturer may not know this until the oil is in his tanks and he has produced several batches of sub-standard product.

7.3.2 Current methods to test for authenticity and their evaluation

Authenticity testing

Oils and fats that are mis-described, insufficiently described or adulterated, or deceptive by any combination of these factors can all be detected potentially or in theory by analysis and comparison of the resulting data with those from an authentic population. In this exercise, it is important to ensure that the 'authentic population' is representative. One of the problems with Codex Alimentarius specifications[28] in the past was that the data was derived as a result of an extensive literature survey. While the obvious errors of unsound data could be anticipated and measures taken to evaluate the reliability of the literature, it was not so easy to screen the reliability of the samples chosen for analysis. It was discovered that a high proportion of literature reports, understandably perhaps, related to obscure varieties of oilseed that had been analysed simply because they were botanical cutiosities, or in some other way unusual. In other cases, samples for analysis had been 'hand-picked' and were superior examples of the variety under study. The full stock of literature information was, as a result, not related to the vegetable oils traded in bulk for use as food on world markets. While vegetable oils, like other natural products, vary from batch to batch as a result of climatic or other variation, this variation is less marked when the oil is available in a single lot of (say) 5,000 tonnes. Furthermore, in contrast to many other natural products, a tank of 5,000 tonnes of liquid vegetable oil is likely to be more or less homogeneous, giving fewer sampling problems than arise with honey, rice, wheat or fish for example. However, this also means that any accidental impurity or fraudulent adulteration will be distributed throughout the whole lot.

The provision of a databank of authenticity criteria is also a problem with branded frying oil products. The manufacturer of a branded frying oil will maintain his declared specification, but not necessarily maintain the same oil blend for all time. He may change the oil blend to take advantage of market circumstances while still maintaining his published quality criteria. This makes it almost impossible for anyone other that the analysts employed by the maker of the branded product to confirm its purity and authenticity.

Authenticity criteria

Authenticity criteria were obtained at the Leatherhead Food RA over a period of years in a project[7,8,11,12,19,22,23,33–41] jointly funded by the (UK) Ministry of Agriculture, Fisheries and Food, the Federation of Oils Seeds, and Fats Associations Ltd. (FOSFA International), and The Leatherhead Food RA. Additional funds were also provided by the Egyptian Government and several companies in membership of the Leatherhead Food RA, in particular CPC Ltd., Unilever and Nestle. Over 600 samples of oilseed were obtained from trade sources, manually sorted to ensure absence of foreign seeds or impurity, and conformity with the declared seed type. Oil was extracted from the seed in the laboratory by the method described in ISO 659, and analysed by appropriate techniques.

In order to compensate for worries about sample reliability, work carried out at the Leatherhead Food RA, and which is described in more detail elsewhere,[7,8,11,12,19,22,23,33–41] concentrated on collecting a set of authentic vegetable oilseed samples that were representative of world trade. Efforts were made to avoid botanical curiosities, unusual or experimental agricultural varieties and handpicked specimens. Oilseeds were collected from the international trade, representing all major production areas. In all frying oil cases except safflowerseed oil, over ten different geographical production areas were sampled, and rapeseed and groundnut oils seeds from 17 different origins were studied. The seeds were normally of known origin and post-harvest history and on receipt in the laboratory were hand-sorted to remove foreign material, ensuring purity and authenticity of the seeds. These were then used to prepare the analytical vegetable oil samples by solvent extraction in the laboratory.

Fatty acid compositions were determined by converting the oils to methyl esters according to ISO 5509 and analysis of the methyl esters by ISO 5508 to give the values shown in Table 7.2. Iodine values were in the main calculated from fatty acid composition by American Oil Chemists Association (AOCS) method Cd 8-89, and occasionally by titration according to ISO 3961 (carbon tetrachloride solvent). Good agreement between the two methods (less that 2 units, generally less than 1 unit, difference) was obtained. Tocopherols (see Table 7.6) were determined by HPLC with fluorescence detector according to IUPAC method 2.432, and sterol compositions by ISO method 6799 (see Table 7.7). There is no standard method for the determination of Stable Carbon Isotope Ratio (SCIR), which was measured by the technique described elsewhere.[39]

These purity criteria are reported in Leatherhead Food RA Research Reports in the series entitled 'Authenticity of Edible Vegetable Oil and Fats', and became the basis of the compositional criteria of the FOSFA Guideline specifications[37] and the Codex Alimentarius Worldwide Oils and Fats Standards.[42]

Table 7.6 Ranges (means) of tocopherol and tocotrienol levels in vegetable oils[35,43] (mg/kg)

Tocol	Relative retention time	Oils analysed											
		Palm kernel[a,b]	Coconut[a]	Cottonseed	Soya-bean	Maize	Groundnut	Palm	Sunflower seed	High erucic rapeseed	Low erucic rapeseed	Rice bran[41]	Sesame[38]
αT	1.0	ND–44	ND–17	136–543 (338)	9–352 (99.5)	23–573 (282)	49–304 (178)	4–185 (89)	403–855 (670)	39–305	100–320 (202)	334.1–424.6 (385.9)	ND–3.5 (0.2)
βT	1.5	ND–248	ND–11	ND–29 (16.9)	ND–36 (7.7)	ND–356 (54)	0.41 (8.8)	–	9–45 (27)	24–158	16–140 (65)	– –	– –
γT	1.8	ND–257	ND–14	158–594 (429)	409–2 397 (1 021)	268–2 468 (1 034)	99–389 (213)	6–36 (18)	ND–34 (11)	230–500	287–753 (490)	54.5–68.3 (64.3)	499.9–1037 (761.7)
δT	2.7	–	ND–2	ND–17 (3.3)	154–932 (421)	23–75 (54)	3–22 (7.6)	–	ND–7 (0.6)	5–14	4–22 (9)	227.7–333.1 (267.6)	1.4–21.5 (12.8)
αT_3	1.1	ND–tr	ND–5	–	–	ND–239 (49)	–	4–336 (128)	–	–	–	200.7–285.1 (255.3)	– –
βT_3	1.7	–	–	–	–	ND–52 (8)	–	–	–	–	–	tr–22.9 (18.7)	– –
γT_3	1.95	ND–60	ND–1	–	–	ND–450 (161)	–	42–710 (323)	–	–	–	388.1–488 (416.0)	tr–19.1 (4.4)
δT_3	3.2	–	–	–	–	ND–20 (6)	–	tr–148 (72)	–	–	–	ND–69.5 (21.0)	ND–tr (tr)
Total		ND–257	tr–31	410–1 169 (788)	575–3 320 (1 549)	331–3 402 (1 647)	176–696 (407.4)	98–1 327 (630)	447–900 (709)	312–928	424–1 054 (766)	1 265–1 660 (1 428.8)	516–1 056.1 (779.1)

[a] = Means are negligible.
[b] = High values may be due to migration of some palm oil into the palm kernels before separation.
Means are shown below ranges. T = tocopherol; T_3 = tocotrienol.
ND – Not detected. tr = trace (<0.05mg/kg).

Table 7.7 Ranges (means) of sterol compositions in vegetable oils (percentage of sterol fraction; totals in mg/kg)[35,43]

Sterol	Oils analysed										
	Palm kernel	Coconut	Cottonseed	Soya-bean	Maize	Groundnut	Palm	Sunflowerseed	Rapeseed	Rice bran[41]	Sesame[38]
Cholesterol	1.0–3.7	0.6–3.0	0.7–2.3	0.6–1.4	0.2–0.6	0.6–3.8	2.7–4.9	0.2–1.3	0.4–1.3	0.4–3.7	tr–0.4
	(1.7)	(1.7)	(1.0)	(0.9)	(0.4)	(1.5)	(3.5)	(0.5)	(0.7)	(2.2)	(0.2)
Brassicasterol	ND–0.3	ND–0.9	0.1–0.9	ND–0.3	ND–0.2	ND–0.2	ND	ND–0.2	5.0–13.0	0.1–0.4	ND–0.5
	(0.1)	(0.5)	(0.3)	(0.1)	(0.05)	(0.0)	(ND)	(<1)	(9.6)	(0.3)	(0.1)
Campesterol	8.4–12.7	7.5–10.2	7.2–8.4	15.8–24.2	18.6–24.1	12.3–19.8	20.6–24.2	7.4–12.9	18.2–38.6	18.4–22.9	7.5–24.2
	(10.0)	(8.7)	(7.9)	(19.5)	(21.3)	(17.0)	(23.0)	(9.5)	(34.1)	(20.4)	(16.8)
Stigmasterol	12.3–16.1	11.4–13.7	1.2–1.8	15.9–19.1	4.3–7.7	5.4–13.3	11.4–11.8	8.6–10.8	ND–0.7	8.6–12.9	2.9–7.5
	(13.7)	(12.5)	(1.4)	(17.5)	(5.5)	(8.7)	(11.7)	(9.4)	(tr)	(10.8)	(5.6)
Sitosterol	62.6–70.4	42.0–52.7	80.8–85.1	51.7–57.6	54.8–66.6	48.0–64.7	56.7–58.4	56.2–62.8	45.1–57.9	31.4–38.3	40.0–75.4
	(67.0)	(46.7)	(83.2)	(54.3)	(63.4)	(58.5)	(57.7)	(59.9)	(49.9)	(34.8)	(58.6)
Δ^{5-}	4.0–9.0	20.4–35.7	1.9–3.8	1.9–3.7	4.2–8.2	8.3–18.8	2.1–2.7	1.9–6.9	ND–6.6	2.6–3.8	4.5–14.5
Avenasterol	(6.2)	(26.6)	(2.6)	(2.4)	(5.7)	(12.3)	(2.5)	(3.4)	(4.5)	(3.2)	(8.5)
Δ^{7-}	ND–2.1	NS–3.0	0.7–1.4	1.4–5.2	1.0–4.2	ND–5.2	0.4–0.8	7.0–13.4	ND–1.3	7.5–163	0.2–15.1
Stigmastenol	(0.6)	(2.4)	(1.0)	(2.9)	(1.8)	(2.1)	(0.7)	(10.4)	(0.5)	(11.6)	(3.5)
Δ^{7-}	ND–1.4	0.6–3.0	1.4–3.3	1.0–4.6	0.7–2.7	ND–5.5	0.4–2.5	3.1–6.5	ND–08	1.9–2.8	0.5–3.3
Avenasterol	(0.1)	(1.1)	(0.7)	(0.6)	(0.5)	(1.2)	(1.0)	(4.8)	(0.4)	(2.4)	(1.7)
Others	ND–2.7	ND–3.6	ND–1.5	ND–1.8	ND–2.4	ND–1.4		ND–5.3	ND–4.2	7.6–23.7	0.1–26.5
	(0.7)	(1.1)	(0.7)	(0.6)	(0.5)			(2.0)	(0.5)	(14.4)	(4.3)
Total	792.1 187	470–1 110	2 690–5 915	1 837–4 089	7 931–22 137	901–2 854	389–481	2 437–4 545	4 824–11 276	24 332–47 259	4 506–16 257
(mg/kg)	(1 025)	(807)	(4 490)	(3 199)	(13 776)	(1 575)	(446)	(3 387)	(7 516)	(36 758)	(10 141)

ND = Not detected; NS = Not separated; tr = Trace.

Use of authenticity criteria

In the investigation of a suspect oil or fat, it is usually the fatty acid composition that is the first criterion studied. This is because the fatty acid compositions of the different oils are easy to determine and are usually sufficiently different to clarify the majority of uncertainties. Thus, in the case of groundnut and sunflowerseed oils suspected of contamination with soya or rapeseed oils, as mentioned above, the linolenic acid (C18:3) content of the oil is a good indication of purity. As shown in Table 7.2, sunflowerseed and groundnut oils contain less than 0.1% of this acid, whilst rapeseed and soya-bean oils contain around 10%. A 5% contamination with rapeseed or soya oils will therefore increase the C18:3 content to 0.5%, an easily determined increase. If confirmation is needed, the concentration of the fatty acid at the triglyceride 2-position can be used to distinguish between these two candidate contaminants.

It was found[34] that the fatty acids at the triglyceride 2-position could be used as useful purity criteria. The acids at the 2-position come from a different biochemical pool from those at the 1- and 3-positions, and have a different composition. They can be determined by the lipase hydrolysis technique described in IUPAC Method 2.210. Although the acids have different compositions, they are present in an oil type in a constant ratio to the concentration overall.[34] Thus, if the concentration of an acid at the triglyceride 2-position in palm oil samples is plotted against the overall concentration for that acid, a straight-line graph is obtained. What is more, the slope of this line varies from oil to oil. For example, a plot of the C18:3 concentration at the 2-position against the overall C18:3 concentration for rapeseed oil gives a line with a slope quite different from that of soya-bean oil. As a result, if a sample of sunflowerseed or groundnut oil, which should contain no C18:3, is found to contain this acid, and as a result it is deduced that it is probably contaminated with either rapeseed or soya-bean oil, determination of the ratio of 2-position C18:3 to the overall concentration of this acid (the so-called enrichment factor)[22,34] will help to clarify which of these two oils is present.

For confirmation, tocopherol analysis can be useful (Table 7.6). While it is true that tocopherols are natural antioxidants and that their concentrations can fall as an oil ages, the concentrations cannot rise, so any sunflowerseed or groundnut oils having concentrations of these specific tocopherol compounds above the established maximum values are clearly contaminated. For instance, if soya-bean oil is suspected as a contaminant in sunflowerseed oil, Table 7.6 shows that the latter contains less than 40 mg/kg of gamma-tocopherol, while soya-bean oil contains up to 2,400 mg/kg of this compound. If a sunflowerseed oil sample contains more than 40 mg/kg, contamination with soya-bean oil is therefore indicated. The detection of soya-bean oil in groundnut oil by this means is not quite so sensitive as groundnut oil contains up to 390 mg/kg gamma-tocopherol. However, in this case, the concentration of delta-tocopherol can also be used, as groundnut oil contains less than 22 mg/kg in contrast to the concentration of up to 932 mg/kg in soya. If concentrations of both tocopherols are enhanced, soya-bean oil contamination is indicated, showing that tocopherol analysis is still a useful test

In the case of suspected rapeseed oil contamination, confirmation can be obtained by determination of the sterol concentrations (Table 7.7). Sunflower-seed and groundnut oils contain no brassicasterol (i.e. less than 0.2 mg/kg) whereas in contrast rapeseed oil contains over 100 mg/kg, making it relatively simple to detect as little as 5% rapeseed oil by this means.

In cases where there is still doubt, it is worthwhile calculating the iodine value from the fatty acid composition and comparing this with the established range. This approach is, sometimes wrongly, discounted in favour of the fatty acid composition, but it is a fact that a fatty acid composition of a contaminated oil can lie within the ranges for the oil declared on the label, while the iodine value can nevertheless be outside the corresponding range, demonstrating impurity. This is because the concentrations for the unsaturated acids, which contribute to the iodine value, are never all at the same end (high or low) of their ranges at the same time in any pure sample. Consideration of the iodine value removes this uncertainty.

As explained earlier, palm oil can sometimes be contaminated with the hard or soft fractions (stearins or oleins, respectively) produced by commercial fractionation. These can usually be detected by plotting values of iodine value against slip melting point, since it has been established that authentic samples give points lying in a characteristic area of the plot.[22,34,44] An alternative approach also developed[34] at the Leatherhead Food RA uses the product of palmitic acid enrichment factor and the C48 (tripalmitin) triglyceride carbon number concentration. The problem of adulteration of cottonseed oil by palm olein can often be clarified by reference to the fatty acid composition, since cottonseed oil has a palmitic acid content lying between 21.4 and 26.4%, whereas palm olein contains 36.8–43.2 (see Tables 7.3 and 7.4).[33,34] A further distinction is in the triglyceride carbon number composition, since cottonseed oil contains 12.6-19.9% C50 (palm olein 37.7%)[33,34] and 32.2–43.6% C54 (palm olein 12.8%),[33,34] the two ranges showing a clear distinction.

Rice bran oil is unusual in containing a high level (3–7%)[45] of unsaponifiable matter, comprising mainly wax esters,[46] sterols and tocopherols. In fact, rice bran oil is one of the few oils containing appreciable concentrations of tocotrienols,[41] the only other common oil containing these powerful natural antioxidants being palm olein[10] (see Table 7.6). The oil also contains relatively high concentrations[41] of avenasterol isomers (Table 7.7), materials said to have antioxidant properties at frying oil temperatures.[5,17] It also has a low level of polyunsaturated fatty acids. It therefore has the attributes of a premium frying oil and is incorporated into the branded product 'Goodfry'.[4,13]

Table 7.2 shows that rice bran oil has almost equal amounts of oleic (40–44%) and linoleic (35–37%) acids but a very low level of linolenic acid (1.1–1.7%). It thus resembles groundnut oil and to a lesser extent maize oil. Groundnut oil contains very long chain fatty acids, with up to 24 carbons, and any contamination of rice bran oil with groundnut would be shown up by this means. Maize oil in comparison has a higher level of linoleic acid (C18:2) which would reveal any contamination with this oil. Furthermore, rice bran oil has a high total tocol content and contains all the tocol compounds in moderate amounts, with the exception of beta tocopherol. Most other vegetable oils, except palm, contain appreciable concentrations of ß-tocopherol and any increase in the concentration of ß-tocopherol, or distortion of the ratios of concentrations of the other tocols, can therefore be used to substantiate deductions about impurity drawn from the consideration of fatty acid compositions.

Sesame seed oil also has about equal concentrations of oleic and linoleic acids but a low concentration of linolenic acid and therefore also resembles groundnut oil. It has low concentrations of saturated long chain fatty acids, with up to 24 carbons, enabling simple detection of any groundnut oil contaminant. The concentrations of palmitic and linoleic acids cover narrow ranges also enabling detection of oils such as palm, cotton rapeseed, soya and maize by means of the fatty acid composition. It has high concentrations of the natural antioxidant γ-tocopherol and only low concentrations of α- and ß-tocopherols. Sunflowerseed, soya-bean, groundnut, low-erucic rape and maize have much greater levels of these tocopherols enabling detection of the presence of these oils by means of tocopherol analysis. The presence of rapeseed can be confirmed by the brassicasterol content while an overall increase in the sterol content is indicative of maize oil contamination.

Sesame seed oil also contains the unusual compounds sesamin, sesamolin and sesamol, phenolic compounds that also serve as natural antioxidants. These compounds are of course responsible for the colour-forming reactions employed in the sensitive Baudouin and Villavecchia tests for the detection of sesame seed oil in other oils. The combination of tocopherols and sesamin related compounds confer a good resistance to oxidation on sesame seed oil making it an excellent frying oil. It is included in the branded frying oil 'Goodfry'.

In the case of maize oil, the fatty acid composition lies between those of other oils in such a way that it is possible to add a blend of different oils to maize oil without changing the fatty acid composition. This is also the case with the tocopherols. The sterol concentration of maize oil is much higher than that of other oils so that in any contaminated maize oil it is still the maize sterols that predominate. This therefore caused a problem in the authentication of maize oil until work at the Leatherhead Food RA[22,39] established that the stable carbon isotope ratio (SCIR) of maize oil is quite different from that of other commercial oils. Measurement of the SCIR therefore enables the identification of as little as 7% impurity in maize oil.[40] This technique was used in a recent MAFF survey of single-seed vegetable oil on sale in the UK, where it was found that maize oil was the most often contaminated oil.[32]

As mentioned earlier, rapeseed and soya-bean oils, although both cheaper than other oils, nevertheless command different prices, depending on the time of year. There are worries that they may sometimes be comingled, e.g. in order to alleviate a short-term stock shortage. The composition of such a blend can be determined initially by the fatty acid composition, since rapeseed oil contains about 50% oleic (C18:1) acid in contrast to the concentration in soya-bean oil of about 20%. Conversely, soya oil contains about 54% linoleic (C18:2) acid in contrast to the level in rapeseed oil of about 20% (see Table 7.2). Deductions derived from the fatty acid composition can be confirmed by measuring the enrichment factor as mentioned earlier, and by determination of the brassicasterol and tocopherol concentrations (see Tables 7.6 and 7.7).

The presence of lard in beef tallow is harder to establish, partly because there are wide ranges in the compositions of the individual fats. This is due on the one hand to the varied diets of the animals. As the pig is a monogastric animal, the composition of its depot fat resembles that of its diet; if the diet includes any beef fat, this will be laid down in the depot fat, which will then resemble the beef fat used in the animal's rations. Cattle, on the other hand, are ruminant animals, one consequence of which is that fatty acids become hydrogenated in the rumen of the animal. Thus, although a cow's diet may be strictly vegetarian and contain polyunsaturated fatty acids as part of the grass that it eats, these polyunsaturated acids become converted to acids with a lower level of unsaturation by the bio-hydrogenation in the rumen. Thus lard contains up to 0.5% of eicosadienoic acid (C20:2) whilst most samples of beef tallow are free from this acid. This was at one time thought to be the basis of a method to distinguish the two fats, but in some work at the Food RA one sample of tallow was found to contain 0.1% C20:2 while another sample had a trace (<0.05%) of this acid, showing that the criterion is not reliable. This observation was also reached by Rugraff and Karleskind[47] who analysed 25 commercial beef tallows and four laboratory samples, finding a mean content of 0.05% C20:2 in the beef fat samples.

However, the digestion of fat by animals involves liberation of the fatty acids, absorption of the liberated fatty acids and re-esterification of these by the animal organism. The way in which the fatty acids are re-esterified appears to differ between the pig and the cow as lard has much higher concentrations of palmitic acid at the triglyceride 2-position than beef tallow. Thomas[48] claims that 1% of lard in tallow can be detected by isolation of the triglyceride fraction with one saturated and two unsaturated fatty acid groups, followed by determination of the palmitic acid concentration at the glyceride 2-position by lipase hydrolysis. He reports that determination of the palmitic acid in the whole fat without prior separation of this particular triglyceride group gives a higher (worse) limit of detection.

The problem of detecting refined or hot-pressed oil in products labelled cold-pressed is as yet not fully resolved, although there are now European regulations that incorporate some sophisticated analytical approaches to clarify the composition of olive oil.[49] Some of these tests for olive oil can be applied to other oils but the background criteria need first to be established for each individual oil. Methods for the detection of hot-pressed and/or extracted oil have been suggested by a number of authors but none of them has as yet been fully established.[50,51,52]

The determination of the origin of an oil is still more difficult, mainly because oils from different parts of the world, but of the same type, differ only slightly from one another, usually being from related agricultural varieties that have been selected or optimised for best yield, disease resistance or other feature. For some reason, this is not the case with groundnut oil which does vary depending on origin. For instance groundnut oil from the USA has the lowest C18 content, whilst oils from West Africa and the continent of South America have the lowest and highest contents of C18:2, respectively. This enabled clarification of a suspect USA groundnut oil in the author's laboratory. The oil had the following fatty acid composition:

C16:0	11.9%
C18:0	2.4%
C18:1	41.8%
C18:2	35.7%
C20:0	1.3%
C20:1	1.2%
C22:0	3.3%
C24:0	1.7%
Calculated IV 99.4	

These values are within the worldwide range for groundnut oil (Table 7.2) and the oil was initially considered pure. Closer inspection revealed, however, that it did not conform to the criteria for USA oil as the C16:0 content was a little high, and the IV also too high (USA oil should have less than 11.8% C16:0, and an IV in the range 93.4–97), the C18:1 content was too high (USA oil has <46.5%), and the C18:2 content was too high (USA is <31.9%). Further inspection of the fatty acid composition data for groundnut oils from different origins revealed that the suspect oil did not correspond to oil from any single location. The C18:2 content was too high for West African oil, the C18:1 content too high for Argentine or Brazilian oil, and the C18:0 content too low for oil from Paraguay, Asia or the Nile Valley. It was concluded that the oil was slightly adulterated. It could have been a blend of Chinese and USA groundnut oils, a possibility that was discounted on commercial grounds by the client. Another possibility was groundnut oil contaminated with either 2–3% soya, 5% cotton or 10% maize oil, a conclusion that would not have been reached without consideration of the compositions of authentic groundnut oils from each of the different production areas.

Rossell *et al.*[22,39,40] demonstrated that maize oils grown in the Southern Hemisphere have a different stable carbon isotope ratio from those grown in the Northern Hemisphere. The iodine values of the oils grown in each hemisphere also differed. The reasons for these differences are not known but they give the opportunity to clarify whether a maize oil was harvested north or south of the equator.

Breas *et al.*[53] studied the ratio of oxygen isotopes in a range of olive oils of various geographical origins by continuous pyrolysis/isotope ratio mass spectrometry and plotted these against the stable carbon isotope ratio results obtained independently by standard techniques.[22,39,40] They found that the plot demonstrated a clear geographical dependence, one that had been previously found with wine.[53] Lower O^{18} contents were observed for the oils from northern regions whilst higher values were observed for oils harvested in more southerly areas. This was attributed to evapotranspiration, which creates an enrichment of the heavier isotopes of both oxygen and hydrogen in plant water, an effect that will be more extreme in a warmer climate. There was also a significant variation in the carbon isotope ratio, thought to be due to environmental conditions (water stress, atmospheric water and temperature), leading to both axes of their plot showing a geographic dependence. The isotopic ratio technique therefore shows promise as a method of distinguishing the geographical origin of a vegetable oil.

7.3.3 Potential future approaches

Many of the problems of identifying more than 10% adulteration or contamination of the bulk edible oils have been clarified. The present difficulty is that some of the methods, such as sterol analysis, are long and tedious, and therefore expensive. Others, such as stable carbon isotope ratio analysis, require sophisticated equipment, which is available in only a limited number of laboratories. Challenges for the future are therefore either to simplify the difficult methods used at present, or to replace them with routine tests, and to positively identify lower levels of impurity.

Another factor is the introduction of genetically modified oilseed plants. Sometimes the modification is by traditional plant breeding, as with low-erucic rape or high-oleic sunflowerseed oils. In these cases, the EU legislation[54,55] allows the seeds and oils to be traded without the need to specify them as genetically modified. They are not transgenic. There is, however, the growing belief that modified oils will become commonplace and that this will lead to a large number of oils with related characteristics. There is then a major problem for the analyst as he may have no database of the potential oils that could be present in his blend. A problem of this nature is already at hand with high-oleic sunflowerseed oil, as this has a fatty acid composition so close to that of olive oil that it is impossible to detect it without recourse to the most sophisticated tests, such as determiantion of Δ^7-stigmastenol.

In the case of oils produced by gene transfer there is, at the time of writing, a problem in Europe as the public are not prepared to accept food produced by this technique. They want GMO-free oil. While it is not legally necessary to label a vegetable oil as one that contains GMO,[55] since a refined oil contains no DNA or protein, neither is it permitted to label it as 'free from GMO' (unless of course it can be proved that it is). At present there are no reliable methods to detect GMO oil as a contaminant in oil from non-GMO seeds. The polymerise chain reaction (PCR) test can be used to amplify strands of DNA to the point where they can be easily identified and this gives a method of detecting moderately low concentrations of most GMO foods in non-GMO foods, but it does not work with vegetable oils since they contain no DNA. Unless the public concern about GMO foodstuffs diminishes, this will be a continuing problem for food industry analysts.

The main difficulty in the analysis of sterols is that they need separating from the triglycerides in the oil. This is achieved by saponifying the triglycerides, and extracting the sterols from the aqueous alkaline solution of water-soluble soaps that is formed. Unfortunately, the aqueous soap solution is also a good emulsifier, and formation of an emulsion can impede the extraction. This stage of the test is generally messy and can lead to losses of sterols. It is therefore in the process of being replaced (for instance in ISO 12228:1999) by column chromatographic separation of the sterols, a column of aluminium oxide being used to retain the fatty acid anions. The extracted sterols are then concentrated by thin-layer chromatography, removed from the TLC plate, derivatised and analysed by GLC. One consequence of the approach is that sterol esters (for example sitosterol oleate) are also saponified. A method that could easily separate sterols and sterol esters from each other and from the triglycerides would give a big saving in time and would allow separate analysis of the free and esterified sterols. It would generate additional analytical information, as it is known that free sterols have different compositions from esterified sterols and that the fatty acids esterified with the sterols also differ from the fatty acids in the triglycerides. Suitable techniques for application here might be HPLC or solid-phase extraction.

The study of the hydrocarbon content of oils also merits further consideration. It has been shown by McGill *et al.*[56] that vegetable oils naturally contain low concentrations of n-alkanes with odd carbon number chains ranging from 15 to 33, most having 27, 29 and 31 carbons. However, the relative proportions of the different alkanes varied from oil to oil in a way that could be used to identify the oil. An attraction of this approach is that it will also reveal if an oil has been contaminated with any unnatural hydrocarbon, e.g. from petroleum. This approach can also be used in the detection of processed oil in oil labelled virgin or unrefined since the action of bleaching earth at temperatures of over about 100°C leads to the generation of steroidal hydrocarbons by dehydration of the sterols.

A development of stable carbon isotope ratio analysis is to study the isotope ratios of specific components in an oil. This has already been carried out with respect to the isotope ratio of individual fatty acids in an oil and it has been shown that there are slight differences.[39,40] Any contamination of an oil will upset these slight differences, the nature of the distortion from the established pattern revealing the cause of the impurity. The technique of site specific NMR (also called SNIF-NMR) has been applied to honey, alcoholic beverages and fruit juices.[57] There is every possibility that it may also show advantage in the evaluation of edible oils.

As previously mentioned, Breas *et al.*[53] studied the ratio of oxygen isotopes in a range of olive oils of various geographical origins by continuous pyrolysis/isotope ratio mass spectrometry and plotted these against the stable carbon isotope ratio results obtained independently by standard techniques.[22,39,40] They found that the plot demonstrated a clear geographical dependence. This technique merits further study as it may well open the door to developments in clarifying not only the presence of foreign oils but also false origin declarations.

The technique of pyrolysis MS also merits further study as it is thought to be able to resolve complex mixtures into components. Two problems with this technique are that there is no easily understood link of the output of the pyrolysis MS apparatus with the original chemical composition of the oil, and an interpretation relies heavily on sophisticated statistical interpretation, both of which hinder its acceptability with some workers.

Another not fully resolved problem is in proving the purity of oils labelled 'cold-pressed'. The presence of refined oils can be detected by the techniques outlined above but the detection of hot-pressed oil is at present not fully clarified. Since frying oils are in any case heated during frying the utilisation of 'cold-pressed' oils is uncommon since any flavour advantages with cold-pressed oils are soon lost. This aspect of purity is not therefore of any great significance in the frying industry

7.4 Minor components

There are several minor components in edible oils and fats. Some of these impede the frying process, whilst others are beneficial. The most important minor components of all are of course the flavours, which may enhance, or in the case of off-flavours detract from the quality of the food. It is appropriate to review these various minor components separately. Minor components are also discussed from another viewpoint in Chapter 6.

7.4.1 Minor components that are regarded as detrimental

Although COMA[58] discounted dietary cholesterol as a major factor in heart disease, some consumers nevertheless prefer frying oils with low cholesterol contents. The concentrations of cholesterol in some foods are therefore listed in Table 7.8, olive oil having the lowest level of all. It is ironic, however, that some consumers may ask for fish portions, or a chicken piece, battered with a composition containing egg yolk, to be fried in oil substantially free of cholesterol, when the base food contains much more cholesterol than the cooking oil. It should be noted that vegetable oils are not totally free from cholesterol, but in comparison with animal products contain only trivial amounts. Olive oil has the lowest cholesterol content of all common vegetable oils. When animal fats are used as frying media, these will of course contain much higher levels of cholesterol than vegetable oils, as shown in Table 7.8, and vegetable products fried in animal fats will acquire an increased cholesterol level. However, although dietary cholesterol is not regarded as a problem by nutritional experts,[58] it is regarded as a problem by customers. Although the

Table 7.8 Cholesterol contents (range (mean) – mg/kg)

Safflower oil	6–10(10)	Soya oil	20–35(28)
Sunflower oil	10– 40(17)	Fish oil	2,000–6,000
Maize oil	20–100(50)	Dairy butterfat	2,200–2,800
Groundnut oil	10– 40(30)	Beef tallow	1,000–1,200
Cotton oil	20–100(43)	Egg yolk	*ca.* 12,600
Rapeseed oil	25– 80(50)	Fish	500–7,000
Olive oil	1–24(7)	Chicken, lamb chop	*ca.* 1,000

customers may be scientifically ill-informed, their views must be taken into consideration if we want them to eat the food we produce.

Another result of the presence of minor componets is foaming. Foaming can be a major problem in frying operations, and can cause a fire. Figure 7.1 shows the difference between foaming and bubbling oil. Foaming is when the oil surface becomes covered with a mass of thick bubbles that do not disperse during frying. This is dangerous because the foam can rise over the top of the fryer and cause a fire. Foaming should not be confused with the more open bubbles that appear when the food is first added to the oil and which should disperse. In the latter case, the bubbles are located only above the fried food, and do not cover the whole surface of the oil.

Foaming is usually associated with an increase in the oxidised triglyceride components in the oil, as these are surface-active and stabilise the foam.

Fig. 7.1 Comparison of foaming and bubbling oil in the frying of potato chips (French fries). Courtesy of Craig Millar.

Blumenthall and Stier[59] claim that the large surface area of the foam is a site for further oxidation, releasing oxidised components into the oil when the foam bubbles burst, thus further degrading the oil. Since oxidised triglycerides stabilise the foam, the concentration of non-oxidised triglycerides is therefore a useful criterion of oil quality.

Farkas and Hubbard[60] carried out some interesting trials into the transfer of heat from the oil into the food being cooked. They reached the conclusion that heat flux increases with increased oil degradation through a reduction in vapour (steam) bubble size and increase in bubble frequency due to changes in interfacial properties of the oil. The changes in the bubble size and frequency led to changes in the bulk movement of the oil and thus to increased heat transfer.

The most detrimental minor components are trace metals such as iron and copper. These promote oxidation if present at levels of over 0.1, or 0.01 mg/kg, respectively, and all efforts should be taken to minimise contamination of the oil with these transition metals, especially copper. Table 7.9 illustrates the concentrations of metals that reduce the AOM stability of an oil by 50%.[61] These values are of course relative and should not be taken to mean that an oil is safe if it contains less than the concentration of metal given in this table. It was established that the concentration of copper needed to be reduced to below 0.02 mg/kg to preserve margarine flavour.[61]

Residual solvents will not normally be present; if they are, they will reduce the flash point and give potential fire or explosion danger. Proteinaceous residues may be present in the oil when purchased. This is a more serious problem in unrefined cold-pressed oils such as groundnut oil or beef dripping. Residues of the food fried will also fall into this category, and should be removed by effective filtration or skimming. The hot oil should not continue to flow through the filter bed of hot debris, however, as this will negate the benefits of filtering it off. Filters should therefore be discharged on a regular basis.

Alkaline-reacting materials (ARM) such as sodium oleate are also reputed to cause deterioration of the frying oil,[62] as illustrated in Table 7.10, where it can be seen that the concentration of alkaline-reacting material (as sodium oleate)

Table 7.9 Concentration of metal at which the oxidative stability is reduced to 50% of its original value[61]

Metal	mg/kg
Copper	0.5
Iron	0.6
Manganese	0.6
Chromium	1.2
Nickel	2.2
Vanadium	3.0
Zinc	19.6
Aluminium	50.0

Table 7.10 Alkaline-reacting material (ARM) in restaurant frying oil

Days of use at 180°C	ARM (as sodium oleate)
0 fresh	7 ppm
1	15 ppm
2	18 ppm
3	37 ppm
4 (discard)	43 ppm

From Blumenthal, Stockler and Summers[62]

increased during the use of frying oil in a restaurant. When the level of ARM had risen to 43 mg/kg (as sodium oleate) it became necessary to discard the oil.[62] Irwandi *et al.*[63] claim that natural antioxidants were effective in reducing the ARM content of palm olein in a five-day frying trial. It must be added, however, that the view that alkaline-reacting materials are a major cause of oil deterioration is not widely held.

Other minor components that may be considered to be harmful to the quality of the fried food may be present in an oil. Lecithin and partial glycerides cause excessive foaming if they are present in moderate or high concentrations. Free fatty acid will cause smoke when present at levels of over 2%. This influence depends on the volatility of the free fatty acid, being more of a problem with the shorter-chain acids present in palm kernel or coconut oils, as discussed earlier. Table 7.11 gives a general summary of the detrimental minor components that may be found in a frying oil.

7.4.2 Minor components that are regarded as beneficial

As mentioned above, some minor components are beneficial to performance during frying. Tocopherols are natural, phenolic antioxidants, and are present in all vegetable oils.[7,22,33,34] Processing of the oil, and its subsequent storage and handling, should be arranged so that the natural tocopherols are preserved for as

Table 7.11 Detrimental minor components

Lecithin – foaming if over 100 ppm
Partial glycerides – foaming
Free fatty acids – smoke if over 2.0%
Trace metals – (oxidation). Cu <0.01 ppm Fe <0.1 ppm
Residual solvents – reduces flashpoint-fire, explosions
Alkaline materials – sodium oleate <40 ppm
Cholesterol (risk of promoting CHD?)
Oxidised triglycerides (promote foam, and perhaps fire)
Solvents (fire risk)
Proteinaceous residues, food scraps (cause darkening of oil and burnt flavours)
Alkaline reacting materials (cause oil degradation)

long as possible, or that excessive quantities are in part removed to an optimum level as recommended by Frenkel.[64] High levels of tocopherols, e.g. over 1,000 mg/kg, can promote oxidation, and suitable processing can reduce excessive levels present in some natural oils giving an improved stability.[64] For this reason, additional quantities should not normally be added. Any tocopherol additions should therefore be adjusted to replace tocopherols unavoidably lost during frying, or to achieve some agreed optimum level bearing in mind the intended use. Animal fats do not contain natural tocopherols, and, with these frying media, addition of moderate amounts of tocopherol can give an improvement. Some sterols – for example, Δ^5-avenasterol, also act as antioxidants.[64,65,66] Rice-bran oil contains more Δ^5-avenasterol than most other liquid oils, and this may confer good frying performance on this oil. This has been utilised by the production of 'Good-Fry'®, a carefully prepared blend of rice bran oil and sesame seed oil.[4,13] Irwandi *et al.*[63] studied the influence of rosemary and sage extracts on the frying life of palm olein, while Boskau[67] has reviewed the effects of natural antioxidants on the properties of an oil during frying.

Low levels of phospholipids (lecithin) act synergistically with tocopherols in alleviating risk of oxidation.[68] Concentrations of lecithin should, however, be limited to a maximum of about 100 mg per kg as higher levels cause foaming. Carotenoids can also inhibit oxidation during frying and it is sometimes claimed that an oil lasts longer if small pieces of carrot are added to the frying bath during frying. Various additives can also be used to enhance or retain oil quality. Synthetic antioxidants such as BHT or BHA are permitted at total concentrations of up to 200 parts per million. Although BHT and, to a lesser extent, BHA, are steam-volatile and are lost from the oil during the frying operation, some benefits are passed on to the fried food. Citric acid is also useful because it complexes transition metals such as iron and copper, thus limiting their catalytic pro-oxidant influence.[64] Most suppliers of refined oil will add up to 100 mg/kg citric acid to the oil immediately after the deodorisation if requested. Silicon oil (polydimethyl siloxane) is useful in that it is claimed to form a film on the surface of the frying oil, limiting uptake of atmospheric oxygen, as shown in Table 7.9.[69]

Table 7.12 illustrates the influence of polydimethyl siloxane (PDMS) on the quality of a sunflowerseed oil during different periods of heating, and with different concentrations of PDMS, or silicon oil, anti-foam agent.[69] The beneficial action of the PDMS is clearly illustrated, even when the agent is present at concentrations as low as 0.04 mg/kg. Levels of up to 2 mg/kg are normally recommended, as some of the anti-foam agent is lost on the fried food during processing.

MirOil Life Powder is claimed[70] to enhance the properties of the frying oil. It is a mixture of citric acid and moisture absorbed onto the surface of a food-grade volcanic ash. It is claimed that, when the powder is added to a used frying oil, the citric acid and moisture react with alkaline materials in the oil, such as soaps, precipitating these in a form in which they can then be removed along with the earth during filtration. Any residual iron or copper compounds are doubtless

Table 7.12 Influence of polydimethyl siloxane (% non-oxidised triglyceride in sunflowerseed oil)

Heating time	Silicon anti-foam concentration (ppm)			
at 180°C	0	0.04	0.06	0.10
1(h)	78	77	80	81
2(h)	70	77	77	82
4(h)	63	66	69	75
6(h)	50	54	56	72
8(h)	40	44	51	64

From Freeman, Padley and Sheppard.[69]

removed at the same time, and any suspended particulate matter is absorbed and removed by the earth during filtration. The filtration must be very thorough, however, to prevent carry-over of any powder into the frying bath, as this would contaminate the food with the powder, which, although it may be used as a processing aid, is not an approved food ingredient. Schulz[71] claims, however, to have tested MirOil Life Powder in a large-scale industrial frying operation with no beneficial effect. Table 7.13 lists the minor components that may be considered to be beneficial.

7.4.3 Development of flavour during use

In the majority of cases, industrial frying is carried out with fully refined oils which have no initial flavour. In these cases, any flavour in the oil is a flavour that has developed in the oil during the frying operation. This may be due to oxidation of the oil and the breakdown of the peroxides that formed;[17] this generally leads to unpleasant flavours. In contrast, the interaction between the peroxides and the amino acids in the food is claimed to be responsible for the pleasant fried flavour that develops during the frying process.[72] In a few cases, however, the oil may initially have an attractive natural flavour, which is preferred in high-quality fried foods. This would be the case, for instance, in the use of virgin olive oil in frying.

Table 7.13 Beneficial minor components

Naturally present:	Tocopherols – up to 1,000 ppm Sterols, e.g. Δ^5-avenasterol Phospholipids (lecithin) – up to 100 ppm Carotenoids
Additives:	BHA (BHT) up to 200 ppm Citric acid up to 100 ppm Silicon oil up to 2 ppm MirOil Life Powder

Another important aspect of the food quality that is influenced by the oil is the nature of the flavours that eventually develop during storage of the fried food. Although these may all be described as off-flavours, some are more acceptable than others. Thus coconut oil develops an initial coconut flavour, but, in the presence of moisture, a strong soapy off-flavour can be produced,[27] these soapy off-flavours also being a problem with palm kernel oil. Soya-bean and rapeseed oils develop painty, fishy, or beany off-flavours, while those in corn (maize) are slightly burnt or corny, and those in cottonseed or peanut are often nutty. The off-flavours that may develop in hydrogenated oils are more varied, as there is a wider variety of unsaturated fatty acid isomers in the oil. In some cases, they are not detected when the food is eaten, but lingering cardboardy notes can develop on the palate some time after the food has been consumed. Fortunately, however, this is a rare occurrence, and is probably encountered only with oils hydrogenated under poorly controlled conditions. Blumenthal, Trout and Chang[73] found that the flavours developing in hydrogenated soya-bean oil were less attractive than those developing in hydrogenated corn, cottonseed or peanut oils.

7.5 Quality limits for a fresh (unused) frying oil

The frying oil must also, of course, be fresh and subject to normal quality control criteria, as illustrated in Table 7.14. The peroxide value should be less than 0.4 mEq/kg and the colour should be less than 2 Lovibond red units in a 5.25-inch cell. In many countries, the linolenic acid level must be below 3%. For freshly refined oils, the free fatty acid level should be below 0.05%, but with some natural oils, such as olive oil, or unrefined animal fats (e.g. beef dripping),

Table 7.14 Quality limits for a frying oil

Peroxide value (mEq/kg)	0.4 max.
Para anisidine value	5 units max.
Linolenic acid (%)	3.0% max
Lauric acid	5% max.
Trans fatty acids	20% max.
Iodine value	130 units max. (preferably less than 100 units)
Colour (Lovibond red units, 5.25 inch cell)	2.0 max.
Free fatty acid (%)	0.1% max. (but 0.4% is optimum according to Matz[74] to provide better heat transfer)
Flavour	Bland
Flash point	315°C
Smoke point	200°C
Moisture	0.05% max.
Soap (sodium oleate)	25mg per kg max.
Copper	0.02 mg per kg max.
Iron	0.1 mg per kg max.

higher levels of free fatty acid may be tolerated. In fact, as discussed later, a slightly higher free fatty acid level is said to be beneficial as it promotes good heat transfer between the hot oil and the wet food. However, assessment of an oil intake at a bulk frying installation must require the initial quality to be low acidity and bland, as any development of acidity or of cooked, fried or partially oxidised flavours before the oil is used must denote some deterioration or premature breakdown of the oil, and indicates poor quality.

In addition, the flash and smoke points of refined oils should be above 315ºC and 200ºC, respectively. Unrefined oils such as virgin olive or fresh beef tallow will have flash and smoke points at lower temperatures – of about 285ºC and 169ºC, respectively – owing to the volatility of the free fatty acids in the unrefined oils. Oils with very low flash points can give rise to possible fire or explosion dangers.

The criteria that should be taken into account in selecting a frying oil are therefore price, freedom from foreign oils, fluidity, a low tendency to foam or smoke formation, oxidative stability of the fried oil, especially with products that are eaten after a period of storage, and good flavour stability of the fried product. Where slight off-flavours do develop, they should be acceptable in terms of the food fried. When dietary requirements play a role, the ratio of polyunsaturated to saturated fatty acids (P/S) must be taken into consideration. The cholesterol content of the oil, while not scientifically linked to coronary heart disease, is judged to be so linked by members of the public, and for this reason the cholesterol content of the oil should preferably be as low as possible.

7.6 Transport, delivery and storage

Catering establishments normally receive their oil in drums (or fats in boxes), whereas industrial units usually receive their oils by bulk road tanker. Melting of boxed fat needs care – an often-overlooked risk can be to add lumps of solid fat to the fryer in such a careless way that a bridge forms, leaving some of the heating surfaces bare. They can then overheat, causing burning of any traces of fat, burnout of the heater, or even fire.

For the bulk liquid oils, the average vehicle carries up to 25 tons of fat in a tank of cylindrical or near cylindrical shape, mounted on a four- or six-wheeled rigid chassis or as part of an articulated unit. The tank will be so mounted that it slopes down toward the outlet, or different outlets if divisions exist within the main unit. The tank will be lagged with a layer of mineral wool 50 mm or more thick, the whole being covered with metal sleeving. Fat is normally loaded at a temperature of about 45ºC at the works, and, with the temperature drop not exceeding 1ºC per hour, it can be expected to arrive at its destination at well above the melting point, should this be above ambient. However, delays do occur, giving rise to exceptionally long journey times and, for this reason, steam pipes are normally installed in the bottom of the vehicles' tanks. These can be connected to the customer's steam main on arrival for melting of any solidified oil. Discharge of the vehicle is either by an engine-driven pump or by introduction of compressed air above the oil in the tank. In either case, delivery is by flexible hose, several lengths of which are carried on the vehicle. To allow undivided tankers to deliver part loads, modern vehicles will have reversible pumps or accessory oil lines to clear the intake pipe to the receiving tank, thereby helping to avoid solidification of fat and blockage of the line between deliveries. If it is necessary to reheat the vehicle contents, then this must be done with care to avoid local overheating and damage to the fat, and it is essential that no fat be drawn off until at least 75% of the load is liquid. While some gain in speed may be achieved by drawing liquid oil from the bottom of the road tanker and pumping it in again at the top, this causes a risk of oxidation, and it is therefore safer to allow natural convection to do the work. Oxidation is also inhibited by blanketing the oil with nitrogen in bulk storage tanks, and also in the road tanker, where the agitation of transport might otherwise promote aeration. Work commissioned at Leeds University by Henry Colbeck Ltd[76] demonstrated the advantages of nitrogen blanketing as illustrated in Fig. 7.2.

Oil delivery tankers should, in the UK, be registered with the Seed Crushers and Oil Processors Association (SCOPA), which has a scheme to ensure that only suitable tankers are used, and that their tankers carry no industrial chemical cargoes prior to an edible-oil delivery. The tankers carry a registration number permanently painted on the rear of the tanker and a logbook with the same registration number that shows journeys to and from food plants and cleaning installations. Further details of the scheme are available from SCOPA. A typical road tanker is illustrated in Fig. 7.3.

The storage time involved from refining to use should normally be less than five days, and bulk storage tanks should be sized to this throughout. Care should be taken not to introduce any copper from drums, tanks, funnels or valves. The

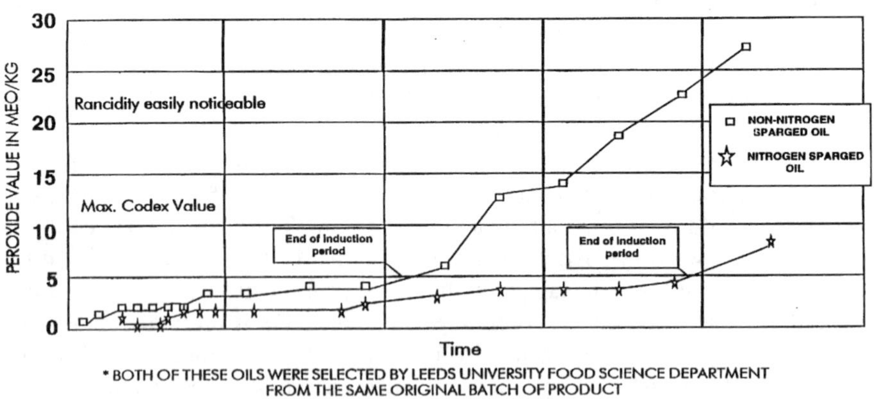

Fig. 7.2 Changes in peroxide value of rapeseed oil when comparing non-nitrogen sparged oil with nitrogen sparged oil (independent analysis from the Procter Department of Food Science, Leeds University).

Fig. 7.3 Bulk liquid oil tanker.

last-mentioned item is of most importance, as copper alloys are very useful in the construction of taps, valves and heating coils, but should be avoided when these are for edible-oil installations.

Oil oxidation is autocatalytic, and an oxidised oil can therefore promote and accelerate oxidation in fresh oil. Care should therefore be taken not to load deliveries of fresh oil on top of old stock, as any peroxides or secondary oxidation products present in the old oil will promote oxidation in the fresh delivery. Care should also be taken not to aerate the oil, and, as mentioned above, if possible, the oil in the tanks should be blanketed with nitrogen. The main cause of aeration is the discharge of fresh oil into a land tank. This may be carried out by discharge of the oil from an open pipe at the top of the tank, but, unfortunately, this is a very effective way of aerating the oil. It is preferable to arrange for the oil to discharge through a bottom outlet, but here again care must be taken. It is common to displace the last few hundredweight of oil from a delivery pipe by blowing the pipe with compressed air. It is easy to allow the compressed air supply to continue for some time after the last traces of oil have been displaced from the pipe, thus blowing air in at the bottom of the tank, and again ensuring full aeration of the oil. Some suitable system of valving and pressure monitoring can be used to avoid this.

Oils that have melting points above ambient should be kept warm, but not overheated. This should be no problem in lagged tanks. Oil should not normally be circulated in the tank, e.g. to promote melting of any solid deposits, as this will promote aeration and oxidation. Transport of oils and fats has been reviewed in depth by Rossell.[77]

7.7 The frying process

Matz[74] claims that a free fatty acid concentration of 0.4% is optimum for heat transfer to the fried food, as it reduces the interfacial tension between the hot oil and the film of surface moisture on most uncooked foods. The low concentrations of free fatty acid in most grades of olive oil are therefore a benefit in the frying process.

Blumenthal and Stier[59] developed this concept further, and claim that one of the most important aspects of frying performance is the influence of surface-active agents in the oil, breaking down interfacial tension between hot oil and wet food. Thus partly oxidised fatty acids, which form in the hot oil, act as surface-active agents and improve frying performance as an oil 'matures'. This is termed 'breaking in' the oil. As discussed later, more recent work has shown that the increased heat transfer with a matured oil is due to bubble size and frequency causing increased oil flow. Another possible advantage in allowing an oil to mature is that some of the fried food flavours can arise as a result of interactions between aldehydes in used cooking oils and proteins in the food. Such flavours may be absent in foods fried in very fresh refined oils.[72,75]

7.7.1 The nature of the food fried

Oil quality may be influenced by the nature of the food fried, especially when the product introduces foreign oils, emulsifiers, trace metals, food scraps, free fatty acids, alkaline-reacting material or other components into the frying oil. Potato 'whiteners' may improve the appearance of fried chips, but they can degrade the frying oil. Smith[78] has reviewed the frying of potato products and the relation to oil quality.

The introduction of foreign oils is most troublesome when fatty fish is fried, as any highly unsaturated fish oil will cause rapid oxidation of the oil. For this reason, it is recommended that fish products should be battered before frying. Chicken fat and dairy fats may also be a hazard, in the latter case because the short-chain acids present in dairy fats may be a cause of excessive smoke. Residual fat from grilled meats such as chops or sausages should not be used to top up frying kettles, as the animal fats are of low stability. This is due in part to the fact that they lack natural tocopherols, but also to the presence of iron from haemoglobin decomposition. In addition, residual meat fats will be already partly oxidised, and for these various reasons will promote oxidation of the frying oil if added to the frying bath.

Emulsifiers are often used in the production of foods, and if these escape into the oil they will cause excessive foaming. Free fatty acids may also migrate from the food product into the oil, and cause fume or smoke formation and in extreme cases also cause foaming. Some products, especially of the snack type, may introduce excessive quantities of trace metals into the oil. This is a particular problem with iron and copper, which will promote oxidation.

7.7.2 The frying operation

The frying operation can extend or shorten the fry-life of the oil considerably. It is accepted[79] that the rate of formation of decomposition products, and indeed the products themselves, vary with the food being fried, the fat being used, the choice of the fryer design and the nature of the operating conditions. It is therefore appropriate to review these influences on the quality of the fat.

Food scraps can become overcooked or char and impart a burnt flavour to the oil. They should therefore be removed from the oil by some form of filtration apparatus, e.g. a continuous filter unit with automatic discharge, or alternatively two batch-type filtration units working in unison. In a restaurant or batch-frying operation, the oil should be skimmed regularly. Furthermore, it is useful to arrange for a cool zone at the bottom of the fryer, e.g. below the heaters, so that scraps can fall into this zone and remain there without damaging the oil. Steps should be taken to ensure that this 'cold zone' does not become overloaded, as accumulated debris may touch the bottom of the heaters and char, developing unexpected off-flavours.

With a continuous filter it should be possible to clean the filter at regular intervals, to discharge the foreign material removed from the oil. If this is not done, the latter can form a bed of hot debris, which continues to deteriorate, damaging the good oil pumped through it. In some cases, this bed of material on the filter may contain high levels of detrimental material, for example iron or copper, previously filtered from the oil, but nevertheless in contact with oil passing through the filter and still able to catalyse oxidation. Furthermore, the filter should be located at the coolest part of the system, not immediately after the heat exchanger, for instance. In order to prevent rapid build-up of deposits in the filter, it is useful to shake or blow loose breadcrumbs or batter from the coated food prior to frying, and thus minimise the risk of their falling off during the frying operation. A careful optimisation of the air flow to the blower on the coating unit can thus improve oil quality – a relationship easily overlooked.

Filtration of the oil was once a considered to be a straightforward task, and none of the above precautions were considered necessary. However, the modern 'light eat' breadcrumb coatings are more friable (as well as being more fryable) and easily break down to give increased amounts of particulate matter in the oil. Factory sites that optimised their filtration system some years ago may therefore need to re-investigate this aspect if the 'life' of their oils appears to have deteriorated. Filtration with filter papers or woven cloths can remove particles down to about 5 μm in diameter and lessens the development of pyrolysis products and burnt flavours. 'Depth filtration' through non-woven fabrics can remove particles down to 1 μm, and may also remove microemulsified water, and thus slow oil deterioration.[59] 'Active filtration' is the technique in which adsorbent materials are used to remove soluble compounds that adversely affect the oil.[59,80] With active filtration, the adsorbent material must be fully removed from the oil before food is fried. Yates[81] has evaluated passive and active filtration. Excess quantities of breadcrumbs in the frying oil are also detrimental for another reason, since flour for breadmaking must (in the UK and several other countries) be fortified with iron for nutritional purposes. The iron in the breadcrumbs will contaminate the oil and catalyse oxidation.

Alkaline-reacting material has been identified as a possible cause of deterioration in frying oils,[62] probably owing to migration of alkali metal ions from the food into the frying oil, where they react to form soaps. This can be a particular problem with salted food, and it should be remembered that food should be salted after frying, and not before. When salt is added to a food before frying, some of the salt will fall into the frying oil, and can build up to high concentrations. Although sodium chloride is a much less active pro-oxidant than iron or copper, it is a pro-oxidant at the concentrations that may be encountered when food is salted before frying. In some work carried out in 1975 at the Leatherhead Food RA,[82] 10 ppm of sodium ion were found to reduce the induction period (IP) of a fully refined palm oil sample from 48 hours to 39 hours, a reduction of 9 hours. Since 5 ppm of copper ion (i.e. half the amount) reduced the IP of the same oil to 2 hours, a reduction of 46 hours, it appears from this rough and ready comparison that sodium has about 10% of the pro-oxidant catalytic activity of copper. Furthermore, the sodium chloride used in frying establishments, although of food grade, will not be of analytical chemical purity, and will undoubtedly contain trace quantities of iron and copper salts, a form in which these metals are very active pro-oxidants.

As mentioned previously, Farkas and Hubbard[60] carried out some interesting trials into the transfer of heat from the oil into the food being cooked. They reached the conclusion that heat flux increases with increased oil degradation through a reduction in vapour (steam) bubble size and increase in bubble frequency due to changes in interfacial properties of the oil. The changes in the bubble size and frequency led to changes in the bulk movement of the oil and thus to increased heat transfer.

Fryers that take solid fats should be provided with a special melting cycle. In some, ‘pulses’ of heat are followed by resting periods to allow solid fat to melt without overheating. In others, a low setting can be used to melt the fat, the higher setting being used only when there is enough oil to cover the heaters. The fryer may be heated by direct gas flames under the bottom of the vessel, but in this case the provision of a ‘cold zone’ under the heaters cannot be easily achieved. Otherwise, electrical resistance heaters or heating coils/ pipes may be installed a few inches above the bottom of the fryer to give the cold zone. Some fryers are equipped with external heat exchangers, in which oil is pumped through the fryer, through a filtration unit, and then through the heat exchanger before return to the frying bath. Rossell[83–86] has drawn attention to the toxicological hazards of any leaks of heat-transfer fluids into frying oils.

It is also important to ensure that recommended cooking temperatures are not exceeded. Table 7.15 lists recommended cooking temperatures during batch-frying operations, although it may be possible to exceed these temperatures in continuous frying equipment when there is a fast throughput of food and a correspondingly large need for topping up with fresh oil. Obviously, small pieces of food cook more quickly whilst larger pieces may need a lower temperature to ensure heat penetration without surface burning. In general, however, the oil will have a better quality, and the food a longer shelf life if the temperatures indicated in Table 7.15 are not exceeded. Blumenthal and Stier[59] point out that food quality and operating efficiency are compromised if a large fryer is used at only partial capacity. The fryer needs to be matched to the throughput. If two fryers are in use in a batch or restaurant situation, it is best to shut one down in slack periods rather than keep both hot but under-utilised. Filtered oil from the less busy unit should then be used to top up the busiest fryer, the former being topped up with new oil. This reduces oil deterioration.

In conclusion, Table 7.16 lists recommendations for processing, and aspects that should be avoided. The oil should be heated slowly in order to avoid

Table 7.15 Cooking temperatures for batch frying

Food	Temperature °C	Temperature °F
Potato chips – blanch	166–188	330–370
– fry	188	370
Potato crisps and straws	175	347
Fish fillets – battered and breaded	170–180	338–356
Fish cakes – breaded	188	370
Meat – battered and breaded cutlets	182	360
– lean chops (uncoated)	177	350
Chicken – large, battered and breaded	163–177	325–350
– small, battered and breaded	170–185	338–365
– pre-cooked, battered and breaded	170–185	338–365
Scampi – battered	180–185	356–365
Doughnuts	190	375
Choux paste	180	356
Onion rings – battered and breaded	180–185	356–365
Fruit fritters – battered	175–180	347–356
Vegetables	165	330
– battered and breaded	185	365
Maize snacks	190	375
Pre-cooked rice	190	375
Croutons	155	310
Nuts – almonds	115	240
cashews	135	275
peanuts	177	350
blanched peanuts	149	300
Chinese noodles – battered	190	375
Croquets – breaded	190	375

Table 7.16 Processing do's and don'ts

Processing do's

Heat up very slowly.
Avoid overheating.
Cool and then cover between batches and at end of frying operation.
Filter/skim to remove food debris from contact with hot oil (i.e. do not pump good oil through a bed of decomposing debris).
Use correct food/oil ratio (1:6).
Top up regularly with fresh oil. Oil turnover rate 5–12 hours.
Clean fryer regularly with strong detergent or boil out with aqueous alkali.
Clean thoroughly with clear water to remove soaps or detergent, and dry.

Processing don'ts

Don't overheat – check thermostat.
Don't allow drip-back from fume hood.
Don't use excessive fume extraction draught.
Don't allow oil filter to block.
Don't interrupt circulation of hot oil.
Don't salt food before frying.
Don't allow any copper metal.
Don't fry wet food.
Don't run plant with oil turnover times of more than 12 hours.

overheating; it should be cooled and covered with a floating lid between frying batches and at the end of the frying operation.[86] If a lid is positioned above the surface of the oil, condensate will drip back into the oil, causing deterioration. For this reason a floating lid is best;[87] it is also more effective in excluding air. The frying kettle should be topped up regularly with fresh oil; the correct food/oil ratio of 1:6 should be used, and the oil should be filtered and/or skimmed regularly to remove food debris. In this last respect, it is important to ensure that the debris does not ignite. The debris will already be warm, will have a large surface area, and will be impregnated with unsaturated vegetable oil. Oxidation of the oil distributed over the large surface area will doubtless take place, and in some circumstances can generate sufficient heat to cause smouldering or ignition, especially if the debris is stored near the cooker, where it will be kept warm. It is possible for the oil to catch fire, e.g. after foaming. Water and forced-jet extinguishers should not be used on such a fire as these will spread the hot oil. It is best to cover the hot oil, e.g. with a fire blanket, which should be kept nearby for the purpose.

Steps should be taken not to overheat the oil, and in this respect it is important to check the thermostat. The fume extraction equipment should also be checked to ensure that there is not an excessive draught of fresh air above the surface of the oil. This will not only cool the oil, but will also promote oxidation. It is also important to ensure that any fumes condensing in the extraction equipment do not drip or run back into the frying kettle. In slack periods, the kettle should not therefore be kept hot and should not be covered with a lid situated several inches above the fat, as this has the same effect. A floating stainless steel lid has been recommended for this purpose[86] in cases where it is necessary to cover the hot oil. Action should be taken to ensure that the circulation of hot oil is not impeded, e.g. by allowing the oil filter to block. If this happens, oil in the heat exchanger, or above the heating elements, will be subjected to excessive temperatures during the period of restricted circulation. Food should not be salted before frying, as this will lead to contamination of the oil by sodium chloride together with any impurities in the sodium chloride, all of which are pro-oxidants. Wet food should not be fried as the additional moisture will cool the oil and cause additional hydrolysis of the fat. In this respect, however, it must be mentioned that the steam evolved during cooking does in fact help protect the oil. The moisture normally present in food therefore has a beneficial effect on the oil. This is because the evolution of steam during the frying process causes a type of deodorisation of the oil, since the escaping steam bubbles carry with them volatile material, such as aldehydes and ketones, which are thus steam-distilled out of the oil. Furthermore, the steam forms a blanket above the surface of the oil, inhibiting access of atmospheric oxygen.

Since the fried food absorbs oil, the level of oil in the fryer will fall. It is usual to replace this with fresh oil. The amount of fresh oil added is related to the turnover rate. Banks[87] has suggested that this may be expressed as a ratio:

$$\text{Oil turnover} = \frac{\text{weight of oil in fryer}}{\text{average weight of fresh oil added per hour}}$$

or, alternatively, as the time taken to replace 100% of the original oil charge. However, it should be remembered that an 8-hour turnover will not correspond to all of the original oil being replaced in the 8 hours, as at each fresh addition the oil already in the fryer becomes diluted by the new oil, and will be at a reduced concentration in the oil removed by the food. Based on fryer design, a 5–12 hour turnover is normal for continuous production. In intermediate frying, with idling periods, longer turnover times are likely. Since idling puts more stress on the oil than active frying, longer turnover times amount to inefficient oil use. Fryer oil quality cannot be maintained with turnover times of 20 hours or more,[88] and after a few days it will be necessary to remove and discard some oil manually to enable a top-up with fresh oil. Oil turnover rate should be a primary factor in any thoughts about changes to fryer use.

A steady state is usually achieved in the fryer after four oil turnovers, and analytical values stabilise. This reflects blending of new and used oil at a constant rate, and uniform take-up by the food being fried. This steady state should be achieved before any finished product evaluations are undertaken.

Finally, stringent measures should be taken to avoid any contact with copper metal. All pipes, especially valves, thermometer pockets, electrical heating elements, and other fittings, should be checked to ensure that they are not constructed from any copper-containing alloy. There have been occasions when batch-frying baskets have developed holes, which have been given temporary repairs with short lengths of twisted copper wire. No doubt this can also happen in continuous frying installations, where the conveyor belt carrying the food into the fryer may develop a fault. It may be expedient to carry out a quick repair with a piece of twisted copper wire, rather than close the plant and call for the maintenance crew, but this must damage the quality of the oil. While a production manager may be able to justify this as a short-term measure, steps should be taken to ensure that a proper repair is carried out with the utmost speed, and the copper wire used initially should be removed. It would, of course, be much better to provide short lengths of soft stainless steel wire in situations where temporary repairs of this nature might become necessary. It is also rumoured[89] that disgruntled employees, such as those recently made redundant, have thrown small copper coins into oil storage tanks. While storage tanks should be frequently checked for cleanliness, this is especially so if a deterioration of oil quality is noted.

It is also necessary to clean the fryer regularly in order to remove traces of polymerised oil, which will promote degradation of subsequent oil batches. This may be achieved by fully draining the fryer, removing any baffles or guide bars for separate cleaning, filling with a 2% alkaline solution, which is then heated to boiling for at least 30 minutes during which time any conveyors used to move food in the fryer should be set in motion to immerse any paddles that are otherwise above the level of the detergent, and use of a long-handle mop to swab all surfaces above the liquid level. The fryer should then be emptied and drained, rinsed with dilute acid to neutralise the alkali, and finally washed thoroughly with clean hot water to remove traces of soap or detergent, and dried prior to refilling with oil. This cleaning treatment should be carried out once a week with fryers subjected to heavy use with unsaturated oils. Failure to remove alkaline cleaning compounds is claimed to be a prime cause of oil degradation.[59]

7.7.3 Evaluation during use

Several methods of evaluating a frying oil have been recommended. These include colour, free fatty acid content, polymerised and/or polar compound level, various colour reactions such as the Rau test and the Fritest, dielectric constant, viscosity, etc. The applicability of these various tests is discussed in depth in Chapter 8.

7.8 Future trends

Future trends relating to clarification on oil authenticity have already been discussed in section 7.3.3. Future trends and developments with frying oil fall into two main fields – improvements to the oil and improvements in the handling of the oil.

Oil improvements will probably come about by plant breeding, This may be via traditional means as it has already with high-oleic varieties of sunflowerseed oil, or it may come about via gene transfer to give oils now labelled as genetically modified organisms (GMOs). The efforts are likely from two approaches, namely increasing the fluidity of saturated fats and by reducing the levels of polyunsaturated fatty acids in liquid oils. In the case of the saturated fats, palm oil is the best candidate as it is already regarded as a good frying oil. The fluidity of palm oil will be improved by increasing its oleic acid content and reducing its saturated fatty acid content, this also making it less suspect to those at risk from CHD. It is more likely that this will be achieved by plant breeding rather than oil fractionation as so many efforts have been made to improve palm oil by fractionation that the best oleins have probably been already produced. However, it is fairly clear that the best palm oil is unlikely to be produced by plant breeding alone and, at least in the short term, candidate palm frying oils will be further improved by fractionation.

There will probably also be efforts to improve on oils such as high-oleic sunflowerseed oil. Although this has an admirable fatty acid composition it does not have the excellence in frying that may be expected of it. This is because it has a limited natural antioxidant content, having only moderate concentrations of alpha-tocopherol. In contrast soya-bean oil has a much higher total content of tocopherols and these include gamma and delta tocopherols, compounds that are much more effective at limiting oxidation at frying temperatures. Palm oil and rice bran oils contain tocotrienols, natural compounds that are even more effective. One approach to the provision of an improved oil would therefore be to improve high-oleic sunflowerseed oil by breeding into it natural antioxidants such as gamma and delta tocopherols together with a proportion of tocotrienols.

Oil handling is likely to be improved by more regular nitrogen blanketing, a technique already shown to have promise (see Fig. 7.2). The removal of food scraps from an oil by modified filtration systems is also likely to provide improvements in oil fry-life and product quality. A recent paper[90] claims that use of ash from rice hulls reduced the FFA and peroxide contents of used frying oils significantly. It is curious that the new branded frying oil 'Good-Fry' utilises rice bran oil[4,13] and the query arises of whether or not there could be some 'magic ingredient' in rice hulls and/or bran (such as perhaps selenium?) which act as an antioxidant in both cases.

7.9 References

1. FEDELI E. (1983) *J Am Oil Chem Soc*, **60**, 404–6.
2. CHAMPAGNE E.T. (1994) in *Rice Science and Technology*, p. 30. Eds. W.E. MARSHAL and J.I. WADSWORTH. Marcel Dekker, New York.
3. ROGERS E.J., RICE, S.M., NICOLOSI, R.J., CARPENTER, D.R., MCCLELLAND, C.A. and ROMANCZYK, Jr., L.J. (1993) *J Am Oil Chem Soc*, **70**, 301–7.
4. KOCHHAR, S.P. (2000) 'Stable and healthful frying oil for the 21st century'. *Inform*, **11,** 642–647.
5. GORDON, M.H. (1989). in 'New Aspects of Dietary Lipids: Benefits,

Hazards and Uses' pp 23–24. Proc. Int. Symp. Organised by Intern. Union of Fd.Sci. and Technol. Goteborg, Sweden (1990).

6. BUTRISS, J. (1998). 'An Update on Peanut Allergy' *BNF Nutrition Journal*, **23**, (85), 237–240. British Nutrition Foundation, London.
7. ROSSELL, J.B. (1998). 'Development of Purity Criteria for Edible Vegetable Oils' in *Lipid Analysis in Oils and Fats*. Ed. R.J. HAMILTON. Blackie Academic and Professional, London and New York.
8. ROSSELL J.B. (1992). 'Chemistry of Lipids' in *The Chemistry of Muscle Based Foods*. Ed. D.A. LEDWARD, D.E. JOHNSTONE and M.K. KNIGHT. Pub Royal Soc. of Chem. Cambridge, UK.
9. INTERNATIONAL OLIVE OIL COUNCIL (IOOC), (1999). 'Trade Standard Applying to Olive oil and Olive Pomace-Oil' Document COI/T.15/NC no. 2/Rev 9 (Madrid, 10 June 1999).
10. MCCASKILL, D.R. and ZHANG, F. (1999, February). 'Use of rice bran oil' *Food Technology*, 53, (2), 50–52.
11. GRIFFITH, R.E., FARMER, M.R. and ROSSELL, J.B. (1997). Authenticity of Edible Oil and Fats. Part XXV. Walnut, Cherry, Passion Flower, Oat, Wheat, Barley, Poppy, Rice Bran, Lupin, Pumpkin and melon-seed oils' Leatherhead Food RA Research Reports No. 744.
12. GRIFFITH, R.E., FARMER, M.R. and ROSSELL, J.B. (1994). 'Authenticity of Edible Oils and Fats. Part XXI. Sesame Seed Oil Samples'. Leatherhead Food RA Research Reports No. 721.
13. KOCHHAR S.P. (2000). 'Sesame Oil – a powerful antioxidant'. *Lipid Technology Newsletter*, **6(2),** 35–39.
14. ABOU–GHARBIA, H.A., SHEHATA, A.A.Y. and SHAHADI, F. (2000). 'Effect of processing on oxidative stability and lipid classes of Sesame seed oil'. *Food Research International*, **33**, 331–340.
15. LIN, S.W. (1999). 'Overcoming the crystallisation of palm olein'. *Palm Oil Technology Bulletin*, PORIM, Malaysia. **5**(6),6–7.
16. BERGER, K.G. (1987). 'Palm oil'. *Chemy Ind.*, 846–51.
17. FRANKEL, E.N. (1998). 'Frying Fats', in *Lipid Oxidation*. Ed E.N. FRANKEL. pub. Oily Press, Dundee, Scotland. 243.
18. BERGER, K.G. (1984). 'Food uses of palm oil'. *Leatherhead Food RA Symp. Proc. No. 29*, pp 78–94.
19. GRIFFITH, R.E., FARMER, M.R. and ROSSELL, J.B. (1997) 'Authenticity of Edible vegetable oils and fats. Part XXIV. High oleic oil samples – hazelnut, apricot, almond, high-oleic sunflower and high-oleic safflower-seed oil'. Leatherhead Food RA Research Reports No. 743.
20. HASTERT, R.C. (undated). 'Partial hydrogenation of soya-bean oil with Nysel DM-3 catalyst and its subsequent winterisation'. Harshaw Chem Co., Cleveland, Ohio, USA.
21. DURKEE INDUSTRIAL FOODS CORP. (1987). 'Durkex 500 – Multi–purpose Ultra-High Stability Oil' Brochure IND-218P, Durkee Export Sales Office, 925 Euclid Avenue, Cleveland, Ohio 44115, USA. p. 3.
22. ROSSELL, J.B. (1991). 'Purity Criteria in Edible Vegetable Oils'. *Fat. Sci.*

and Technol., **93**, 526–31.

23. DOWNES, M.J., SOLLARS, J., JORDAN, M.A. and WHITEHEAD, P.A. (1990). 'Authenticity of Edible Vegetable Oils and Fats. Part XIX. Analysis of Additional Samples of Rapeseed Oils 1986–1988 Crop Years'. Leatherhead Food RA Res. Reps. No. 682.
24. BRITISH NUTRITION FOUNDATION (BNF) (1995). *Trans* fatty acids – Report of the BNF Task Force, July 1995.'
25. VARIOUS AUTHORS (1985). *Leatherhead Food RA Symp. Proc.* No. 31.
26. GLIDDEN CO (1961). UK Patent No. 867,615.
27. ROSSELL, J.B. (1994) 'Measurement of rancidity' in *Rancidity in Foods*, 3rd edn. Eds. J.C. ALLEN and R.J. HAMILTON, pp 22–53. Blackie Academic & Professional
28. FAO/WHO (1982). 'Codex Alimentarius Volume XI. Codex Standards for Edible Oils and Fats'. FAO/WHO, Rome.
29. FAO/WHO (1993). Codex Alinorm, 95/17 Pub FAO/WHO Rome. November 1993, 69–80.
30. LAZARUS ROKK. (1983). 'Cottonseed Rip-Off – Our Refineries Cleared'. *New Straits Times* of Wednesday March 2nd 1983.
31. LAZARUS ROKK. and BALA, K. (1983). '$14m Export Swindle'. *New Straits Times* of Wednesday March 2, 1983.
32. MINISTRY OF AGRICULTURE FISHERIES AND FOOD (MAFF) (1996). 'Authenticity of Single Seed Vegetable Oils' Report of a Working Party on Food Authenticity. MAFF Report reference No. 162.
33. ROSSELL, J.B., KING, B. and DOWNES, M.J. (1983). 'Detection of adulteration'. *Journal of the American Oil Chemists Society*. **60**, 333–9.
34. ROSSELL, J.B., KING, B. and DOWNES, M.J. (1985). 'Composition of oil'. *Journal of the American Oil Chemists Society*. **62**, 221–30.
35. KING, B., TURRELL, J.A. and ZILKA, S.A. (1986) Authenticity of Edible Vegetable Oils and Fats. Part XI. Analysis of Minor Fatty Acid Components by Capillary Column GLC, and of Triglycerides by HPLC (Contains a synopsis of Parts I–X). Leatherhead Food RA Research Report No 563.
36. TURRELL, J.A. and WHITEHEAD, P.A. (1989) Authenticity of Edible Vegetable Oils and Fats. Part XVIII. Analysis of Additional Samples of Sunflowerseed, Groundnut and Maize Germ Oils. Leatherhead Food RA Research Report No 637.
37. FEDERATION OF OILS FATS AND SEEDS ASSOCIATIONS LTD (FOSFA INTERNATIONAL) (1993). 'Guideline Specifications' FOSFA International, 20 St Dunstans Hill, London EC3R 8HL.
38. GRIFFITH, R.E., FARMER, M.R. and ROSSELL, J.B. (1994). Authenticity of Edible Vegetable Oils and Fats. Part XXI. Sesame Seed Oil Samples. Leatherhead Food RA Research Reports No. 721.
39. ROSSELL, J.B. (1994). 'Stable carbon isotope ratios in establishing maize oil purity'. *Fat Science and Technology*, **96**, 304–8.
40. WOODBURY, S.E., EVERSHED, R.P., ROSSELL J.B., GRIFFITH, R.G. and FARNELL,

P. (1995). 'Detection of vegetable oil adulteration using gas chromatography combustion/isotope ratio mass spectrometry'. *Analytical Chemist*, **67** (15), 2685–90.

41. GRIFFITH, R.E., FARMER, M.R. and ROSSELL, J.B. (1997) Authenticity of Edible Vegetable Oils and Fats. Part XXV. Walnut, Cherry, Passionflower, Oat, Wheat, Barley, Poppy, Rice Bran, Lupine, Pumpkin- and Melon-seed Oils. Leatherhead Food RA Research Reports No. 744.
42. CODEX ALIMENTARIUS COMMISSION (1995). 'Report of the Fourteenth Session of the Codex Committee on Fats and Oils'. Alinorm 95/17. FAO/WHO, Via delle Terme di Caracalla 00100, Rome, Italy.
43. ROSSELL, J.B. (1991) 'Vegetable Oils and Fats' in *Analysis of Oilseeds, Fats and Fatty Foods*. Ed J.B. ROSSELL and J.L.R. PRITCHARD. Pub: Elsevier Applied Science, London and New York.
44. TAN, B.K., SIEW, W.L., OH, F.C.H. and BERGER, K.G. (1981). 'Detection of palm Oil in Palm Stearin' Paper T9 given at conference 'The Palm Oil and Its products', Kuala Lumpur, Malaysia.
45. ROSSELL, J.B. (1986) 'Classical Analysis of Oils and Fats' in *Analysis of Oils and Fats*. Ed R.J. HAMILTON and J.B. ROSSELL. Pub Elsevier Applied science. 1–90.
46. KUNDU, M.K. and DEB, A.T. (1981) 'Column Chromatographic determination on unsaponifiable matter in vegetable oils', *Fette Seifen Anstrichmittel verbunden mit die Ernährungsindustrie*, **83**(2), 73–6.
47. RUGRAFF, L. and KARLESKIND, A. (1983) 'Analyse des mélanges de graisses animals. Application au contrôle de l'absence de graisse de porc dans les suifs et subsidiairement dans les produits carnés et derivés crus ou cuits', *Revue fr. Cps gras*, **30** (9), 323–31.
48. THOMAS, A. (1987) 'Fats and Fatty Oils'. *Ullmans Encyclopaedia of Industrial Chemistry*. Vol **A10**, 173–243 esp. Table 22 p 215.
49. Commission Regulation 2568 of 11 July 1991 on the characteristics of olive oil and olive residue oil and the relevant methods of analysis. Official Journal of the European Communities of 5 September 1991, Vol. **34**, l248/1.83.
50. GOMBOS, J. and WOIDICH H. (1987). 'Einfluss von Gewinnung und Verarbeitung auf die Inhalts- und Begleitstoffe der Pflanzenöl'. (Teil 1 und Teil 2). *Ernährung*, **11**, 459–64, and 539–45.
51. GERTZ, C.H. (1991). 'Native und nicht rafinierte Speisefette und Öle'. *Fat Science Technology*. **93**, 545–8.
52. BRUHL, L. and FIEBIG, H.J. (1995). 'Quality characteristics for cold pressed edible oils', *Fett Wissenschaft und Technologie*. **97**, 203–8.
53. BRÉAS, O., GUILLOU, C., SADA, E. and ANGEROSA, F. (1998). 'Oxygen-18 measurement by continuous flaw pyrolysis/isotope ratio mass spectrometry of vegetable oils' Rapid Comm. In *Mass Spectrometry*. **12**, 188–92.
54. Council Regulation (EC) No. 1139/98 of 26 May, 1998.
55. Commission Regulation (EC) No. 49/2000 of 10 January 2000.
56. MCGILL, A.S., MOFFAT, C.F., MACKIE, P.R. and CRUICKSHANK, P. (1993). 'The

composition of alkanes in edible oils'. *Journal of Science and Food Agriculture*, **61**, 357–62.

57. COLQUHOUN, I.J. and LEES, M. (1998). 'Nuclear Magnetic Resonance Spectroscopy' in *Analytical Methods of Food Authentication*, Eds P.R. ASHURST and M.J DENNIS. Pub: Blackie Academic and Professional, London and New York. 36–75.
58. DHSS (1984). 'Diet and Cardiovascular Disease'. DHSS Report No. 28 on Health and Social Subjects by the Committee on Medical Aspects of Food Policy. (COMA Report) HMSO London.
59. BLUMENTHAL, M.M. and STIER, R.F. (1991). 'Optimisation of deep-fat frying operations'. *Trends in Fd. Sci. Techol.*, **2**, 144–8.
60. FRAKAS, B.E. and HUBBARD, L.H. (2000). 'Analysis of Convective Heat Transfer During Immersion Frying' *Drying Technology*, **18**(6) 1269–85.
61. SWERN, D. (1982) *Baileys Industrial Oil and Fat Products*, Vol 1, 4th edn. Wiley, New York.
62. BLUMENTHAL, M.M., STOCKLER, J.R. and SUMMERS, P.K. (1985). 'Alkaline contaminant materials in used frying oils: A new quick test'. *J Am Oil Chem Soc*, 62 (9), 1373–4.
63. JASWIR, I. MAN, Y.B.C. and KITTS, D.D. (2000). 'Effect of Natural Antioxidants in Controlling contaminant materials (ACM) in heated palm olein', *Food Research International*, **33**, 75–81.
64. FRANKEL, E.N. (1998). 'Antioxidants', in *Lipid Oxidation*. Ed. E.N. FRANKEL. Oily Press, Dundee, Scotland. 142.
65. BOSKAU, D. and MORTON, I.D. (1976). 'Effect of plant sterols on the rate of deterioration of heated oils'. *J Sci Fd Agric*, **27**, 928–32.
66. GORDON, M.H. and MAGOS, P. (1983). 'The effect of sterols on the oxidation of edible oils'. *Fd. Chemy*, **10**, 141–7.
67. BOSKAU, D. and ELMADFA, I. (1999). Frying of Food: Oxidation, nutrient and non-nutrient antioxidants, biologically active compounds and high temperatures'. Technomic Publishers.
68. DZIEDZIC, S.A. and HUDSON, B.J.F. (1984). 'Phosphatidyl ethanolamine as a synergist for primary antioxidants in edible oils'. *J Am Oil Chem Soc*, **61** (6), 1042–5.
69. FREEMAN, I.P., PADLEY, F.B. and SHEPPARD, W.L. (1973). 'Use of silicones in frying oils'. *J Am Oil Chem Soc*, **50** (4), 101–47.
70. ANON (1994) 'Frypowder, an oil stabiliser from MIROIL', *Food Technology* (December), 92.
71. SCHULZ, E.W. (1986). Personal communication to Dr J.B. Rossell.
72. GILLATT, P. (1988). 'The interaction of oxidised lipids with proteins'. Leatherhead Food RA Sci. & Tech. Survey No. 163.
73. BLUMENTHAL, M.M., TROUT, J.R. and CHANG, S.S. (1976) 'Correlation of GC profiles and organoleptic scores of different fats and oils after simulated deep fat frying'. *J Am Oil Chem Soc*, **53**, 496.
74. MATZ, S.A. (1976). *Snack Food Technology*, p. 23. AVI Publishing Co., USA.

75. HO, C-T, CARLIN, J.T. and HUANG, T-C. (1987). 'Flavour development in fried foods' in *Flavour Science and Technology*. Eds M. MARTENS, G.A. DALEN and H. RUSSWURM JR., pp. 35–42. J. Wiley & Sons.
76. DUNCAN MCLEAN OF HENRY COLBECK LTD (2000), Personal communication to J.B. Rossell.
77. ROSSELL, J.B. (1998) 'Transport of Oils and Fats' in *Food Transportation*. Ed. R. HEAP, M. KIERSTAN and G. FORD. Blackie Academic & Professional, London and New York, 129–143.
78. SMITH, O. (1987). 'Potato chips' in *Potato Processing*, 4th edn. Eds W.F. TALBURT and O. SMITH. AVI, Westport, Conn., USA. 371–489.
79. STEVENSON, S.G., VAISEY-GENSER, M. and ESKIN, N.A.M. (1984) 'Quality control in the use of deep frying oils', *J Am Oil Chem Soc*, **61**(6) 1102–1108.
80. JACOBSEN, G.A. (1991). 'Quality control in deep-fat frying operations', *Fd Tech.* **45**(2), 72–4.
81. YATES, R.A. (1996) 'Evaluation of active and passive filtration media', in *Deep Frying*. Eds E.G. PERKINS and M.D. ERICKSON, pp 297–310. AOCS Press, Champaign, Il. USA.
82. MEARA, M.L. and WEIR, G.S.D. (1975). 'Components Affecting the Oxidative Stability of Palm Oil'. Leatherhead Food RA Research Reports No. 234.
83. ROSSELL, J.B. (1991) 'Survey of heat-transfer materials in the vegetable oil processing industry'. Leatherhead Food RA Scientific and Technical Surveys No. 171.
84. ROSSELL, J.B. (1993a) 'Heat transfer fluids in the oils and fats industry. I Specification types and toxicity', *Lipid Technol.* **5**, 110–14.
85. ROSSELL, J.B. (1993b) 'Heat transfer fluids in the oils and fats industry. II Acceptable media, characteristics and analysis', *Lipid Technol.* **5**, 134–7.
86. ROSSELL, J.B. (1999) 'Safety of Heat Transfer Fluids (HTFs)'. *INFORM*, **10**(2) 204.
87. WEISS, T.J. (1983). 'Frying shortenings and their utilisation' in *Food Oils and their Uses*. Ed T.J. WEISS p. 163 Ellis Horwood Publishers Ltd., Chichester, England
88. BANKS, D. (1996) 'Food service frying', in Deep Frying Eds E.G. PERKINS and M.D. ERICKSON, pp. 245–57 AOCS Press, Champaign, Il., USA.
89. PETTY, D. (1996). Personal Communication to J.B. Rossell, August 1996.
90. KALAPATHY, U. and PROCTOR, A. (2000) 'A new method for free fatty acid reduction in frying oil using silicate films produced from rice hull ash', *J Am Oil Chem Soc*, **77**(6), 593–598.

8

The measurement of frying oil quality and authenticity

R. F. Stier, Consultant

8.1 Introduction

A number of years ago, the computer industry developed a term called 'Garbage In, Garbage Out', or GIGO. This same statement may also be applied to any industry. It is especially true when processing foods or manufacturing food ingredients. A failure to maintain or a lack of commitment to quality is a sure path to failure. There are reasons why companies like McDonald's are successful. One can walk into a McDonald's operation anywhere in the world and, except for menu items that are a concession to local tastes, the quality of the basic foods and services are consistent. The company strives for consistency and their customers expect the hamburger or french fries purchased in San Francisco, California to be the same as ones purchased in Cairo, Egypt.

The same quality standards must be applied to all aspects of the production of cooking oils and the manufacture of fried foods. Processors involved in the production of crude refined oils need to be committed to the production of quality ingredients. Failure to do so can damage their reputation and their business. They must also be willing to work with their buyers, especially in this day and age. Product quality between companies may not vary a great deal between different operations, but there may be a marked difference in their commitment to service. The companies who will thrive in the new millenium are not just those who manufacture quality foods and ingredients, but those who provide their customers with technical and other services. Mr. Monoj Gupta put this quite succinctly a number of years ago in when he stated, 'Oil processors have always handed down the product (oil) to the snack food industry. The oil has been judged to be good by criteria chosen by the oil processors'.[1] What Mr. Gupta was doing was telling the suppliers that they needed to start listening to their customers, rather then telling them what they need. The bottom line, however, is that vendors and their customers need to work together for the common good. It costs a business five times more to find a new customer than to retain an existing one.[2]

One of the key issues for assuring quality are product specifications. They are an essential part of doing business. Written specifications are something that each and every food processor needs to develop. They should be developed for all food ingredients, raw materials, packaging materials and all elements involved in the manufacture of foods or ingredients. The development of such documents helps to assure that all materials entering the plant meet the needs of the operation. These documents also provide a road map for purchasing and the selection of new suppliers. Without good specifications food safety and quality can be compromised. Here is an example.

The processor, a manufacturer of oil roasted peanuts, had initiated frying and within a few minutes the oil in his fryer began to foam badly. Frying was halted immediately, resulting in the loss of a day's production and the product that had been produced. Investigations of the oil determined that the product had a high level of alkaline materials or soap. This was the cause of the foaming. Upon further investigation, it was discovered that the company had found a lot of oil at a good price. The buyer failed to determine why the price was so good.[3] The good deal ended up costing the company far more than the savings achieved. This leads into the next important point, that is, vendor selection or certification. Selecting vendors who can not only meet your needs consistently but are willing to work with you is, as noted above, an essential element in doing business in today's world.

What all processors and suppliers or food ingredients need to do is focus on the quality of the food being produced. The food is what people eat, not the oil, so that all programs to control quality must emphasize food quality. Utilizing substandard ingredients or frying in degraded oil will compromise food quality, and can erode a customer base. In fact, at a recent meeting of the German Society for Fat Research,[4] the first recommendation from the conference was 'The principle index of deep-fat frying should be the sensory parameters of the food being fried.' To do this, it is essential that processors, foodservice operators and suppliers of fats and oils monitor and maintain the quality of the foods they produce or the oils they supply.

8.2 Maintaining quality during frying

Understanding how a frying oil degrades is an essential part of producing high-quality fried foods. Over thirty years ago, Robertson[5,6] proposed several basic guidelines for assuring the quality of fried foods. These basic criteria may be seen in Fig. 8.1. All are aimed at controlling the rate of oil degradation, producing high-quality fried foods and operating in as effective and economical manner as possible. The objective of both foodservice and industrial frying is to

- Proper design, construction and maintenance of equipment
- Proper operation of equipment
- Properly clean equipment
- Monitor chemical indices of oil degradation
- Minimize exposure to ultraviolet (UV) light
- Keep salt and other sources of metal from oil
- Filter oil regularly

Fig. 8.1 Basic criteria for quality frying.

produce quality foods for consumption. When frying food, the operator must understand that once frying is initiated, the oil begins to degrade and that the process is irreversible.[7] Operators need to understand how the oil degrades for their individual operation to control that process.

It is also impossible properly to evaluate whether a change made to an operation is beneficial if one does not understand the existing system. To do this, operators need to establish baseline data on their products and processes.[8,9] This holds true for both industrial and foodservice operations. Data should be developed on the fresh oil being used for frying, how that oil degrades, food quality and how it changes with oil quality, the frying process itself, and whether there are breaks in the process. The key to developing this data is the development of a good sampling schedule which will allow the necessary data to be collected, which should be part of the experimental design. Characterization of the fresh oils should be the first step, followed by studies to determine how the frying oil degrades over time during normal operations. Figure 8.2 shows the analyses that should be conducted on the fresh oils.

It is essential that several lots of oils be evaluated to be sure that there is no significant variation between lots. Variation between lots may be minimized through the development of a vendor qualification program. This may seem excessive, but the more information an operator has, the more information available for decision making. If such a sampling program of sampling and analysis seems excessive, conduct a cost/benefit analysis. Before beginning any study, whether it is for frying or whatever, a good experimental design needs to be developed. This should include the establishment of the necessary monitoring programs, but the most important point is to set goals, and how those will be achieved. This must include indices of evaluation and what parameters will be employed to demonstrate success or failure. If one doesn't know what one is trying to prove and how to determine it, do not start the study.

- Polar materials
- Polymers
- Soaps
- Iodine value
- Metals
- Anisidine value
- Free fatty acids
- Peroxide value
- Lovibond color
- Mono, Di- & Triglycerides
- Fatty acid profile

Fig. 8.2 Characterization of fresh oils.

- Fresh oil
- Oil as it reaches frying temperature, but before frying is initiated
- Just after frying has been initiated
- One/half hour after frying has been initiated
- At eight-hour intervals (minimum); or 1, 2, 4 & 6 hours after frying has been initiated
- End of shift or end of day
- When the oil has reached its endpoint

Fig. 8.3 Sampling schedule for frying study.

In a frying study, protocols to monitor how much oil is used through the course of the study must be established. In industrial operations, this may entail adding meters to oil lines delivering oil to the fryer. In foodservice operations, oil may have to be weighed or the volume of oil added needs to be monitored. It is also important to monitor the weight of food introduced into the fryer. In foodservice operations, the food/oil ratio is an essential decision-making tool. Fryer temperatures, drops after addition of product and frying times also need to be monitored. Lastly, since these studies are carried out in 'real world' environments, a means for recording process deviations should be in place.

When starting the study, make sure that frying times are confirmed and that the weight or volume of oil in the fryer is known. It is also essential that the fryer be clean. And again, it is absolutely essential that agreement has been reached on how the study will be evaluated. The parameter or parameters that determine the endpoint for the study need to be set. Will a chemical index of oil degradation be used or will the study be based on sensory analysis? Will a combination of factors be used?

Part of the experimental design will be the establishment of a sampling schedule. It is also necessary to define what analyses will be conducted each time a sample is collected. An example of a sampling schedule may be seen in Fig. 8.3.[8,10]

The heated oils may be tested for polar materials, alkaline materials (soaps), free fatty acids, Lovibond color, metals, anisidine value, peroxide value, fatty acid profile, and mono and diglycerides. The more information that is gathered on how an oil degrades, the better the understanding of the system. Merging this information with data on food quality allows researchers to establish a relationship between oil chemistry and the quality of the food being produced in that oil. It will provide the company with a means for determining which index or indices of oil degradation should be used for quality control during actual frying. It may also be useful to monitor certain physical characteristics of the frying operation and the oil. Suspended solids, changes in smoke and flash points during frying, foaming, visual evidence of polymer formation and heat recovery are all tools which may be used to better understand the frying operation.

There are also researchers who collect samples during frying and conduct sensory tests on the degrading oil. Methods which may be used are the AOCS method, the USDA/NRRL method which evaluates room odor or the Rutgers method, which evaluates strength versus pleasantness. Oil odor can be a concern as there are some oils which develop distinct and/or unpleasant odors during frying.

Once baseline data has been developed, an operator is in position properly to evaluate any changes to the system. Changes could include the use of a filter aid, a change in product formulation or the use of a new cooking oil. These studies should be conducted in a similar fashion as described above.[11,12] Using the same format, the cumulative effect on the system can be determined and compared directly to the baseline data for oil and food quality. These studies will also help potential users determine whether changes will be beneficial to their operation.

8.3 Regulatory issues

The role of the regulator throughout the world is to protect the public and assure public health. In the United States, foodstuffs are regulated by the Food and Drug Administration and the United States Department of Agriculture. These agencies regulate the industries within the country and are responsible for assuring the safety and wholesomeness of products entering the country. The focus of these programs should be food safety and economic fraud, not quality issues. What is a safety issue and what is a quality varies between nations. In some places, regulatory programs may be seen by others as protectionist, that is, they are designed to protect their own country from foreign goods or products. Through the work of international bodies such as Codex Alimentarius and other, and through the growing world economy, harmonization of rules and regulations governing trade is the ultimate goal. Regulatory issues have been discussed in Chapters 3 and 4 by R. Fox and D. Firestone, respectively. As these issues relate to quality and safety, it is certainly appropriate to review pertinent issues in this chapter.

One area which has received attention throughout the European community is regulation of frying fats and oils, specifically those used in foodservice or restaurant frying. In the early 1970s, a series of complaints to the German authorities led them to examine foodservice and restaurant operations throughout the country. Responding to these concerns, a detailed survey of foodservice and industrial operations was initiated. Although the survey did not reveal any direct links between food and oil quality and the consumer complaints, it did underscore the fact that the quality of the frying oil had a direct effect on the quality of the food being fried.

The German Society for Fat Science, or DGF, organized a symposium to address this issue in 1973.[13] One of the recommendations that emerged from this symposium was that there was a need to regulate restaurant frying oils. These oils were the ones which were found to have suffered the greatest level of abuse. Based on animal studies and extensive chemical analyses of heated oils, oxidized fatty acids (OFA) was selected as the index of oil quality and safety. The DGF organized a second symposium in 1979.[14] Following this program, the DGF

Table 8.1 Regulation of frying fats and oils

COUNTRY	REGULATIONS: YES/NO)	YEAR ESTABLISH	FED/ LOCAL	POLARS	OFA (1)	ACID VAL /FFA	SMOKE POINT	HEAT LIMITS	DPTG*	RAPID TESTS	APPEARANCE	SENSORY	COMMENTS
AUSTRIA	YES		FED	<27 %	<1%	<2.5 (AV)	<170°C	<180°C			No carbonaceous residue, dark color or foaming	No unpleasant odor or taste	Found in Frying Fats Section of 3rd Edition of Austrian Foodstuffs Book
BELGIUM	YES	1988	FED	<25%		<2.5 (FFA)	≤170°C	<180°C	≤10%				Viscosity not > 37 mPa-sec @ 50oC (fats) or 27 mPa-sec (oils);Frying fats may not have >2% linolenic
CHILE	YES	1996	FED	≤25	<1.0	≤1.0 (ffa)	≤170°C						MAXIMUM LINOLENIC CONTENT OF 2%. ANTIOXIDANTS MAY BE USED AS ALLOWED.
FRANCE	YES	1973 & 1979	FED	25%									Frying fats may not have > 2% linolenic; silicone additives prohibited
GERMANY	YES	1973 & 1979	LOCAL	<24%	<0.7%		>170°C						Regulations enforced by individual states
HUNGARY	NO	1988		<25%									
SPAIN	YES	1989	FED	<25%									Frying fat must not alter quality of food & may not be sold for use in preparing foods.
ISREAL	NO												Use guidelines published by Swedish National Food Administration for restaurant operations
ITALY	YES	1993	FED	<25%				<180°C					
JAPAN	NO					<2.5 (av)	>170°C						No regulations, but use guidelines. Also, carbonyl value <50
NETHERLANDS	YES		FED	<27%		<4.5 (AV)			<16%			Sensory qualities can result in rejection.	
PORTUGAL	YES	1993	FED	<25%									
DENMARK	NO												
NORWAY	NO												
FINLAND	NO												
SWEDEN	NO												
SWITZERLAND	YES		LOCAL	<27%			>170°C			FRITEST	COLOR	Sensory qualities can result in rejection.	SILICONE ADDITIVES ARE FORBIDDEN; ENFORCEMENT BY CANTONS
USA	NO					2% (FFA)					SMOKING, COLOR		U.S.D.A. GUIDELINES FOR MEAT & POULTRY OPERATIONS

- -Dimeric & Polymeric Triglycerides
- (1) - Oxidized Fatty Acids

proposed that total polar materials (TPM) be employed as the standard for oil quality. One of the reasons for the change was the difficulty involved in testing for oxidized fatty acids. The rationale for these regulations was that abused frying oils contained compounds that could have an adverse effect on human health.

The proposed German regulations started a trend that has spread throughout most of Europe.[15] There are a number of nations who have established regulatory limits or guidelines for frying oils (Table 8.1).[16] As may be seen in the table, monitoring oil quality by the regulatory bodies throughout Europe entails the use of both laboratory analyses and rapid tests.

8.4 Quality measurements for refining operations

Before a fat or oil reaches a food processor or a foodservice operator, it must be extracted and/or refined for use. Fats and oils for frying are derived from animal or vegetable sources. Some operations may use animal/vegetable (A/V) oil blends, but the use of such oils for frying has almost disappeared in the United States due to concerns about saturated fats in products of animal origin. The refining process for crude oil varies slightly for each product, but the basic process is as follows:[17]

Degumming⟶Alkali Refining⟶Bleaching⟶Hydrogenation⟶
Winterization⟶Deodorization⟶Storage⟶Packaging⟶Shipment

Some oils undergo a process called interesterification. This process allows fatty acids to be rearranged or redistributed on the glycerol backbone. It is usually done at high temperatures in the presence of a catalyst, but there are enzymatic procedures that may be used. Interesterification is, however, seldom used in the production of industrial frying oils. As with any food process, there is a need to monitor quality parameters throughout the operation. These quality operations help to assure that the finished products meet the needs of their users. The basic need is, as noted above, the product of quality fried food. The product must also allow the user to operate efficiently and economically.

The production of crude oils is done by pressing or crushing oil seeds or rendering animal fats. The seeds are then extracted with a solvent to extract the oil. The quality of the crude oil is a direct function of the quality of the seedstock. Damaged or poor quality seeds affect the quality of the crude, so sorting will help enhance oil quality. During the refining process, operators need to monitor the process to assure that they are meeting finished product quality specifications. They also need to assure that refined oils meet the needs of their clients. Of particular importance is to monitor the alkali refining process. Failure to remove the soaps that are formed can severely compromise oil quality and performance.[18] Tests that are conducted on oils during processing include free fatty acids, color, peroxide value, moisture, chlorophyll, iodine value, active oxygen method, solid fat index, flavor, odor and appearance.

As Gupta[1] observed, oil refiners and suppliers should seek to produce oils that meet the needs of their buyers. They should not impose products on them, but work with them to determine what product or products best meet their needs. Customer service will be the key to doing business in the future. An example of a product specification sheet for a fresh oil may be seen in Fig. 8.4.[30] One of the most important parameters on this specification sheet that is often ignored in many specifications is anisidine value.[18] This value measures overall oil abuse from processing. As an example, a refiner could reduce the % free fatty acids in his oil by re-refining the product. If a buyer only specified free fatty acids, this oil would meet the established specification, but due to the extra processing might not perform as it should in the fryer.

The supplier's job does not end with production of the oil, however. The operator must takes steps to protect the oil.[19] Areas that the operator needs to address are storage after deodorization, how trucks or railcars are loaded, transportation, receiving and unloading, storage after unloading and utilization. Of course, all tank trucks or rail cars used to transport foodstuffs should be cleaned before use, and should be used only for transporting food materials.

INGREDIENT Palm frying oil	**SUPPLIER** Quality fat importers
ADDRESS 44 Importer Road, New York, N.Y. 10055	**TEL** (212) 555-4444 **FAX** (212) 555-3333
DESCRIPTION/CONTAINER - Bulk tanks; 5,000 gallon tankers	
CAS# - A-816	**SPECIFICATION #** - D-0041
MSDS SHEET REC'D - Yes	**LOCATION** - QA Lab, office, & employee break area
SPECIAL HANDLING REQUIRED/DESCRIBE: - Check % FFA, % soaps, & PV upon receipt. If target specifications are not met, reject lot.	
CHARACTERISTIC/TEST	**SPECIFICATION**
% FREE FATTY ACIDS (A.O.C.S # Ca 5a-40)	≤ 0.1%
PEROXIDE VALUE (A.O.C. S.# Cd 8-53)	< 10 meq
IODINE VALUE (A.O.C.S # Cd 1b-87)	50-55
VISUAL COLOR	Clear, golden yellow.
SOAPS (A.O.C.S. # Ce 17-79)	< 1 ppm
POLAR MATERIALS (A.O.A.C. #982.27 15th Ed)	< 2%
POLYMERS (A.O.A.C. #977.17 15th Ed.)	< !%
ANISIDINE VALUE (A.O.C.S.# Cd 18-90)	< 5
ODOR, FLAVOR	Bland, flavorless
DEVELOPED BY: John Doe, October 15, 2000 (NAME/DATE)	
APPROVED BY: H. Ross Parrott, October 21, 2000 (NAME/DATE)	
COMMENTS:	

Fig. 8.4 Bakery ingredient specification sheet.

8.5 Developing purchasing specification and certifying vendors

Every operation has different supplier programs. There are many, however, who have initiated programs to certify the companies who supply them. These operations only buy well-defined materials from recognized or approved vendors, or certified suppliers. The process for approval or certification is generally a time-consuming and expensive proposition, but these operations believe it to be worth their while. The reasons for certification are quality and economics. Such operations want products that provide satisfactory performance and are readily available. They usually have multiple suppliers.

The delivery of quality goods or services leads to economies of operation in a number of areas. What these savings are must be determined for each operation, but they do exist. Campbell's has developed a select supplier program as part of their corporate Total Quality Management philosophy. Their data shows that for each dollar they spend as part of this program they get a return of $20.00.[20] Campbell's staff have estimated that it takes between 4–8 years to get a supplier to the desired level.

The first step in selecting a vendor is to know needs. When purchasing frying oils, technical considerations are product flavor, texture, mouthfeel, aftertaste and storage stability. The buyer must have a product specification to take to the supplier. In fact, sending a specification sheet to a potential vendor is a necessary first step. Send the vendor the specification and request a sample of oil. If the sample that the vendor submits does not meet the specification, the potential vendor may be dropped right then and there. Inclusion of anisidine value as a parameter for oil quality is an excellent tool for evaluating a potential supplier's understanding of their own process and your needs. There are a number of criteria that one should look for in a vendor. Each of these will have an influence on whether a processor will want to enter into a relationship with that company. The first is, obviously, that they can produce the material in which you are interested at an affordable price.

It is also imperative that the buyer determine whether the packer is financially sound and whether they have a reputation for quality and good service. This information can be discovered without contacting the company. Financial information can be obtained through any number of financial services. In the United States, the Freedom of Information Act makes it possible to determine whether FDA inspections have turned up any adverse findings.[22] Whatever the case, once preliminary talks are initiated, it is up to the buyer to evaluate that packer or supplier. It is essential that your quality requirements and/or policy towards suppliers be brought up immediately. Safety and quality should not be negotiable.

As part of the evaluation process, your technical staff should look over the potential vendor's operation. This audit may be by one person or a team. The team might consist of persons from quality assurance, engineering, food safety/sanitation, and R & D. Areas that the technical people should look at include:[21]

- Production capabilities – can they meet your needs? Some operations simply do not have the production capacity to supply certain operations.
- Record keeping – does the vendor have the basic HACCP prerequisites of product identification and the ability to trace and recall product?
- Sanitation/GMP compliance – is there a commitment to Good Manufacturing Practices and hygienic operations? Is the staff trained in this area?
- Food safety programs – does the facility have a working HACCP program?
- Laboratory/quality staff – are the staff well-educated and is the laboratory well-equipped?
- Overall commitment – the technical staff can often provide you with the best insights into an operation's commitment to quality operations. Management can promise, but they may not quite understand the realities of delivery. When visiting a facility, be sure to watch and, if possible, talk to the workers. They are often the best indicator of management's commitment to safety and quality.

Many companies have a policy whereby they, the purchaser, make a commitment to help the vendor meet their needs. Buyer and seller become partners. As a buyer who espouses this philosophy, the following tasks are among those which can help build a partnership:

- working with the supplier's laboratory staff to upgrade their analytical capabilities
- conducting training programs with management and staff in SPC, TQM, or HACCP
- inviting the supplier's management and staff to your facility to see how you operate
- assisting in the purchase of new equipment to assure a more efficient supply
- assisting the supplier in upgrading existing programs.

Vendor selection is really the sum of its parts, as opposed to a group of distinct steps. The actual selection process and, perhaps, the certification process will vary between operations, but they could well flow as follows:[19]

1. Establishment of performance characteristics or specifications by management from both supplier and buyer – this is what the buyer expects the product or service to be, and how it will be measured.
2. Review of analytical methods to determine the necessary data to be submitted and the format that the data should take – many provide their suppliers with data acquisition or collection sheets. This makes the processor's life easier, especially if the buyer is dealing with many different suppliers. Submission of the methods that are used is also suggested. If buyers and vendors do not use standardized methods, there can be real problems. This issue is addressed in greater detail later on, however.
3. Measure key attributes and conduct collaborative studies to assure correlation between supplier and buyer laboratories – this seems simple,

but can be a major headache if different methods are used. If a supplier says something is within specifications and the buyer says it is not, things can get bloody.

4. Test runs – test runs of the material(s) are conducted in the presence of both the vendor and the buyer. This provides both parties with a greater understanding of how they operate.
5. Systems implementation – the supplier should begin to implement quality or control systems required by the buyer. These may be SQC, TQM, or HACCP.
6. Performance monitoring – criteria for evaluating product will be established. As a relationship builds, monitoring frequently is decreased as the buyer becomes more confident in the supplier.

The ultimate goal is to improve supplier performance so that buyers not only receive quality goods, but that the supplier can reduce his costs. If he can operate more efficiently (and at a lower cost), the chances of a price increase drop.

8.6 Quality control during frying

There is a direct relationship between the quality of the food being produced and the oil in which it is being fried. As an oil degrades, the food passes through various stages from literally undercooked to greasy and unacceptable. Operators, whether they are working in an industrial operation or a restaurant, monitor both food and oil quality, and usually discard the oil before it begins producing food that will be rejected by the consumer.

Dr Michael Blumenthal published what he called a Frying Oil Quality Curve to explain oil degradation in 1987.[23] He was the first to try to relate the engineering aspects of oil degradation to changing oil chemistry and the effect of the oil on the food. Subsequent publications[10,24] further explained the Surfactant Theory of Frying. This theory described how changing oil chemistry, particularly the formation of surfactant materials, was responsible for how food fried in a degrading oil. This figure may be applied to all frying systems with slight modifications. Pokorny[26] helped to validate the surfactant theory and the Frying Quality Curve when he published data plotting changes of total polar material content versus flavor perception.

To explain how an oil degrades, the French fry or chip was selected. When frying is initiated, the oil in the fryer should be fresh and the fryer clean. French fries produced in that oil are raw or uncooked, light in color, and do not have the rich smell expected of a fried potato. This 'break-in' oil has little or no surfactancy, so the heat is not being transferred to the food. Blumenthal has reported that there is little contact between these 'break-in' oils and the food. Contact time may be as little as 10% of the actual immersion time, the water being pumped from the food as steam pushes the oil away from the surface of the food. Without direct contact there can be no heat transfer by conduction. The surface remains uncooked, starch does not properly gel, and there is little oil pickup by the food.

As frying continues, the degree of surfactancy increases resulting in improved food quality. These stages are known as the 'fresh' and 'optimum' stages. The French fry produced in an 'optimum' oil will be golden brown in color, it will be fully cooked, there will be a minimal crust, there will be an acceptable level of oil absorbed by the food, and the product will have the desired odor. Part of optimizing the frying process is maintaining the oil in this optimum stage for as long as possible. This is easier in industrial frying where oil is continuously replenishing that which is removed by food than in the batch type fryers used in restaurant operations. During the 'fresh' and 'optimum' stages, the contact time between the oil and the food will be approximately 20 and 50%, respectively.

Once an oil begins to break down, the process is irreversible. An operator cannot maintain the oil in the 'optimum' stage indefinitely. It continues to degrade entering what are referred to as the 'degrading' and 'runaway' stages. During these periods contact time between the food and oil increase to 80 and 100%, respectively. Foods produced in these oils are of poor quality. The surface of the French fry will be dark and spotted and it will be too oily. Excessive contact with the surface rapidly dries and thickens the surface trapping moisture in the food and inhibiting heat penetration by conduction to the center. The French fry may not be fully cooked in the center, despite increased contact time with the oil, or it may have no center at all. Foods produced in these damaged oils may have obvious off-flavors and if they are to be stored, will have a reduced shelf life. This is true even with frozen foods as oxidation reactions will occur, albeit slowly in the freezer. Oils that have reached the runaway stage not only produce poor quality food, but are potentially hazardous as the smoke/flash point has been lowered.

It is up to the fryer operator or his research and development staff to establish a relationship between food and oil quality. Quality staff will monitor one or more chemical indices or markers during the frying operation in an effort to maintain optimum product quality. These values should be determined through the use of the controlled type studies described above and the use of sensory testing.[26] With products such as snack foods, it is essential that shelf-life studies be conducted. The product, such as a potato chip (or crisp in the UK), may be acceptable immediately after cooking, but it is meant to have a shelf-life of 40 days or more. That shelf-life should be determined by testing. The product characteristics need to be defined by marketing, but R & D must determine the processing parameters that allow the production of foods that meet those standards. The processor must, therefore, know the product and understand the process. It is necessary to identify when the product becomes unacceptable to consumers and the mode of failure. Expert sensory panels who have been trained on the product must then identify the off-flavor notes in the product. The group must then determine the chemical marker or markers that caused the product to become unacceptable.

The most common chemical test conducted by fryer operators is free fatty acids. Many operators use this test only because it is traditional. They have not taken the time to establish a relationship between oil and food quality. Free fatty acid values that are used as endpoint indicators vary with the product being produced. Snack food manufacturers are more conservative than those producing breaded products or French fries. Other markers include peroxide value, active oxygen method (AOM), oxidative stability index (OSI), total polar materials, polymers, viscosity, color, soaps and anisidine value. The value of some of these analyses with frying fats is debatable, and care must be taken in both their use and in analysis of the results. There are others who use the different rapid or quick tests on the market. These are described later in the chapter. There are some operators who have taken advantage of new technology such as Near Infrared Spectroscopy (NIR) for on-line monitoring of oil quality. The bottom line is that quality is something that each and every operator must determine for themselves.[26] What works for one fryer operator may be inappropriate for another.

8.7 Adulteration of fats and oils

Adulteration of food products, be they fats and oils or other, may be intentional or unintentional. In either case, the results can be devastating. The adulteration with dioxin of an animal fat destined for use in animal feeds resulted in a massive recall of feedstocks, the destruction of countless animals, the loss of millions of dollars and the criminal prosecution of the manufacturer of the product. Reuters[27] reported that insurance companies guaranteed some $664 million dollars US in loans to Belgian farmers to cover their losses. That incident has since resulted in the passage of more stringent regulations in the European community and around the world to assure the safety of the food supply. Perhaps the most tragic example of adulteration of oils occurred in Spain. Rapeseed oil was treated with aniline to render it unsalable for food. The adulterated product was then re-refined and sold to an unsuspecting public by unscrupulous vendors. This resulted in the hospitalization of over 13,000 people and killed some 500. The incident aroused public indignation throughout the country.[28] In this case, someone's desire for greater profits turned out tragically. There is another form of adulteration that is of concern to many around the world. This is adulteration of high-quality oils, such as olive or high oleic sunflower, with lower-quality oils. Again, unscrupulous processors stand to increase profits significantly if they can dilute high-value oils with lower-quality products, and still command the premium prices. The primary reason for intentional adulteration is, therefore, economic. Oil authenticity is discussed in greater depth in Chapter 7, however.

One concern that most oil users and processors do not have is pesticides. The deodorization process effectively eliminates pesticides from RBD or refined, bleached deodorized oils. Processors who use impeller or press oils need to monitor the source and should follow the guidelines for supplier selection described earlier to assure that pesticides are not an issue. Fortunately, impeller or press oils tend not to be used in either foodservice or industrial frying as their stability is poor. IUPAC's method 2.641 has been developed to detect organochlorine pesticides.

Methods for detection of adulterated products have been developed the world over. The International Union for Pure and Applied Chemistry[29] and the Association of Official Analytical Chemists[30] have published a number of these methods. IUPAC method 2.611 describes how mineral oil may be detected in animal or vegetable oils, but the method's detection limit is 200 ppm. There are other methods that have not yet become official that utilize GC/MS which can expand the detection limits to as little as one ppm. As noted above, this kind of adulteration would be unlikely for oils purchased for use in industrial or foodservice frying.

Adulteration of high-value oils is a concern the world over. The oil that is most frequently associated with adulteration is olive oil. It is of a high value and is recognized as a 'healthy' oil due to its high content of oleic acid. It is also an integral part of the 'Mediterranean Diet', reputedly one of the healthiest diet regimens. Work to detect olive oil adulteration has been conducted throughout the world, especially in the principal olive oil producing nations, such as Spain, Portugal, Italy and Greece. International trade in adulterated oils could damage their reputation and businesses. The United States Food & Drug Administration has also actively conducted investigations with European nations, and the International Olive Oil Council (IOOC) to develop methods to detect adulterated oil, and arrest the trade in the illegal product. The IOOC provisions for olive oil have been part of Codex and EU provisions for since 1991. These provisions are designed to safeguard the purity of olive oil and discourage adulteration of high grades of olive oil with lower-quality oils.

8.8 Tests for frying fats and oils

The previous sections have discussed or alluded to many methods for testing fats and oils used for deep-fat frying. The following tests may be used for monitoring refining operations, setting specifications, monitoring oil quality during frying operations and conducting research on fats and oils used in frying.

8.8.1 Chemical methods

Polar materials (TPM)

These materials are those which remain on the column after the first elution when a heated oil is tested for polar materials using AOAC method 28.074 (also, AOCS method Cd 20.91 and IUPAC 2.507) They include all non-triglycerides and particulates in the oil. Polar materials may also be considered to include all non-triglyceride materials soluble in, emulsified in, or suspended in the frying oil. A fresh oil typically contains 2–4% non-triglycerides. Once an oil is exposed to frying temperatures and food product, conversion of the triglycerides is initiated. Polar materials, therefore, can be used to measure cumulative degradation of the frying oil. There are many who consider the polar materials measurement to be the single most important test for degrading oil. Regulatory agencies in Spain, Portugal, France, Germany, Belgium, Switzerland, Italy and The Netherlands have established regulatory limits for polar materials in frying oils.[15,16] The main concern with the use of polar materials in regulatory and quality operation is that the official method is time and solvent intensive. And of course, the use of flammable solvents should never be used near a fryer, much less in a food processing facility. Polar materials are an excellent predictor of food quality for many operations. Pokorny[25] has demonstrated that increases in the polar fraction resulted in an eventual decline in food quality.

Free fatty acids (FFA)

Free acidity is determined using a titration procedure (AOCS method Ca 5a-40,[31] IUPAC 2.201, AOAC 940.28) but has a difficult end-point with highly colored or emulsified oils. It is frequently the only frying oil marker used by many frying operations. Free fatty acid (also expressed as Acid Value outside North America) is a measure of the amount of fatty acid chains hydrolyzed off the triglyceride backbone. It used as chemical marker for the frying oils and may also be used to evaluate oils absorbed onto the surface of a fried food. Although free fatty acid or acid value is used in quality monitoring of frying operations, researchers[32–34] have stated that there is no direct relationship between percent free fatty acids and the quality of a used frying oil. Reasons include the transience of free fatty acids. They both volatilize and are converted to other decomposition products even as they form. Free fatty acids also do not affect frying. Good quality foods can be fried in pure free fatty acids under laboratory conditions.[23] Lastly, free fatty acids can be diluted out using fresh oil.

Soap

Alkaline soaps are an excellent marker of oil degradation. They are determined using AOCS method Cc 17-79. Alkaline soaps (sodium oleate, etc.) are formed by the reaction of metals and free fatty acids in the presence of water. They are fairly stable once formed. Precursors are derived from residual caustic cleaners, from coatings or breading containing metals left over from leavening agents (in cake doughnuts, for example), from rupturing plant and animal cells, and from animal blood and bone cells. As the concentrations of soaps and other surfactants increase, they affect heat penetration, oil incorporation into the crust of a food, and the makeup oil required to replace oil ad/absorbed by a food. Commercial filtration systems have been developed specifically to reduce or eliminate surfactants and water. Fresh oils should have less than 1 ppm soap.

Oil color

This has been used by many as a quality index over the years. Many believe that dark oil will produce bad or dark colored food. High quality food may be produced in dark oil, and poor quality food in apparently clean oil. Lovibond Color (AOCS Cc 13E-92) may be used when developing quality standards for fresh oils or in frying research. Researchers may want to use Lovibond Red, Yellow, and Blue[8] in frying work. Red color in finely filtered oil loosely correlates to combined oxidized fatty acids and pyrolytic condensation products. The Yellow color may relate to the combined peroxides and aldehydes in an oil. Blue color is related to the haze created by water and fine particulates suspended/emulsified in the oil. Oil color has not been found to be an effective index for oil quality or for discard. It is too subjective and really bears no relation to the ability of that oil to produce quality food, although many do use it.

Peroxide value (PV)
Peroxides are unstable radicals formed from triglycerides. They are determined using AOAC method 965.33, AOCS method Cd 8b-90 or IUPAC 2.501. This is a standard test for fresh oils, but has limited value for frying oils. One reason is that the test itself is very sensitive to oil temperature. Users must take great care to standardize their sampling and testing procedures. Peroxides are destroyed at frying temperatures, but begin to reform during cooling and 'staircase' upwards with each additional temperature cycle. Fritsch[31] has stated that this is not a good test for measuring the degree of abuse of a frying oil. One area where this test has found value is in snack foods, where processors extract oil from the fried snack and determine peroxide value as an indicator of shelf life. This is one method that has been modified due to environmental concerns.

Anisidine value
Aldehydes are products of the decomposition of peroxidized fatty acids. The Anisidine Value quantifies important aldehydes (AOCS method Cd 18-90 and IUPAC method 2.502). The aldehydes can be used as a marker to determine how much peroxidized material has degraded. In conjunction with peroxide levels, the past and future degradation profile of an oil can be determined. This test is especially useful for detection of re-processed 'fresh' oils.[18] Anisidine values should be employed as a parameter when establishing specifications for fresh oils. There are suppliers who believe that oils with high free fatty acid levels can be improved by additional processing. Such an activity will be reflected by an increase in the anisidine value. Anisidine value might be used more if it was not such a difficult method to conduct. There are some potato chip (crisp) manufacturers who have employed Near Infrared Spectroscopy (NIR) to monitor changes in anisidine value.

Polymers
These are usually the largest single class of degradation products in a frying oil. They are determined using AOAC method 993.25, AOCS METHOD Cd 22-91 or IUPAC method 2.508. They include dimers, trimers, tetramers, etc., and may be formed through oxidative and thermal reactions. They are an excellent chemical marker of oil degradation, but not applicable in food-quality

monitoring due to the time involved in the analysis. The official method utilizes high pressure liquid chromatography (HPLC). Polymers formation is not something that can be readily measured as part of a quality control program, but understanding the rate of polymer formation is extremely important towards understanding how an oil degrades and optimizing the system. Polymers are used as a regulatory index in certain nations in Europe. Increases in polymer levels also result in an increase in the viscosity of the oil.

Iodine value (Wijs Method)
The iodine value is commonly used as quality index for fresh oils. It is used to measure the average degree of unsaturation of an oil as determined by titration with active iodine. The greater the degree of unsaturation, the higher the iodine value and the greater the susceptibility to oxidation. This is determined using AOAC method 993.20, AOCS method Cd 1-15 or IUPAC method 5.205. The AOAC method utilizes a cyclohexane-acetic acid solvent system, which is a more 'environmentally friendly' system than the traditional system that used carbon tetrachloride. This is one of the trends that oil chemists will continue to see through the coming years.

8.8.2 Rapid tests

Rapid tests may be used to provide a quick and easy means to monitor the quality of products. With oils, the object might be to 'clear' incoming lots or test a frying oil to determine whether it is still acceptable for frying, that is, it will produce good food. Stier and Blumenthal proposed criteria for the ideal frying oil quick test.[35] These criteria can in fact be expanded to cover any rapid test (Fig. 8.5). Rapid tests must provide good value for their cost. Any test used in foodservice operations must be of especially good value. The test must be both inexpensive and also be easy to use, that is, relatively 'idiot-proof'. Figure 8.6 summarizes the different rapid tests used for fats and oils.[36] Quick tests may be used for quality monitoring, troubleshooting and other applications.[37]

Physical tests (no instrument)
There are several simple physical tests used by the industry. These are used to 'measure' foam, oil color and oil clarity. Degrading oils begin to foam due to the formation of polymers, soaps and other surfactant materials. In continuous industrial processing, foaming is rarely a concern. It is a much greater problem in foodservice operations, doughnut frying or operations using batch fryers. There are operators who discard oils when the foam reaches a certain height. Some fryers have a 'grid' or ruler (much like the depth markers on the bow of a ship) etched onto the rear panel which is used for this purpose. Others insert a marker or ruler in the oil to measure foam height. This is a simple and inexpensive test, but there is no real correlation between foam and food quality. No fryer wants his oil to foam, however, as foaming can result in a 'boil over' which will create a fire hazard.

- Correlate with official/recognized methods
- Provide an objective index
- Easy to use
- Safe for use in food preparation/production area
- Correlate to food quality/safety
- Field rugged
- Allow for remote reinspection

Fig. 8.5 Ideal rapid test.

Oil clarity is another 'quick test' which is used. Some operators feel that as oils become darker and less clear, it is no longer fit for use. A clarity test, called the visibility tester, is used by Kentucky Fried Chicken (KFC). They use a stainless-steel rod attached to shiny silver disk or plate. The rod has a scale showing three indented marks which are used to judge shortening quality. The shiny disk is lowered into the oil with the depth of insertion monitored on the rod. A penlight is held next to the rod, turned on and the disk observed. If the disk is invisible at or above a certain depth the oil is discarded.[38]

The third physical test is oil color. Oil suppliers provide users with a dark-colored wand. Oils which reach the color of the wand are considered unfit for use and should be discarded. To conduct the test, an operator removes a sample of oil from the fryer with a dropper. The color of the oil in the dropper is compared to the wand and a decision is then made. These tests have no real relationship to actual oil quality and are very subjective. A good oil can be dark

Type of test	Manufacturer
Physical tests – no instrument	
Foam height	
Color wands	
Clarity	
Physical tests – instrument	
Viscosity meter	Leatherhead Fri-Check
Chemical tests – instruments	
Coat	Lanson Industries
Ruler	Lanson Industries
Near infrared	Several suppliers
PCT 120	3M
Food oil sensor	Northern Instruments
Chemical quick tests	
Oxifrit	Merck
Fritest	Merck
ACM/PCM	Mir-Oil
TPM, FFA, WET	Test Kit Technologies
AV-Check	Advantec
Shortening monitor	3M

Fig. 8.6 Rapid tests for fats and oils.

and a bad oil light in color. Perhaps the greatest argument against color is that it has not been adopted as a regulatory tool by any of the nations who have established regulations for restaurant frying oils despite its ease of use.[1,3]

Physical tests (instrument)

GEC Marconi of Chelmsford, England developed, with assistance of the scientists at Leatherhead Food Research, an instrument to measure viscosity.[39–41] The instrument was designed to measure viscosity of used frying oils based on the formation of 'polymerized and oxidized matter' or POM. The inventors selected viscosity as the parameter to measure because they felt the formation of degradation products during frying would result in an increase in this particular index. These researchers determined that an oil in excess of 15% POM was unfit for human consumption. Within the instrument itself were two short vibrating steel tubes which are excited by internally mounted piezocrystals. The mode of operation has been compared to that of a tuning fork. Dampening depended on the viscosity of the surrounding fluid and resonance depends on density. The probe provided continuous readings of these two parameters. The probe also contains an internal thermocouple to compensate for changes in viscosity due to temperature. A common cut-off for the probe was defined for which the reference value fell between the limit values for this reference value to indicate the discard point. The developers felt that the instrument could be used in both industrial and foodservice frying operations. They felt that it could be modified to allow a digital readout of results (industrial frying) or red, yellow and green lights for restaurant operations. The latter instrument would have been considerably less expensive. The unit never achieved commercial success.

Dr Christian Gertz[42] recently reported on the development of a new unit for monitoring oil quality. It is currently being marketed under the name Fri-Check (Fig. 8.7). The unit measures viscosity, which Gertz determined could be correlated with the polar content of the degrading oil. The Fri-Check unit consists of an electronic box and removable steel tube. Oil may be added directly from the fryer or at room temperature. No filtering is required. When the sample has equilibrated, a piston-like body is dropped through the tube and its falling time is taken to correlate with the percentage of polar materials. Results are obtained in 5–7 minutes. The unit has been designed to help users conform with European guidelines for regulations governing food safety and hygiene.

Chemical tests (instrument)

There are several companies who have developed instruments for measuring oil quality which have been designed to measure chemical components in degrading frying oils. Lanson Industries developed two instruments called COAT and RULER, and Northern Instruments of Lino Lakes, Minnesota manufactures and markets the Food Oil Sensor (FOS).

The COAT (Cooking Oil Analysis Technique) and RULER (Remaining Useful Life Evaluation Routine) sprang from technology developed at the University of Dayton and Wright Laboratory at Wright-Patterson Air Force base

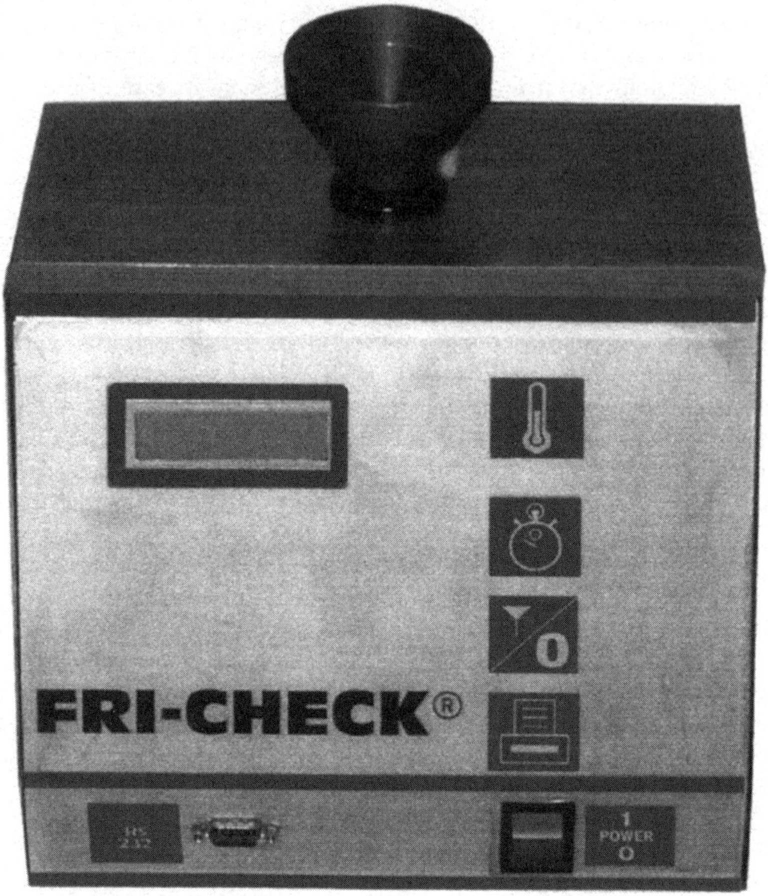

Fig. 8.7 Fri-check unit.

to monitor the condition of engine oil lubricants. The researchers felt that the technology could be expanded to be used as a monitor for cooking oil quality. The RULER was developed for determining the presence of antioxidants, such as BHA, BHT, TBHQ and tocopherols) and was designed to be used 'off-line'. Lanson personnel felt that this instrument was more suitable for industrial operations. COAT was also known as the 'Smart Dipstick'. It was a simple instrument which used solid state technology that yields an oil quality value when the oil probes on the instrument were inserted in hot oil. COAT represented an overall oil quality value. In-house tests conducted by Lanson staff demonstrated good correlation with free fatty acids, oil color and smoke point. The instrument was also reputedly able to detect when fresh oil was added to fryers. The manufacturers implied that use of their instrument would prevent premature disposal of cooking oils, leading to overall cost savings and that their instrument will provide more consistent end-point readings than other methods.[43]

The Food Oil Sensor (FOS) has been on the market since the late 1970s (Fig. 8.8). This sensor monitors the change in dielectric constant of degrading frying fats. A sensitive bridge circuit detects minute dielectric changes in the oil.[44] Operating the instrument is a two-step protocol. In the first step, calibration, the instrument is turned on and the sensor cup should be thoroughly cleaned. The cup must then be filled with fresh shortening of the same make and type used in the fryer. Turn the instrument on and holding down the test switch, 'zero' the instrument. To run a test, place the test shortening in the sensor cup. When the 'green' light goes on, depress and hold the test switch down. Read the shortening status number.

This instrument has been evaluated by many workers the world over. It is, in fact, used rather extensively in Europe as both a research tool and in quality monitoring. Fritsch *et al.*[45] acknowledged that the FOS provided a good measure of oil deterioration, but acknowledged that there were concerns, such as with calibration. Studies by Croon *et al.*[46] found that the FOS correlated well with polar materials (the best index of oil quality) and triglyceride dimers as a means of evaluating oil quality. Saltmarsh[47] described other concerns with the instrument including sensitivity to air currents, long warm-up period, possible adverse effects of water on results and effects of oils leaching from the foods on readings. Schwarz[48] recently affirmed the value of the food oil sensor in a comparative study with other rapid test methods. To overcome the problems with the unit, the manufacturer now suggests that the instrument be 'zeroed' with fresh oil and that the sensor cup be heated to 50ºC when solid fats are used for frying. Filtration may also be employed to remove suspended solids and entrained water which will adversely affect the readings.

3M recently introduced an instrument called the Polar Compound Tester or PCT 120.[49] This unit (Fig. 8.9) allows users to evaluate % polar materials in degrading oils. According to the manufacturer, calibration of the unit allows accurate determination of polar compounds in the samples. The user places a treated strip in the metallic well of the instrument and pours oil through a funnel so it contacts the strip. The operator pushes the 'Start' button and reads the result when a flashing green light indicates that the sample is ready. Results are read using an embossed scale on the lid. 3M believes that the unit is easy to use, precise and will help users assure that they satisfy European regulations on restaurant frying oils.

Chemical tests (quick tests)

The Merck Company of Darmstadt, Germany developed these oil quality tests in response to regulatory concerns regarding frying oil quality.[13] Among the recommendations proposed by the DGF was that oxidized fatty acids (OFA) be used as an endpoint index for restaurant frying oils. In 1979,[14] the DGF changed direction and suggested that total polar materials be used instead of the OFA test. Merck developed their Oxifrit and Fritests to provide operators and regulators with a quick and easy means of monitoring oil quality. The Fritest measured the alkali color number and the Oxifrit Test oxidation products. Both

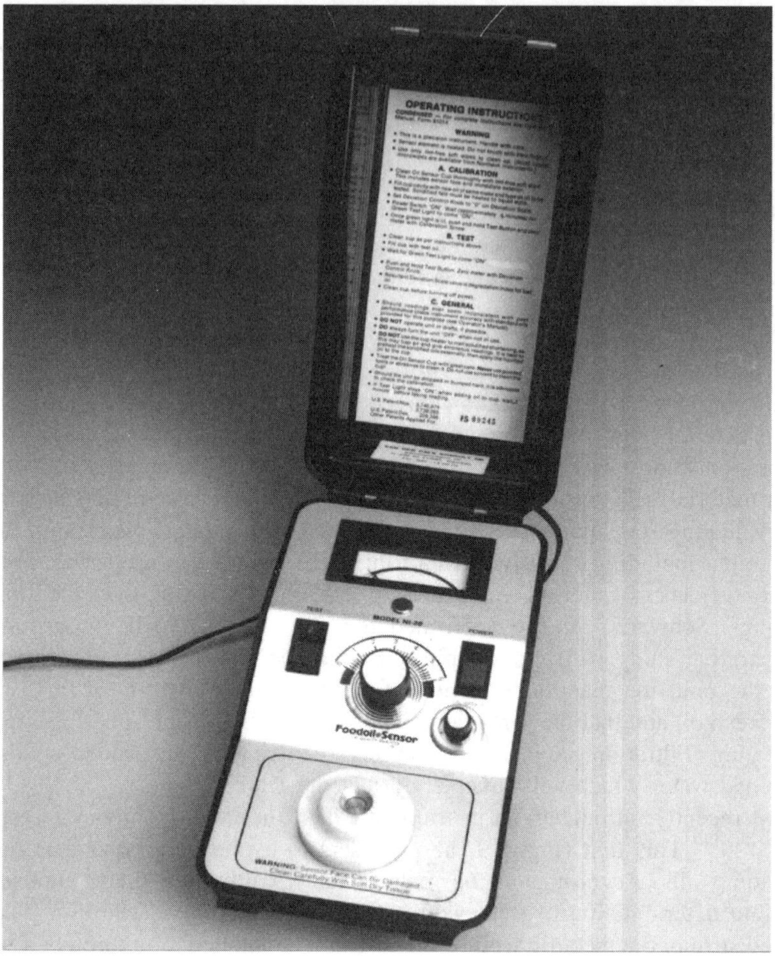

Fig. 8.8 The Food Oil Sensor.

tests are colorimetric tests that use a solvent-based reagent system. The tests are being or have been used by regulators in Switzerland, Finland, Denmark, Austria, Luxembourg, Portugal, Norway and Sweden as a monitoring tool.[15,16] To use the Fritest, the test reagent is placed into a sample tube to the fill line and hot oil is added to a second line. The tube is stoppered, gently shaken and read against the color scale after one minute. The Oxifrit Test is slightly different in that five drops of a second reagent are added to the sample tube before the addition of oil. The colors for each test are used to judge the quality of the oil. The Fritest has a four-color scale and the Oxifrit three. For Fritest, Merck recommends that the oil be changed at color three and that it has 'gone off' at color four. The three-color Oxifrit test recommends a change at color two (green) and that at color three (olive) the fat is unfit for use.

Fig. 8.9 Polar Compound Tester or PCT 120 (3M Corp).

The ACM and PCM tests were developed and patented by Libra Laboratories (50). Patent and distribution rights are now owned by Oil Process Systems of Allentown, Pennsylvania (Fig. 8.10). The ACM test measures alkaline contaminant materials, which include soaps. Oil Process Systems sells a filter aid that is designed to reduce the ACM load in used cooking oils, so this test was developed to demonstrate efficacy of their treatment system. The PCM measures polar contaminant materials (accumulated polar) and was developed as an endpoint test. These tests are colorimetric solvent based tests. To conduct a test, tubes are placed on a flat surface and hot oil is added to a fill line using a glass dropper. The appropriate solvent is then added to the second line and tube is capped and shaken. The colors which form are then compared to a color card. The color changes which occur with the ACM are very dramatic going from yellow to dark blue.[36] Adoption of these tests have been hampered by the fact that the solvent system is flammable and, therefore, not something that operators wanted to use around a hot fryer.

The Shortening Monitor was developed by the 3M Corporation of St Paul, Minnesota. The test consists of a white paper strip measuring 1.0 × 10 cm (0.3 × 3.75 inches) which has four blue bands across it. They are used as a dip test to measure accumulated free fatty acids, similar to commonly used pH papers. The tests were developed to provide users, especially those in the fast-food industry with an inexpensive means objectively to measure cooking oil quality.[51] To conduct a test, the operator grips the strip at the top and immerses it in the hot oil for 1–2 seconds. All four bands must be covered. The strip is removed and excess oil is alowed to drain into the fryer. After 15 seconds the strip is held up to the light and the number of bands which have changed to yellow are counted. The approximate free fatty acid (FFA) values for the bands are >2.0, > 3.5, >5.5 and >7.0%. The test, therefore, would have little value in a chip operation in which operators generally strive to maintain fatty acid levels at or below 0.50%, but would be applicable in operations that use higher levels for control.

Advantec manufactures a test similar to the 3M test strip. The test is distributed in Asia by Ajinomoto and is called AC-Check. This test is used to monitor acid value. The operator places a plastic strip with an indicator on the tip into the cooking oil. Based on the degree of abuse, the color indicator changes from dark blue to a light olive green. The dark blue color indicates that the acid value is '0', whereas the light green is 4.0. The color chart goes as follows: 0, 0.5. 1.0. 2.0, 3.0 and 4.0. Corresponding colors are blue, dark green, light green, olive, light olive and very light olive. Both the Advantec and Shortening Monitor are designed to measure free fatty acids, which are not good indices of oil quality as has been explained earlier. The value of these two tests is, therefore, hindered by this issue.

Test Kit Technologies of Metuchen, New Jersey, USA, manufactures and markets rapid tests for fats and oils. Their first-generation test kits utilized the solvent technology and color cards employed by the Mir-Oil and Merck tests.

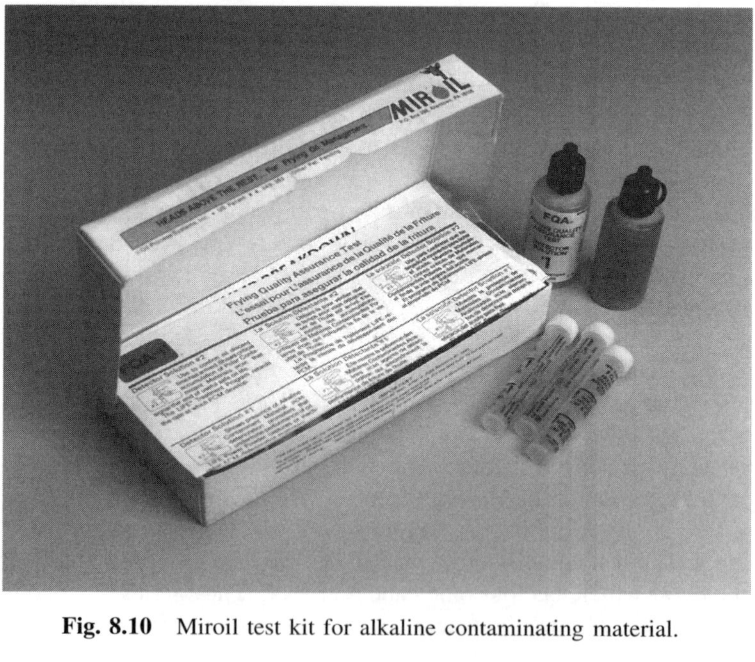

Fig. 8.10 Miroil test kit for alkaline contaminating material.

Since that time, the company has developed and patented a system called GiTIC or Gel-in-Tube Instant Chemistry. With the GiTIC technology, hot oil samples are collected from the fryer, filtered and added directly to the gel. The oil melts the gel and the components of the gel react with the oil to produce a color. While still hot the tube is placed in a small colorimeter and a reading is obtained. The reading may be directly related to oil quality and used for quality monitoring. Tests are available for total polar materials (TPM), free fatty acids (FFA) and surfactant materials or WET for water emulsion titratables. The TPM test is a single-phase test whereas the FFA and WET tests are two-phase, producing a colored bottom layer, all of which are read by the instrument. Tests may be used with a color card instead of the meter.

8.9 The future for monitoring oil quality

Oil quality monitoring and other analytical procedures for fats and oils will continue to evolve if the recently concluded 3rd International Symposium sponsored by the Deutsche Gesellschaft für Fettewissenschaft in Hagen, Germany is any indication. This meeting saw the introduction of several new rapid tests, and discussions on others. Dr Christian Gertz and Dr Jack Truong spoke of the new systems that they had developed.[42,49] Other systems that were discussed at this meeting were the use of the electronic nose and the use of near infrared spectroscopy as tools for rapid analysis of fats and oils. The reports on the electronic nose system[52] showed promise, but at this stage, high costs would render the unit impractical for quality operations. Near infrared spectroscopy also has a potential application in frying operations. There are reports that it is already being used for on-line monitoring and control. Given the way NIR operates, this system could become an extremely powerful tool for on-line monitoring and control.

The future will see more quick tests being developed for use in the industry. Those destined for use in foodservice and restaurant operations, a potentially huge market, will need to be small, rugged and inexpensive. They will also have to be safe for use in a restaurant or foodservice operation. Safety meaning both food-safe, that is without solvents or toxic materials, and fire-safe. Those targeting industrial operations will need to be able to be used on-line, such as NIR or at-line. With the movement towards implementation of the Hazard Analysis and Critical Control Point system worldwide, a system that emphasizes process control and monitoring, technologies like this will be a natural fit.

There will also be a trend towards developing test methodologies, or modifying existing ones, that are more environmentally friendly. The work conducted through the AOAC in 1993 that demonstrated that a cyclohexane-acetic acid solvent system could be substituted for carbon tetrachloride[53] in the test for iodine value (IV) is an example of this evolutionary process. Fats and oils chemists will continue to look for methods that use less solvent or more 'friendly' solvent systems. There are both environmental and economic issues at play here. Solvent and reagent disposal costs will continue to rise, so it behoves the analytical community to find ways to reduce costs. As computer and chip technology continue to become more sophisticated and the costs for manufacture drop, the potential for improvements in analytical techniques are limitless. The future for oil testing will read 'fast, safe and environmental.'

8.10 References

1. GUPTA, M.K., (1994), 'Letter to the Editor', *Inform*, 1994, 5:5, 647–648.
2. TECHNICAL ASSISTANCE RESEARCH PROGRAMS INC., 'Report of the Technical Assistance Research Programs', Technical Assistance Research Programs, Inc., Washington, DC, 1987.
3. STIER, R.F. and BLUMENTHAL, M.M., 'Quality in Frying', *Baking & Snack*, 1994, 16:11, 38–42.
4. Unpublished, 'Proceedings of the 3rd International Symposium of Deep-Fat Frying – Final Recommendations', Sponsored by the German Society for Fat Research, (Deutsche Gesellschaft für Fettewissenschaft (DGF)), Hagen-Westphalia, Germany, March 20–21, 2000.
5. ROBERTSON, C.J., 'The Practice of Deep-Fat Frying', *Food Technology*, 1967, 21:1, 34–36.
6. ROBERTSON, C.J., 'A Review of Deep Frying Current Practices Point the Way to New Horizons', *Canadian Institute of Food Technologists Journal*, 1968, 1:3, A66–75.
7. CHOW, C.K. and GUPTA, M.K., 'Treatment, Oxidation and Health Aspects of Oils', Chapter 11 from *Technological Advances in Improved and Alternative Sources of Lipids*, KAMEL, B.S. and KAKUDA, Y.S. Eds, Blackie Academic & Professional, London, Glasgow, New York, Melbourne & Madras, 1994.
8. STIER, R.F. and BLUMENTHAL, M.M., 'Quality Control in Deep-Fat Frying', *Baking & Snack*, 1993, 15:2, 67–77.
9. STIER, R.F., 'Chemistry of Frying and Optimisation of Deep-Fat Fried Food Flavour', Presented at the 3rd International Symposium of Deep-Fat Frying – Final Recommendations', Sponsored by the German Society for Fat Research, (Deutsche Gesellschaft für Fettewissenschaft (DGF)), Hagen-Westphalia, Germany, March 20–21, 2000.
10. BLUMENTHAL, M.M., 'A New Look at the Chemistry and Physics of Deep Fat Frying ', *Food Technology*, 1991, 45:2, 68–71, 94.
11. STIER, R.F. and BLUMENTHAL, M.M., 'Filtering Frying Oils' *Baking & Snack*, 1991, 13:5, 15–18.
12. STIER, R.F., 'Oil Filtration and Treatment: A Key Element in Quality Food Service and Industrial Operations', IFT sponsored short course, Science & Technology of Frying, Hayward, California, December 1–3, 1999.
13. DGF, 'Meeting Summary, DGF (German Society for Fat Research), Symposium on Frying and Cooking Fats', *Fette Seifen Anstrichm.*, 1973,

75:49.

14. DGF, 'DGF Symposium on Frying and Cooking Fats', *Fette Seifen Anstrichm.*, Special Issue, 1979, 81:483.
15. FIRESTONE, D., 'Regulation of Frying Fat and Oil', Chapter 19 from *Deep Frying: Chemistry, Nutrition and Practical Applications*, PERKINS, E.G. and ERICKSON, M.D. eds, Pp. 323–334, American Oil Chemists Society (AOCS) Press, Champaign, IL, 1996.
16. STIER, R.F., 'Are Fried Foods and Frying Oils Good for You?', IFT sponsored short course, Science & Technology of Frying, Hayward, California, December 1–3, 1999
17. ERICKSON, D.R., 'Production and Composition of Frying Fats', Chapter 1 from *Deep Frying: Chemistry, Nutrition and Practical Applications*, PERKINS, E.G. and ERICKSON, M.D. Eds, pp. 4–28, American Oil Chemists Society (AOCS) Press, Champaign, IL, 1996.
18. GUPTA, M.K., (1999), 'Qualification of Oil Suppliers for Frying', IFT sponsored short course, Science & Technology of Frying, Hayward, California, December 1–3.
19. GUPTA, M.K., 'Handling of Oils After Processing', IFT sponsored short course, Science & Technology of Frying, Hayward, California, December 1–3, 1999.
20. BECHTOL, L., 'Supplier Partnership: An Integral Ingredient of Total Quality Management in the Manufacturing Process', from Total Quality Management a Short Course Sponsored by the Institute of Food Technologists, Berkeley, CA, February 2–4, 1993.
21. STIER, R.F. and M.M BLUMENTHAL, 'Vendor Certification: Is It Worth the Hassle', *Baking & Snack*, 1994, 16:9, 66–70.
22. STIER, R.F. and M.M. BLUMENTHAL, 'Plant Self Inspection', *Dairy, Food & Environmental Sanitation*, 1995, 15:9, 549–553.
23. BLUMENTHAL, M.M., *Optimum Frying: Theory and Practice*, Libra Laboratories, Piscataway, N.J., USA, 1987.
24. BLUMENTHAL, M.M. and STIER, R.F., 'Optimization of Deep-Fat Frying Operations', *Trends in Food Science & Technology*, 1991, 2:6, 144–148.
25. POKORNY, J., 'Flavor Chemistry of Deep-Fat Frying in Oils', Chapter 7 from *Flavor Chemistry of Foods*, SMOUSE, T. and PERKINS, E.G., eds, American Oil Chemists Society (AOCS) Press, Champaign, IL, 1989.
26. GUPTA, M.K., 'Quality Control in Deep-Fat Frying Operations', IFT sponsored short course, Science & Technology of Frying, Hayward, California, December 1–3, 1999.
27. ANONYMOUS, 'Insurers to Cover $664 Mln Belgium Dioxin Loans', Reuters, 1999, August 11
28. GORDON, R.S., 'Oleoanilides and Spanish Oil Poisoning', *Lancet*, 1981, 2:1171–1172
29. INTERNATIONAL UNION OF PURE AND APPLIED CHEMISTRY, *Standard Methods for the Analysis of Fats, Oils and Derivatives*, 7th Edition, 1st Supplement, Blackwell Scientific Publications, Oxford, UK, 1992.

30. ASSOCIATION OF OFFICIAL ANALYTICAL CHEMISTS, *Official Methods of Analysis of the AOAC International*, 16th Edition, AOAC International, Gaithersburg, Maryland, USA, 1997.
31. AMERICAN OIL CHEMISTS SOCIETY, *Sampling and Analysis of Commercial Fats and Oils*, American Oil Chemists Society (AOCS) Press, Champaign, IL, 1999.
32. FRITSCH, C.W., 'Measurements of Frying Fat Deterioration' *JAOCS*, 1981, 55:10, 718–727.
33. MANKEL, A., *Fette Seifen Anstrichm.*, 1970, 72:677.
34. CASTANG, J., *Ann. Fals. Exp. Chim.*, 1981, 74:701
35. STIER, R.F. and M.M. BLUMENTHAL, 'The Use of Rapid Methods for On-Line Monitoring', *Baking & Snack*, 1992, 14:5, 30–35.
36. STIER, R.F., 'Quick Tests for Fats and Oils'. *Baking & Snack*, 1996, 18:9, 62–66.
37. ANONYMOUS, 'Oil: Quick Testing for Quality', *Snack Professional*, 1995, 3:1, pp.16–19.
38. ANONYMOUS, 'Viscosity Tester Method', KFC Corporation R & D/QC, 1995.
39. GILLATT, P., KRESS-ROGERS, E. and ROSSELL, J.B., 'A Novel Sensor for Measurement of Frying Oil Quality', *Lipid Technology*, 1991, July–September, 78–82.
40. KRESS-ROGERS, E., GILLATT, P.N., and ROSSELL, J.B., (1990), 'Development and Evaluation of a Novel Sensor for In Situ Assessment of Frying Oil Quality', *Food Control*, July, 163–178.
41. ANONYMOUS, 'Oils and Fats Research at Leatherhead Food RA', *Inform*, 1992, 3:5, 586–593.
42. GERTZ, C., 'Chemical and Physical parameters as a Quality Indicator of Used Frying Fat', Sponsored by the German Society for Fat Research, Deutsche Gesellschaft für Fettewissenschaft (DGF), Hagen-Westphalia, Germany, March 20–21, 2000.
43. LANSON INDUSTRIES, 'Smart Dipstick Takes Guesswork Out of Oil Changing', Press Release, Sept. 24, 1992
44. GRAZIANO, V.J., 'Portable Instrument Rapidly Measures Quality of Frying Fat in Food Service Operations', *Food Technology*, 1979, 33:9, 52–57.
45. FRITSCH, C.W., D.C. EGBERG and J.S. MAGNUSON, 'Changes in Dielectric Constant as a Measure of Frying', *JAOCS*, 1979, 56:8, 746.
46. CROON, L.B., A. ROGSTAD, T. LETH and T. KIUTOMO, 'A Comparative Study of Analytical Methods for Quality Evaluation of Frying Fats', *Fette Seifen Anstrichm.*, 1986, 88:3, 87–91.
47. SALTMARSH, M., 'Rapid Analysis of Fats and Oils: Thermal Degradation', Presentation at Western Food Industries Conference, UC Davis, March 1990.
48. SCHWARTZ, K., 'Quick Tests Used for Fats and Oils', Sponsored by the German Society for Fat Research, Deutsche Gesellschaft für Fettewissenschaft (DGF), Hagen-Westphalia, Germany, March 20–21, 2000.

49. TRUONG, J.G., 'A Universal Easy to Use and direct Method for Measuring % Total Polar Compounds in Degrading Oils', Sponsored by the German Society for Fat Research, Deutsche Gesellschaft für Fettewissenschaft (DGF), Hagen-Westphalia, Germany, March 20–21, 2000.
50. BLUMENTHAL, M.M., J.R. STOCKLER and P.J. SUMMERS, 'Alkaline Contaminant Materials in Used Frying Oils: A New Quick Test', *JAOCS*, 1985, 62:9, 1373–1374.
51. ANONYMOUS, 'Test Strip for Frying Fats', *JAOCS*, 1986, 63:7, 838–839.
52. DEMISCH, U. and MUHL, M., 'Electronic Nose for Detection of Deterioration of Frying Fat Comparative Studies for a New Quick Test', Sponsored by the German Society for Fat Research, Deutsche Gesellschaft für Fettewissenschaft (DGF), Hagen-Westphalia, Germany, March 20–21, 2000
53. FIRESTONE, D., (1993), 'Determination of the Iodine Value of Oils and Fats: Summary of a Collaborative Study', *The Referee*, 1993, 17, July, AOAC International.

Part III

Improving product quality

9

The manufacture of pre-fried potato products

M. J. H. Keijbets, Aviko BV, Steenderen

9.1 Introduction

Manufacturing pre-fried potato products, although originating from small-scale traditional methods some fifty years ago, is now a complex, highly developed industrial food production activity. The full manufacturing process comprises many individual processes or unit operations, from raw material reception and peeling to freezing, packing and storage.

Modern state-of-the-art manufacturing lines for pre-fried potato products, i.e. French fries (chips in the UK), are continuous processing lines with a high capacity per hour. In order to save running costs the capacity of French fry lines has gradually risen from an output of 3–5 tons per hour to 15–25 tons per hour. This means that such lines have to be fed with 30–50 tons potatoes per hour, the approximate yield of a pre-fried French fry line being about 50%. These modern lines are thus highly capital intensive, fully automated but low in their use of labour.

In this chapter a short description is given of what pre-fried potato products are (Section 9.2) and which products exist in the marketplace (Section 9.3). In the following sections an overview is given of the key requirements for their production (Section 9.4), the key manufacturing processes for French fries and what they mean for the product (Section 9.5) and the key manufacturing processes for 'formed' pre-fried potato products (Section 9.6). Storage and distribution to the consumer of the finished pre-fried potato products is dealt with in another section (Section 9.7). Major quality determining factors during manufacture are treated (Section 9.8) and an outlook on future developments in manufacture is given (Section 9.9). The chapter finishes with cross-references to key books to consult and other sources of valuable information in the field of pre-fried potato products (Section 9.10).

9.2 What are pre-fried potato products?

Pre-fried food products are products that are pre-fried in the manufacturing factory and that need to be finished before consumption by the end-user i.e. the consumer. Traditionally, this finishing process during preparation in the kitchen was by means of finish-frying but, due to the desire to develop more low-fat and healthier food, oven- or even microwave-finishing has become an alternative way of preparation. The frying process is thus partly carried out during the industrial manufacturing and partly in the kitchen. Pre-fried potato products are the largest and may be the most popular representative of this food group in the world. They form a major menu item of the widely spread and still rapidly growing fast food chains all over the world, even in places where potato products did not belong to the traditional food consumption pattern (Asia, Africa).

Frying in fact is a very effective cooking process; cooking in oil. Pre-frying in the factory means that the potato product is partially cooked and that it is convenient and rapid to prepare in the kitchen before consumption. For instance, pre-fried deep-frozen French fries may be prepared by after- or finish-frying in a deep fat fryer in about three minutes. Frying imparts very characteristic taste and texture properties to a food material. The frying process is a cooking process in oil at relatively high temperatures (140–180ºC). It leads to changes in the structure and taste that are recognised and beloved as the typical fried food characteristics. This in broad terms means that the potato gets a crispy crust with a cooked inside and a fried potato flavour.[1,2] At the same time the potato product takes up oil during the frying process. Oil reinforces the fried taste and the pleasant feeling of the fried product in the mouth. At the end of the manufacture after pre-frying, the potato products are chilled or frozen, packed, stored and distributed to wholesalers or retailers and via them to the consumer.

9.3 Range of pre-fried potato products and their use

Pre-fried potato products are manufactured in a wide variation. World wide, French fries is by far the main product in terms of volume. French fries are manufactured in many forms (cut size, type of oil or fat, coated or non-coated, frozen or chilled). They are produced for retail or food service, including fast food chains (McDonald's, Burger King, etc.). The major manufacturers of French fries are based in North America (Canada, USA) or Europe. Production also occurs in Australia, Argentina, China, Egypt, Indonesia and South Africa. Chilled French fries are produced only in Europe and mainly used in traditional food service.

Other pre-fried potato products are cut from potatoes the same way French fries are or formed from mashed potatoes. Cut pre-fried products are for instance slices, wedges, cubes, small whole potatoes (France: 'pommes parisiennes'), hash-browns ('rösti's') and pancakes (Germany: 'Reibekuchen'). Mainly in

Europe a variety of formed potato products exists, e.g. croquettes, 'pommes duchesses' (duchesse potatoes), 'pommes noisettes' (small potato nuts), potato waffles, 'pommes dauphines' (France: potato puffs), 'tortilla patata' (Spain: potato egg cake), potato letters, potato strip figures, etc.

Pre-fried potato products are finish-fried in the kitchen by deep-frying in oil or fat. This is the case for food service operations and for countries where deep fat frying is a common kitchen practice in the household (Belgium, France and the Netherlands). In other countries finishing in the oven (Germany, Scandinavia, UK, USA) or shallow pan frying for some products ('pommes noisettes' in France) is more common in the household. Microwave heating is another now widespread kitchen tool. If other finishing methods than deep fat frying are required, the manufacturing process for pre-fried potato products has to be adapted. This will be discussed in Section 9.5.

9.4 Key requirements for pre-fried potato products

Key requirements for pre-fried potato products with emphasis on French fries are:

- raw material selection: potato and frying oil
- primary processes before frying
- frying process
- chilling and freezing process
- packing process
- storage and distribution.

Although the frying process is the key and distinguishing process for any pre-fried food, raw materials selection and pre- and post-frying processes are also vital for manufacturing high-quality pre-fried French fries.

The selection of proper potato raw material is very essential for pre-fried French fries production. Important characteristics of the potato raw material are:

- variety: purposely bred for French fry processing, suited for growing in normal agricultural practices, resistant against plant pathogens, black spot and bruising, prolonged storage
- shape: big tubers, elongated form, shallow eyes (peeling)
- dry matter content: moderately high 20–24%, uniform distribution in the tuber
- reducing sugar content: low on prolonged storage, maximum 0.5%
- good taste and texture.

Traditionally, varieties as Russet Burbank and Shepody are used in the USA, Maris Piper, Pentland Dell (both UK) and Bintje in Europe. In Europe new French fry varieties are emerging, mainly because of increasing processing and environmental demands e.g. Agria, Asterix, Felsina, Victoria (more disease resistant, less use of chemicals during growing).

Frying oil selection depends on frying characteristics (stability, solidifying behaviour) and desired nutritional value (more unsaturated fat). Traditionally, hardened palm and soybean fat are major frying fats for manufacturing French fries. Palm oil, sunflower oil and canola oil (zero erucic acid rapeseed oil) are now common in the industry as well.

After selection of the right raw materials, correct pre-processing converts the potatoes into strips of the desired dimensions and shape, well sorted, blanched and dried before entering the industrial fryer. These pre-processes in combination with frying are vital to obtain the desired colour, texture and taste of French fries. After frying the potato strips are rapidly cooled down, chilled or frozen and packed, mainly to preserve the characteristics obtained during further storage and distribution.

9.5 Key manufacturing processes for pre-fried French fries (see Fig. 9.1)

9.5.1 Peeling

After transport of the potatoes to the factory, raw material quality control and washing to remove adhering soil, peeling with steam is the first unit operation. Steam peeling is carried out at 14–18 bar steam pressure. It is a very efficient peeling method using a combination of short steam pressure cooking to loosen the potato skin followed by a subsequent peel removal operation. Peeling losses are considerably lower than in mechanical peeling, averaging 10–20 mass %. Peel removal is done with minimal water use by a belt brushing machine or a centrifugal peel remover (Europe) or by a rotating drum with water spraying (USA). Peeling is not necessary when French fries with skin on are produced.

9.5.2 Cutting

Cutting is the next unit operation. This determines the shape of the French fries. Originally, mechanical cutters were used in which potatoes were first sliced and then cut into strips. This method is still in use for manufacturing wave-like shapes ('crinkle fries') and for special shapes, e.g. waffle fries. For straight French fry shapes the water knife cutting system has been developed.[3] In this

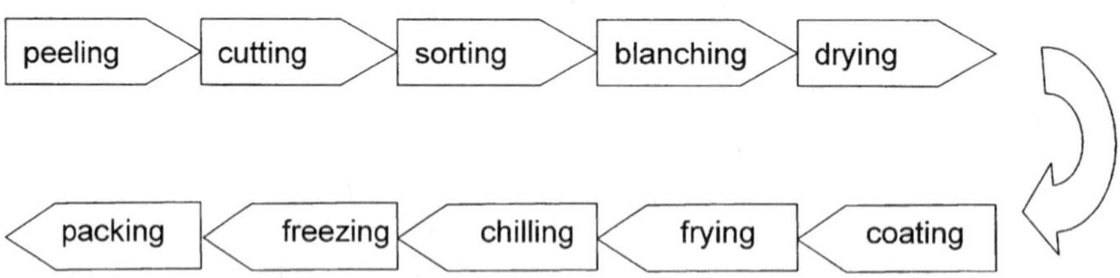

Fig. 9.1 Flow diagram of manufacture of pre-fried French fries.

cutting system the potatoes are transported by water at high speed (about 30 km/h) through a tube in which a knife assembly has been placed. Due to the high water speed the tubers are oriented in the length direction and pushed through the knife assembly. As a result maximum length of potato strips is obtained from the tuber. Length of French fries is an important quality characteristic, particularly in the fast-food business. The design of the water knife cutting system, including supply and discharge pipe, knife assembly construction, knife material, knife exchanging mechanism, etc. is a major issue in improving cutting quality and preventing cutting losses[4] which may be as high as 4–10 mass %.

9.5.3 Sorting and grading

Sorting and grading together with peeling and cutting determine the outer appearance of pre-fried French fries. Two types of sorting/grading operations are in use. The first is the mechanical removal of thin and short strip cut sizes that originate from cutting whole potatoes into strips. Roller sorters remove too thin strips. The strips are transported over rotating rollers that are placed at a predetermined distance depending on the cut size. Thin strips fall between the rollers and are removed. Too short strips are removed on shaking sieves with sieve configurations depending on the final product quality required. The short strips fall through the openings of the sieve.

The second sorting/grading operation is the optical sorting.[5] Optical sorters remove potato strips with deviating flesh colour i.e. brown, black, green, etc., so-called 'defects' caused by damaged and diseased potato material, but also strips with peel residues. The principle of optical or 'colour' sorting is based on making pictures of the potato strips with a video colour camera while these are transported in aligned form on a white belt. These pictures are digitised and analysed with the aid of a computer program in order to decide that the potato strip is acceptable or has to be rejected. Potato strips to be rejected are localised and removed at the end of the sorter belt by blowing them out in free fall. An additional piece of optical sorting equipment is the so-called 'defect-cutter'. The principle is the same as that of the optical sorter. However, the defective parts of the potato strip, mainly residing at the end of the strip, are cut off from the strip at the end of the sorter belt by plastic knives, mounted on a rotating drum and directed by the computer program. Optionally this defect-cutter may be placed behind the optical sorter. The strips with defects blown out by the first optical sorter are transported through the second one, the defect-cutter, the defect is cut off and the good part brought back in the processing line increasing total yield of good potato strips. Total sorting and grading loss may be as high as 20 mass %.

9.5.4 Blanching

Together with the subsequent drying and pre-frying processes the blanching process determines the major external and internal eating characteristics of pre-fried French fries; colour, texture and taste. During blanching in water several physical and chemical processes occur simultaneously in the potato strips.

First of all, reducing sugars glucose and fructose diffuse from the outer layers into the surrounding blanching water.[6] The intensity of the Maillard reaction between reducing sugars and amino acids decreases. As a result the brown reaction colour of finish-fried French fries will remain within acceptable and agreeable borders. The diffusion rate of sugars is mainly determined by the concentration difference between potato tissue and surrounding water and water temperature. The extent of leaching out of sugars is rather time dependent.

Secondly, heating of potato tissue during blanching inactivates enzymes.[7] This inactivation is necessary to prevent biochemical changes at prolonged storage of deep-frozen French fries. Inactivation of the enzyme polyphenol oxidase (PPO) is necessary to prevent enzymatic grey discoloration, a major quality defect of pre-fried French fries.

Thirdly, heating changes texture and taste of French fries. The raw potato strips have to be converted into palatable fries with a desired, well-cooked texture at the inside. Because pre-frying also cooks the potato tissue, the total cooking 'load' of the French fry manufacturing process (blanching and frying) has to be controlled during blanching in such a way that the inside of the fry is well-cooked but firm and not-overcooked.[8] Overcooking results in disappearance of the inside or an empty filling within the French fry crust. Undercooking, on the other hand, leads to a too firm texture and a raw, unpalatable potato taste. By choosing the right blanching conditions (temperature and time) and designing a blanching process in various consecutive steps, the desired colour, texture and taste are obtained.

At the end of the blanching process some ingredients and additives or processing aids may be added. This is done in a separate, small blancher, in a dip tank or transport flume connecting the last blancher to the drier. Glucose is added for reaching the desired colour after finish frying. When manufacturing oven and microwave fries addition of glucose is particularly necessary to obtain the desired golden brown fry colour at final preparation. Salt (NaCl) may be added for taste. Sodium acid pyrophosphate (SAPP) ($Na_2H_2P_2O_7$) is needed to complex iron (Fe^{2+}). In this way SAPP prevents that iron in the potato reacting with chlorogenic acid at the heating processes. The Fe^{2+}-chlorogenic acid complex is oxidised by oxygen from the air into a greyish coloured substance that causes the so-called 'after-cooking' grey discoloration.[9]

As a result of all these considerations, blanching is frequently carried out in the industry in various steps in water of 60–90ºC. Total blanching time may vary between 5 and 60 minutes. In order to firm the inner texture, the pre-blanch-cool-blanch process from the potato flake and granule manufacture is sometimes applied. After pre-blanching at about 70ºC, the potato strips are cooled in cold water until about 20ºC and thereafter blanched again at 70ºC. Steam blanching is rather rare in the French fry industry, having as main disadvantage that no sugar leaching takes place. Several types of blanching equipment are in use. Screw blanchers filled with water are the predominant types in Europe. The potato strips are transported through the blancher by the blades attached on the turning screw axis. Special devices are constructed for in- and out-feed of the strips. The water is circulated through the screw, partly refreshed to maintain the sugar concentration difference and heated by direct steam injection. In the USA more expensive belt blanchers are in use. The potato strips are transported on a wire or plate belt through a huge water bath. Instead of transport through the water, water may be sprayed over the strips lying on the belt. An advantage of belt blanchers is that less breakage of the rather vulnerable potato strips may occur. This is especially true for thin cut sizes, e.g. 7 to 8 mm, the preferred cut size for fast-food French fries.

9.5.5 Drying

In the drying process water is evaporated with the aid of hot air at the outside of the potato strips (10–30%). As a result the outer part of the strip will be drier than the inside leading to crust formation. Due to air drying before pre-frying a thicker and firmer crust develops than without. Originally, the drying process was developed for French fry manufacture in the 1970s in order to be able to produce French fries with improved ovenability.[10] Side effects of the drying operation are that less water has to be evaporated in the fryer, increasing the capacity of the fryer and the possibility to recover heat from the fryer exhaust vapour through a heat exchanger. In this way a more efficient use of energy is made during processing and emission of frying vapours and smell is reduced.

Crust formation through drying is of major importance when manufacturing French fries for the fast-food restaurants. One of the main quality requirements of fast-food French fries is the so-called 'holding time'. After deep fat finish frying fast-food fries should keep their crispiness for 5–10 min. under a heat lamp. In order to serve their clients at high speed, in particular the drive-in ones, the waiting time should be at a minimum. For that reason fast-food restaurants hold a part of their finish-fried fries under a heat lamp. Apart from other contributing raw material and processing factors, severe crust formation during drying is vital for fulfilling holding time requirements.

French fry air dryers essentially are belt dryers. The potato strips are mechanically de-watered before they enter the dryer. In the dryer hot air of 70–120°C is circulated and blown downside up through the transport wire belt. Depending on the desired uniformity of drying the dryer consists of several sections with their own belts. At entering into another section the potato strips are turned around. In the 1980s 'air radio frequency assisted' (ARFA) dryers were introduced into the French fry industry. These dryers, which used radio frequency radiation at 27 MHz, would dry more uniformly and at shorter transport length thus saving space. Due to running problems and high costs they failed. A more recent breakthrough in drying technology in French fry manufacture, is the high-speed air suction dryer. In this dryer air is sucked at much higher speed through the belt than possible with blowing, resulting in more uniform and shorter drying time. When a high mass loss at drying is required as for fast-food French fry production, the dried potato strips must be equilibrated before frying in order to re-distribute the moisture in the strips. When this does not take place, blisters may be formed in the crust at pre-frying.

9.5.6 Coating

Coating with a batter after drying is a relatively new unit operation in French fry manufacture. The goal of coating French fries can be both increasing crispiness and creating a vehicle for adding new taste experiences to French fries or similar pre-fried potato products (e.g. wedges) or a combination of these functions. Through coating, a new crispier crust than the original potato one is created. When more crispiness is the goal, French fries are coated with a thin, clear-coat batter. When new flavour and taste are added, more visible, thicker tempura-like batters including spices and colouring materials are used.

Clear-coat batters are widespread in the USA for fast food manufacture. A clear-coat batter should be rather invisible but have high functionality; a low batter pick-up during coating and a significantly improved crispiness after deep fat finish frying or oven re-heating. The main advantage of these clear-coated French fries is a better holding time of fast-food fries. This is due to better holding characteristics of the new crust in combination with a slower cooling down under the heat lamp. Increase of ‘holding time’ for 10–20 min. is the prerequisite. Clear coats are also applied for oven fries to increase their ovenability.

Clear-coat batters may be flour type batters, wheat flour and modified starch being the main components. More modern types are composed of a modified (potato) starch, a (potato) dextrin, rice flour, salt, glucose, rising agents and gums.[11] New and maybe better recipes certainly will be developed.

At coating of French fries dried potato strips are coated with a batter suspension before pre-frying. The strips are aligned after the dryer and transported on a belt through a batter bath, under a batter curtain or both. Batter uptake will typically be between 10 and 20 mass %. The batter is set in the first fryer. A second fryer is necessary in order to create the possibility to break clumps originating from the first fryer on a vibrating shaker after lifting the fries from the first fryer and to finish the pre-frying process. Investment costs in a coating line including batter make-up equipment, batter bath and recovery, second fryer and auxiliary equipment are high. Together with the high costs of batter dry matter these result in a considerable cost increase for manufacturing.

9.5.7 Frying

Frying is the ultimate key process when manufacturing pre-fried French fries. In this unit operation, in combination with finish frying, the ultimate characteristic features of French fries come into being. Industrial deep fat frying in oil is carried out by transporting potato strips from the dryer or the coating line on a wire belt into a bath with hot oil at 140–180°C. Due to the high frying oil temperature water evaporates at the potato surface at an explosively high speed. From the inside of the potato strip water diffuses to the outside. At the outside a dry, porous potato layer is formed: the crust.[1,12] Due to the porosity and dryness of this – thin, 0.3–0.8 mm thick – layer, the consumer experiences this crust at eating as crispy (Fig. 9.2). At the same time the tissue at the inside of the potato strip will be cooked. Cooking progresses at higher speed than during blanching because of the high temperature (100ºC). Ultimately, a finish-fried French fry has a crispy crust and a well-cooked inside.

At the end of the frying process the pre-fried potato strips take up frying oil when they leave the fryer. The oil or fat content mainly depends on length of frying time and surface area (cut size). Frying time is determined by the amount of moisture to be evaporated and frying temperature. The moisture content of the incoming potato strips depends on dry matter content of the potato used and the drying loss in the dryer. The moisture content of the pre-fried French fry is determined by the specifications set for the particular product. For instance, ovenable fries that are prepared for consumption by re-heating in the oven at about 220ºC are pre-fried to a lower moisture content (and a higher oil content) than fries that are finish fried in deep fat. Thin cut sizes have lower (relative) moisture content than thicker cut sizes. Microwavable French fries have a very low moisture content, being already almost finish-fried.

During frying the Maillard reaction between reducing sugars and amino acids takes place. Apart from the desired brown colour formation, rather characteristic Maillard flavour products arise in the French fry. At the same time frying oil hydrolyses and oxidises to some extent and generates typical lipid degradation

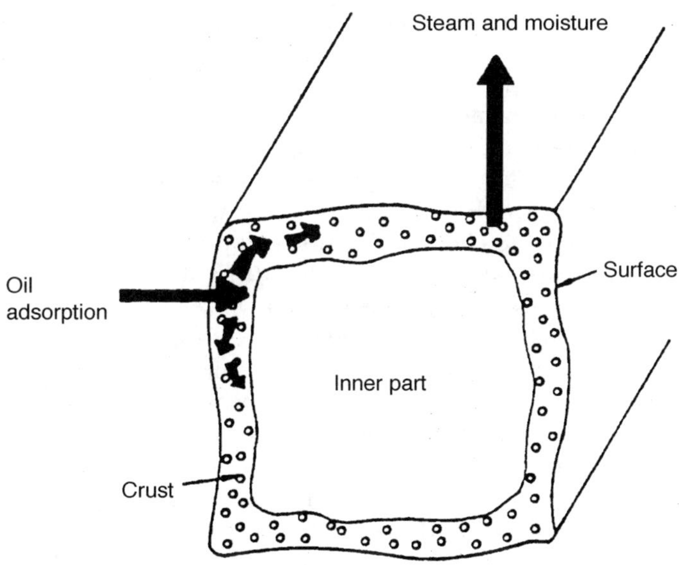

Fig. 9.2 Crust formation and oil adsorption at frying French fries (adapted from Guillaumin[12]).

products as aldehydes, etc. The extent of oxidation and the type of lipid breakdown products arising is very much dependent on the type of frying oil and the frying conditions, i.e. entrance of oxygen, etc. More unsaturated types of oil such as sunflower oil will be more liable to oxidation than more saturated ones as (hydrogenated) palm oil. This means that each frying oil imparts its typical characteristic flavour to pre-fried French fries. Altogether Maillard reaction and oil oxidation in combination with the frying oil uptake in the fry are responsible for the very characteristic flavour and taste and the pleasant feeling of French fries in the mouth.[2,13]

Degradation of frying oil by hydrolysis and oxidation, followed by polymerisation, may lead to adulteration of frying oil. An important quality type of specification in the French fry industry is that the content of free fatty acids (FFA) should remain within certain limits e.g. maximum 1%. Legal limits are set to frying oils in several countries, e.g. for FFA content, content of polar compounds (PC) and of dimeric and polymeric triglycerides (DPTG).[14] The design of industrial fryers and their use during manufacture determine oil quality. Essential features are that the oil volume of the fryer is as small as possible, no overheating of frying oil occurs, entrance of oxygen is prevented as much as possible and proper oil filtration equipment is installed. A small oil volume results in high turnover rates of frying oil. Typical values in the industry are 7–12 hours. External heat exchangers where the oil is heated rapidly with overheated steam, limit the oil volume of the fryer. Heating of frying oil over 180ºC will severely accelerate oil breakdown. A slight overpressure under the fryer hood prevents entrance of air under the hood coming into contact with the oil. Removing small potato particles by filtration prevents charring them and their contribution to oil degradation.

9.5.8 Chilling and freezing

After the heating processes, the pre-fried French fries have to be cooled down before packing them. This cooling process can be split up in chilling and freezing. When pre-fried French fries are distributed as chilled product, cooling remains limited to chilling. The first part of cooling down is chilling. The hot fries are transported on a wire belt from the fryer, via a shaking vibrator to remove excess fat, through a tunnel while air is blown through the layer of fries on the belt. The air is circulated in the tunnel by means of high-capacity ventilators. In French fry manufacture chilling was done in the past with outside air. Due to more strict environmental regulations limiting odour emission, outside air chilling is no longer possible in many countries. Nowadays the chilling air is internally circulated in a closed cooling tunnel system and cooled down in heat exchangers with outside air or mechanically cooled down with the use of refrigeration equipment. The latter, in fact, is the same system as used for freezing. So, for manufacturing chilled French fries, the product is cooled down to about 0ºC. Deep-frozen French fries are cooled down to at least −18ºC.

Chilling and freezing are common processes in the food manufacturing industry. Rapid freezing of French fries below −5ºC is important to maintain the structural characteristics. During cooling down the frying oil or fat at the outside of the fries will start to solidify. When the fries are pre-fried in a fat with a relatively high melting point, e.g. hardened palm or soy oil, the fat starts to solidify at about 40ºC. At 20ºC the fat has about 75% solid parts. It is important that the fries are turned on the cooling belt at this point to avoid clumping together. Turning is accomplished by using more than one belt in the cooling tunnel.

After the chilling and/or freezing process, French fries may be graded again. Grading is necessary when the length of the fries is not yet in accordance with the length specifications set. For that reason the fries are transported over shaking sieves and too short fries fall through as in the sorting operation before blanching (see Section 9.5.3). French fry manufacturers may sort their fries in A-, B- and C-quality streams with different length quality specifications. The very short fries are rejected as waste.

9.5.9 Packing

Packing is an essential unit operation in industrial food processing. The food is protected against spoilage and abuse and the packing material helps to sell the food. Deep-frozen pre-fried French fries are packed in polyethylene (PE) bags, normally on automatic, continuous working vertical form-fill packing machines. Bags are formed and sealed from a PE roll on these machines. After filling the bag with the predetermined mass of fries, the last seal is closed. Weighing of the portion to be packed – weigh and metal checks are part of the packing process. The bags with frozen fries are packed in carton cases on a case packer. Cases are transported and manually or automatically palletised before they are stored.

Traditionally, chilled French fries are directly packed in cartons of 10kg and palletised. Gas-packed chilled French fries are packed in a similar way as deep-frozen fries. The only difference is that before closing the last seal, the bag with the fries is flushed with gas until a sufficiently low concentration of oxygen (O_2), e.g. under 1%, has been reached. For gas-packing laminates with sufficiently low gas permeability have to be used, e.g. nylon-PE laminates. The packing gas is a combination of carbon dioxide (CO_2) and nitrogen (N_2), e.g. 70% CO_2 + 30% N_2. CO_2 limits growth of aerobic spoilage micro-organisms of pre-fried French fries, thus increasing shelf life.[15] N_2 is added to the packing gas in order to avoid the emergence of a 'pseudo-vacuum' in the bags. Due to uptake of CO_2 by the fries an under pressure emerges soon after packing. This vacuum might damage the rather fragile fries.

9.6 Key manufacturing processes for 'formed' pre-fried potato products

When manufacturing pre-fried slices, wedges, cubes, small whole potatoes, etc. the key processes are similar to those for pre-fried French fries. The only difference is another cut size. The manufacture of 'formed' pre-fried potato products, i.e. products formed from potato dough, differs in many respects. Key processes for these formed potato products are mashing/separating, mixing, forming and breading. The majority of the volume of formed products is manufactured from thin, short and defective potato strips sorted out during manufacturing French fries (Section 9.5.3). These potato strips mostly undergo a blanch-cool-cook process as for potato flakes or granules.

For the manufacture of hash-browns ('röstis') the potato material may be submitted to a cutting step after cooking. The next step is that the small potato strips are fed into a mixer in which potato dough is made by adding ingredients for binding and taste (herbs and spices). This potato dough is transported to a forming machine in which hash-browns of the desired form and dimensions (rounds, triangulars, etc.) are formed. The hash-browns are pre-fried, frozen and packed.

Potato products that are made from potato mash are manufactured in a similar way. The cooked potato material, however, is mashed in a mashing and

Fig. 9.3 Formed pre-fried potato products (courtesy Aviko BV).

separating machine. The mash is mixed with ingredients in a mixer and the dough is formed into croquettes, duchesses, noisettes, waffles, etc. in a forming machine (Fig. 9.3). After forming (and breading) these potato products are pre-fried in a similar way as French fries, frozen and packed. Freezing of these formed pre-fried potato products normally is done in spiral freezers and not in horizontal tunnel belt freezers. Belt length in these spiral freezers is 500 m or more. Due to their thickness (15–30 mm) these formed pre-fried potato products require a much longer freezing time to reach a core temperature of −18°C. Average freezing times are 40–60 min.

9.6.1 Mashing/separating

Mashing cooked potato strips is carried out in mashing and separating machines. The cooked potato strips are pressed through a screen with small holes (about 1.2 mm) resulting in production of mash. At the same time defective potato material containing discoloured, hard pieces does not pass the screen and collects on the screen. This defective potato material is continuously removed from the screen and transported to the waste. Suppliers of mashing/separating machines apply the same principles but in different technical constructions, in particular concerning the transport and pressing of the mash through the screen and removal of the defective material. This may lead to a difference in quality of the mash. The more gentle the mashing step the better the textural quality of the mash in terms of stickiness, dryness and looseness.

The mashing step is essential for the textural quality of the potato mash. In fact the quality of the mash is a compromise between mashing and separation in the same machine. The better the separation action of the equipment, the more damage may occur to the texture of the mash. The textural quality of the potato mash depends on the characteristics of the potato material, some potato varieties being more suited for mashing than others, a correct performance of the blanch-cool-cook process before mashing and the mashing operation itself. These three factors determine the level of cell wall damage of the potato material, due to which the potato cells during mashing remain intact. Once cell walls are damaged or broken, starch will leak out of the potato cells causing stickiness of the mash or undesirable quality deterioration.[16] The small holes of about 1.2 mm in the screen of the mashing and separation machines favour the risk of cell damage but at the same time improve the separation process by preventing the passing of bigger defective particles. During manufacturing dried potato flakes the size of these holes in the mashing sieve is about 8 mm diminishing the risk of cell damage. Furthermore, it is very important for potato mash quality that the potato strips are well cooked during the pre-processing. Overcooking leads to too much weakening of potato cell walls, undercooking to a middle lamella cell wall layer between two adjacent cells that is too resistant to the mashing force resulting in rupturing of cell walls.

9.6.2 Mixing

In the mixing equipment the cooked and cut potato strips (for hash-browns) or the potato mash are mixed with ingredients necessary to form a potato dough with the desired texture and taste characteristics. Both batch and continuous mixers may be used. It is rather essential again that the mixing operation does not damage the vulnerable potato dough increasing stickiness. On the other hand uniform distribution of the ingredients in a short time is necessary for an efficient mixing operation. This depends very much on the design of the mixer, in particular the configuration of the mixing blades, and residing time.

9.6.3 Forming

During forming the potato dough is converted to the ultimate form of the end product for the consumer. Several types of forming machines are in use in the potato manufacturing industry. In the machines with forming moulds, the mould with the particular shapes is filled with potato dough from a hopper. The potato shape is ejected from the mould on a transport belt by a device with the same shape. Moulding machines are used for hash-browns, waffles and special strip or animal characters. Low-pressure extrusion machines are used for croquettes, etc. In these extruders potato mash is pressed through small pipes and the resulting cylindrical 'sausage' is cut into croquettes. Special forming heads are needed to produce pommes duchesses and pommes noisettes. These are very much related to bakery machines, using rotating nozzles (duchesses) or diaphragms (noisettes). A third type of forming machine uses rotating drums in which the shape is moulded e.g. for small hash-browns.

9.6.4 Breading

Products such as potato croquettes are breaded with crumb before frying in order to increase crispiness during consumption. After forming croquettes, these are transported through a batter containing a protein or flour suspension. The batter improves the adhesion of the crumb on the croquette. Consequently, the battered croquettes are transported on a wire belt through breadcrumbs and take up the desired amount of crumb. Crumbs are applied in a wide variation of particle size, crispiness and colour, depending on the required characteristics of the final product.

9.7 Storage and distribution

Pre-fried deep-frozen French fries and other potato products are stored and distributed at −18°C. Shelf-life of frozen pre-fried French fries may last up to 24 months. Due to this long shelf-life frozen pre-fried French fries may be distributed world-wide from Europe and the USA. From European manufacturing sites, mainly situated in north-western Europe, trucks with refrigeration equipment distribute the French fries throughout Europe including Russia and Turkey. French fries for export to Asia, Latin-America, etc. are shipped overseas in containers. From the USA French fries are shipped to Asia, Japan being an important market, and Latin-America. Frozen fries normally are carried from the frozen storage at the manufacturing site to distribution centres from which they are further distributed to retail outlets.

Chilled French fries are stored and distributed at 0–4°C. Chilled pre-fried French fries are only manufactured in Europe, mainly in the Netherlands, Belgium and Germany. They have a limited shelf life but the image of a fresh product. The traditional chilled fries, directly packed in cartons, have a shelf life of only 10 days at 0–4°C. Due to the development of modified atmosphere packing technology in the late 1970s, a considerable and increasing part of the chilled fries are gas-packed. These have a shelf life of about 21 days at 0–4°C. These chilled French fries with extended shelf-life chilled are now distributed in refrigerated trucks all over Europe. Contrary to frozen French fries chilled fries are frequently distributed direct from the store at the manufacturing site or via distribution centres to the food service outlets, many of which are rather small businesses that do not possess large storage facilities. Traditional chilled fries may be distributed twice a week, gas packed fries once a week or less. Maintaining the chilling chain at 0–4°C is essential for reaching the guaranteed shelf-life of 10 or 21 days. In that sense distribution of chilled French fries is much more complex and vulnerable than that of deep-frozen fries.

9.8 Major quality-determining factors during manufacture of pre-fried French fries

Quality of pre-fried French fries is specifically described by the following attributes:

- colour, texture including holding time, taste/smell (sensory attributes)
- dry matter and fat content
- length and number of defects (outer appearance)
- microbiological quality (chilled fries).

Raw material quality and all manufacturing processes influence these quality attributes. Major influences are exerted by potato raw material (colour, length and defects). Cutting also influences length as does breakage during further processing. Sorting/grading is essential for outer appearance (length and defects). Blanching, drying and frying are the decisive processes for the sensory attributes of pre-fried French fries. Colour, texture (crispiness of the crust and well-cooked inside) and the very characteristic French fry flavour and taste are formed during these key processes. A new process, coating before frying, assists in particular in determining crispiness and 'holding time'.

Microbiological quality of French fries depends on recontamination of fries with spoilage micro-organisms after frying during the chilling and the packing process. Processing hygiene including thorough cleaning and disinfection of the processing line and control of temperature during chilling are the major factors determining shelf-life of chilled French fries.

9.9 Future trends

Manufacturing pre-fried potato products, in particular French fries, is now a highly sophisticated industrial food production activity. A high output per line in order to save costs will have much attention in coming years. Major technological breakthroughs would not be expected, the technology being rather mature at the year 2000. The key processes of peeling, cutting, sorting, blanching, drying, coating, frying, freezing and packing will not change, but individual unit operations (processes) will further be developed in a continuous process of improving technology and equipment. In this way a new generation of steam peelers (static peelers) is emerging.[17] Cutting technology is steadily improving. Colour sorting is a high-tech operation where new functions such as length measuring and data generation for process control arise. Coating technology, being a new technology for French fries, certainly will develop to improve coatings at lower costs.

PC-based process line control and data process management is a development already starting as in other manufacturing operations. In-line real-time measurements of quality parameters will be developed and used to adjust processes on a continuous basis. They will replace traditional, quality control of end products. Further development of process improvement in the potato pre-frying industry will be based on increasing knowledge of the underlying mechanisms of the processes. Knowledge development thus is the key issue. Process models based on this knowledge may be used to control processes.

Development of potato varieties that are more suited for French fry processing will be of major importance. These varieties have the right shape (long oval) and length, high dry matter, low reducing sugar at cold storage and are not susceptible to bruising and plant diseases, in particular *Phytophtora infestans*. Genetic modification might be a very helpful tool for traditional potato breeding in order to change variety characteristics. However, variation in potato quality is inevitable due to the changing weather conditions during and between the growing seasons. It will be a challenge to the manufacturing potato industry to process this variable potato raw material to a constant end-product quality. Analysing the incoming potato for key parameters in a more proper way and adjusting major process conditions on the basis of that information could be another major knowledge-based development in this industry.

Further development of the 'potato chain' from breeding to manufacturing will result in increasing ties between potato growing and processing. Raw material demands for manufacturing will more and more be specified and used to adjust growing at an early stage. Due to increasing consumer awareness organic potato growing and processing will dramatically increase, requiring breeding of suitable organic potato varieties.

9.10 Sources of further information and advice

9.10.1 Books

BOSKOU D, and ELMADFA I, *Frying of food.* Lancaster/Basel, Technomic Publising Company, 1999.

BURTON W G, *The potato*, Harlow, Longman Scientific, 1989.

GOULD W A, *Potato production, processing and technology*, Timonium, CTI Publications, 1999.

LISINSKA G and LESZCZINSKI W, *Potato science and technology*, London/New York, Elsevier Applied Science, 1989

PUTZ B, *Kartoffeln Züchtung – Anbau – Werwertung*, Hamburg, Behr's Verlag, 1989.

RASTOVSKI A and VAN ES A, *Storage of potatoes*, Wageningen, Pudoc, 1987.

TALBURT W F and SMITH O, *Potato processing*, New York, Van Nostrand Reinhold Company, 1987.

VARELA G, BENDER A E and MORTON I D, *Frying of food*, Chichester/Weinheim, Ellis Horwood/VCH Verlagsgesellschaft, 1988.

9.10.2 Organisations

ATO, Agrotechnological Research Institute, Wageningen University and Research Centre, P.O. Box 17, 6700 AA Wageningen, the Netherlands (e.g. potato storage and processing research).

EAPR, European Association of Potato Research, P.O. Box, 20, 6700 AA Wageningen, the Netherlands (European scientific association of potato researchers o.a. potato utilisation).

FNK, Potato Processors' Research Group, Waalsdorperweg 80, 2597 JD Den Haag, the Netherlands.

UEITP, Union Européenne des Industrie de Transformation de la Pommes de terre, Von-der Heydt-Strasse 9, 53177 Bonn, Germany.

9.11 References

1 KELLER CH, *Fritieren in der Lebensmittelverarbeitung. Untersuchungen am Beispiel der Herstellung von Pommes frites*, Dissertation ETH Zürich nr. 8674, 1988.

2 KEIJBETS M J H, 'Vet, smaakmaker in gefrituurde aardappelprodukten', *Voeding*, 1992 **53**(2) 50–53.

3 LISINSKA G and LESZCZINSKI W, *Potato science and technology*, London/New York, Elsevier Applied Science, 1989 (chapter 4, 'Manufacture of potato chips and French fries')

4 MENDENHALL G A, 'Converging tube assembly for hydraulic food cutter', *United States Patent* 1998 nr. 5806397.

5 HEBEL R J, FAZZARI R J and SKORINA F K, 'A high speed mass flow food

sorting apparatus for optically inspecting and sorting bulk food products', *PCT Application* 1997 nr. WO97/09689.

6 GEKAS V, ÖSTE R and LAMBERG I, 'Diffusion in heated potato tissue', *J Food Sci* 1993 **58**(4) 827–31.

7 FRETZDORFF B, BERGTHALLER W and PUTZ B, 'Untersuchungen zur Inaktivierung einiger Enzyme durch Wasserblanchieren bei der Pommes frites-Herstellung. Publication no. 5564, 1987, Bundesforschungsanstalt für Getreide und Kartoffelverarbeitung, Detmold, Germany.

8 CANET W and HILL M A, 'Comparison of several blanching methods on the texture and ascorbic acid content of frozen potatoes', *Internat J Food Sci Technol 1987* **22** 273–77.

9 BURTON W G, *The potato*, Harlow, Longman Scientific, 1989 (chapter 11, 'Cooking and processing quality')

10 MORRIS C R, 'Process for preparing frozen French fried potatoes', *UK Patent* 1975 no.1411838.

11 ROGOLS S, WOERMAN J H and KUNERTH W H, 'Improved French fry formulations containing rice flour', *European Patent Application* 1998 no. EP 0898902.

12 GUILLAMIN R, 'Kinetics of fat penetration in food' in VARELA G, BENDER A E and MORTON I D, *Frying of food*, Chichester/Weinheim, Ellis Horwood/VCH Verlagsgesellschaft, 83–89, 1988.

13 MELA D J, 'The basis of dietary fat preferences', *Trends Food Sci Technol* 1990 **1** 71–3.

14 WARENWET (DUTCH FOOD LAW), 'Warenwetbesluit bereiding en behandeling van levensmiddelen', 1992.

15 CERNY G and GRANZER R, 'Schutzgasverpackung von Pommes frites. 1 Mitteilung: Grundlegende Untersuchungen zur mikrobiellen Stabilisierung vorfritierter Pommes frites durch CO_2-Begasung', *Verpackungs-Rundschau* 1984 **35**(8) 49–52.

16 WILLARD M J, HIX V M and KLUGE G, 'Dehydrated mashed potatoes – Potato Flakes' in Talburt W F and Smith O, *Potato processing*, New York, Van Nostrand Reinhold Company, 557–612, 1987.

17 VAN DER SCHOOT P W C, 'A steam peeling apparatus'. *European Patent Specification* no. 0182434, 1989.

10

Managing potato crisp processing

R. M. Bennett, Consultant

10.1 Introduction

In order to gain consumer acceptance a product must consistently offer quality and value equal or better than that of its competitors. Application of the basics described in this chapter will enable a processor to meet this objective.

10.1.1 The product[1]

A potato crisp is a thin slice of potato, either flat or wavy, fried in vegetable oil and salted or seasoned to taste. A quality crisp is a whole chip having a saddle-shaped curl, light golden in color and having no blemishes. When placed in the mouth, it is crisp and tender, has a slight potato flavor, is properly salted or seasoned and leaves a pleasant aftertaste in the mouth. From the minute a package is opened until it is completely eaten, a consumer can only use his eyes and mouth to determine if a crisp lives up to his expectations. These may be based upon product purchased in the past or, perhaps, from impressions gained from other sources such as the media. He will look at them in three ways:

1. *Appearance* Does it look appetizing and make me want to eat it?
2. *Texture* – When I put it in my mouth, is it crisp, goes down easily and leaves no residue in my mouth?
3. *Flavor* – Does it taste good? Does it have a slight potato flavor? Is the salt or seasoning right? Does it make me want to eat more?

To process a crisp these perceptions must be expressed in quantifiable characteristics that can be used to evaluate raw materials, set process parameters,

Table 10.1 Finished crisp characteristics

By analysis	Visually
Color	Peel
Moisture	Greening
Salt	External blemishes
Seasoning	Internal blemishes
Breakage	Seasoning coverage
FFA	Blisters
PV	Foldovers
	Oil soaked crisps
	Soft centers/clusters

monitor the process and evaluate finished product. Measurable characteristics commonly used in crisp processing are listed in Table 10.1.

10.1.2 The process

The potato crisp process is a dynamic system of interdependent steps linked together to convert a potato into a crisp that will meet the expectations of the consumer described above. Each step in the process, known as a control point, is designed to perform a specific function producing a desired characteristic in the finished product or enabling a subsequent control point to perform its function. The steps in a standard potato crisp process are pictured in Fig. 10.1.

10.2 Oil and fat management

Fats and oils serve an important function in our daily diets. They supply a concentrated source of energy, supply essential fatty acids, serve as carriers for

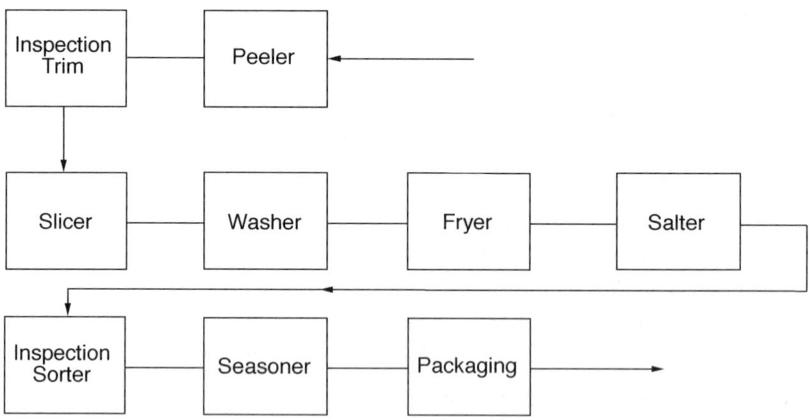

Fig. 10.1 Potato crisp process steps.

fat soluble vitamins, make foods more palatable and give satisfaction after eating. Slightly over one-third of a potato crisp is oil. In this discussion both oils and fats will be referred to as oil. In general, the term 'oil' refers to a material that is liquid at room temperature and 'fat' to a material that is solid or semi-solid at room temperature.

10.2.1 Oil and fat characteristics

The more highly polyunsaturated oils are usually used in crisp processing, with cottonseed and soybean oils the most common. While the highly unsaturated oils are much more desirable from a nutritional point of view, they require careful handling in storage and use. Their polyunsaturated structure makes them quite susceptible to degradation due to oxidation, hydrolysis or polymerization when exposed to air, water or heat. The more saturated oils, while more stable, tend to leave an unpleasant 'tacky' or 'greasy' feeling in the mouth when consumed due to the fact that their melting points are usually above normal body temperature. Table 10.2 shows the relative degree of saturation for some of the most common vegetable oils and animal fats.

When an oil comes into contact with air, oxygen in the air will react with the fatty acids in the oil to form unstable peroxides to produce the undesirable odor and flavor associated with rancidity. The rate of oxidation is accelerated by temperature and light. Although they are not currently used by the major crisp processors, anti-oxidants such as butylated hydroxytoluene (BHT), butylated hydroanisole (BHA), propyl gallate (PG) and tertiary butylated hydroquinon (TBHQ) may be added to retard oxidation.

When an oil comes in contact with water, water replaces the fatty acid in the oil to yield 'free fatty acid', FFA, to create an unpleasant flavor in a crisp. This reaction is accelerated by temperature, contact with copper-bearing metals and 'hot spots' in a fryer due to inadequate cleaning. Free fatty acid is controlled by

Table 10.2 Relative degree of saturation of common oils

Type of oil	% of total fatty acids		
	Saturated	Mono unsaturated	Poly unsaturated
Coconut	92	6	2
Beef tallow	51	46	3
Palm	50	40	10
Lard	41	47	12
Cottonseed	27	19	54
Peanut	20	48	32
Olive	17	72	11
Soybean	15	24	61
Corn	14	28	58
Sunflower	12	18	70
Safflower	9	13	78

continually adding fresh oil to a fryer. Experience has shown that oil in a fryer system must turn over in eight hours or less to avoid free fatty acid build-up.

Polymerization occurs when a polyunsaturated oil is exposed to a high temperature over a period of time. A 'gummy' film forms on exposed surfaces that are not cleaned on a regular basis. As a rule, they will not pose a problem if proper cleaning practices are observed.

10.2.2 In-plant storage of cooking oil

Because oil is so susceptible to oxidation, care must be given to minimize contact with air and temperature at all stages of an operation starting with the arrival of a shipment of oil at a plant. Upon arrival at a plant site, a shipment should be tested for the following: free fatty acid (FFA), peroxide value (PV), presence of foreign material, anti-oxidants, flavor and odor. Prior to unloading it may be necessary to heat the oil because of low temperatures experienced in shipment. The heat is usually applied by passing steam through coils present in the shipping tank. If steam is used, the steam pressure should not exceed 25 psi to limit the temperature the oil is exposed to on the heating coil surface. Oil should not be heated to more than 10ºC above its melting point prior to unloading or stored at temperatures below 45ºC.

Care must be taken when transferring oil in every step in an operation to make sure that pumps do not 'suck in' air and disperse it throughout the oil. When oil is transferred from one point to another, it is important that it is discharged below the oil level of the receiving vessel to prevent splashing and foaming. Further protection against exposure to air may be achieved by blanketing the headspace in the storage tank with either nitrogen or carbon dioxide. Oil should not be recirculated in a storage tank. Oil inventories should be managed so that an oil does not remain in storage for more than three weeks. If a King gauge or similar type gauge that uses air in its operation is used to measure the amount of oil in a tank, it should remain off at all times except when taking an actual measurement. Exposure to light is usually not a problem in the storage and handling of cooking oil.

10.2.3 Handling cooking oil in processing

Oil degradation in processing can be much more of a problem than in storage, primarily because of exposure to high temperatures. A crisp oil should not be exposed to a temperature of more than 205ºC for more than 5 minutes. This may be prevented by installing a high-temperature cut-off switch and setting it at 205ºC. Operating a fryer above its rated design capacity can have a negative impact on oil quality because of the higher oil temperature required to achieve the additional output.

'Hot spots' may occur in the heat exchanger if the small particles of potato are not removed from the cooking oil during processing. They may become 'carbonized' and will tend to adhere to the surface of the heat exchanger tubes.

A fines removal system is provided in most fryers to remove these particles and should be in operation at all times during operation. It is equipped with air jets to remove the fines from the filter screen to keep them from falling back into the oil.

The paddle wheels and the submerger continually whip air into the oil even when not producing crisps. This situation is further compounded by the frequent starting and stopping of a fryer to balance processing with packaging. The paddle wheels and submerger should not be operated at an oil temperature below 125°C. When a fryer is not in use, the oil should be removed from the system and held in a clean, dry and preferably air-free holdover tank.

Particles of oil carried up the fryer stack by the steam developed during the frying process will break down upon exposure to heat and air. This is called 'stack drip back' and will cause 'off' flavors if allowed to fall back into cooking oil. A special bonnet is placed at the base of the stack to collect this material and drain it outside of the fryer. Periodic checks of the drip back lines should be made to assure free flow.

A fryer system should be cleaned and sanitized on a regular schedule. This is usually done by boiling out the system with a caustic solution. It is imperative that the system be thoroughly rinsed after boil out to make sure no cleaning solution remains. Inadequate removal of the cleaning solution will have an adverse effect on flavor and may contribute to oil breakdown.

10.2.4 Monitoring oil quality

Since the flavor of a finished crisp is so dependent upon oil quality, it is essential that oil quality be evaluated at each step in the operation where it may be exposed to degradation. Tests commonly used to evaluate oil quality are:

- Active oxygen method (AOM): used to determine oil suitability for crisp processing.
- Peroxide value (PV): used to determine degree of oxidation during storage and handling. Oil should be tested prior to start-up and heating.
- Free fatty acid (FFA): used to determine degree of oil degradation during processing due to hydrolysis and exposure to certain metals. Oil should be tested daily at start-up.
- Total polar materials (%): used to determine degree of oil degradation during processing. Oil should be tested daily on start-up.
- Polymeric material (%): used to determine degree of oil degradation due to continous exposure to heat and improper cleaning. Oil should be tested daily at start-up.

The last two tests are recommended by the German Society of Fat Research.[8]

10.3 Raw material management

The principal raw materials used in crisp processing are potatoes, cooking oil, salt and flavorings. Storage conditions are critical to the production of a quality crisp and the control of processing costs. Proper procedures for storing and handling cooking oils have been described in the previous section. Storage of salt and flavorings only requires space that is dry, temperatures in the range of 10°C and 30°C and contamination free. Potato storage is much more critical and requires special facilities and controls.

10.3.1 Potato requirements for crisp production

Not all varieties of potatoes will produce quality crisps. In selecting a potato for crisp processing, it must be able to produce a crisp that will meet consumer expectations described in Section 10.1.1, offer value to the consumer and profit to the processor. Of particular importance is the sugar content which, if too high, will cause dark color in the finished crisp. Potato properties important in crisp processing are listed in Table 10.3.

10.3.2 Long-term storage of potatoes

In most areas potatoes are harvested only once a year. Potatoes required for processing during the harvest season can be delivered directly to the processing plant and, although they must be handled gently, they only need be protected from extreme environmental conditions prior to use. On the other hand, a higher percentage of the harvest must be stored for use until the next harvest becomes available. Because of the limited storage in the modern processing plant, off-site facilities must be equipped for long-term storage and be able to recondition potatoes prior to processing.

The potato is a living body up until it enters the slicer. It breathes, gives off carbon dioxide, requires nourishment, is sensitive to light and bruises easily just

Table 10.3 Potato properties important to crisp processing

Characteristic	Impact on process
Sugar content	Finished crisp color, flavor
Solids content	Slicing, oil usage, yield, texture
Size	Slicing, bag fill, breakage
Shape	Slicing, waste
Mechanical damage	Appearance, waste
Wet breakdown	Waste
Greening	Flavor, appearance, waste
External defects	Appearance, waste
Internal defects	Appearance, waste
Foreign material	Edibility, waste

like a human being. As a result, it must be treated with tender loving care until it enters the crisp process.[2–4]

During long-term storage potatoes are cooled to around 5°C to reduce their respiration rate and subsequent weight loss. Under these conditions a potato draws on its reserve food supply and converts starch into sugar. This will result in dark-colored crisps if the potatoes are not reconditioned prior to processing. In addition to a controlled temperature environment, adequate ventilation is required to provide fresh air and remove carbon dioxide. Exposure to visible light must be avoided because it will cause surface greening which, if not removed in trimming, will give the finished crisp an undesirable appearance and leave a bitter taste.

Potatoes held in long-term storage must be reconditioned or 'cured' before shipment to a processing plant. The reconditioning process, which takes from two to five weeks, will reverse the starch to sugar reaction by increasing the potato's respiration rate resulting in improved color in the finished crisp. This is accomplished by raising the storage temperature to a range of 15–20°C in a high-humidity area to prevent moisture loss. As in the case of long-term storage the area must be adequately ventilated with no light.

10.3.3 Receiving potatoes at the plant

Because of limited storage space at a processing plant, potatoes must be able to be processed when received. Verification of their 'fry on arrival' capability must be made prior to unloading. They also must be checked for shipping damage and to provide information to set slicer heads and to man the preparation steps of the process. The common checks made on an incoming shipment of potatoes are: tuber temperature; tuber size; shipping damage; tuber defects; specific gravity; fry sample. It is not uncommon for major processors to develop control procedures with their suppliers to become 'self certified'. The benefits of this program will be to assure quality of incoming shipments, reduce plant inventories and reduce in-plant testing.

After determining a shipment will meet processing requirements, it is unloaded and placed in storage. Care must be used in this operation to prevent damage to the potatoes and to minimize loss. Since 'fry on arrival' potatoes are stored only for a limited time, anywhere from a few hours to several days, they may be placed in large storage bins until used. However, even for the short time they are in storage, environmental conditions, such as temperature, humidity, air circulation and avoidance of light must be carefully controlled.

The final step in making potatoes process ready is to run them through a destoner to remove stones and other heavy objects that may cause damage to processing equipment. Potatoes must be supplied in sufficient quantity to operate the process at its designed production rate.

10.4 Managing the processing operation

Processing plays a significant part in the successful marketing and sale of a product but very often its role is not fully appreciated. While marketing and sales create interest in a product and obtain distribution channels, processing must offer a product of value to aid in the initial placement and deliver consistent quality to attain repeat sales to build market share and provide for the continuing success of a company.

10.4.1 Processing objectives

Processing philosophy should be directed to optimize quality and productivity at all times. To optimize quality, a process must be run in statistical control and centered at the specification target at all times maximizing the amount of heart cut product delivered to the consumer. To quote Dr W.E. Deming,[5] 'This is the best you can do'. Unless a product has some unique characteristic, it must have value equal or better than that offered by competition. This is achieved through maximum utilization of raw materials, manpower and equipment. It requires producing no reject product and as little waste as possible while operating the process at its designed production rate with minimal downtime.

10.4.2 Process control[6]

A process is a complex and dynamic system with many elements at work. In the operation of a process there are only a limited number of adjustments an operator can make while the process is running. These are the controllable variables or operating controls and are worked together to center the process at its specification mid-point, sometimes called its aim value. Once this balance has been established, the process will run, centered at its aim value, until the balance between the operating controls is upset. These control settings should be monitored periodically to make sure they do not change.

Since there are many variables that an operator can do little or nothing about, such as variation in raw material, machinery conditions, testing methods, environmental conditions, personnel, etc., variation will occur in the finished product. This variation is called random variation and, even though a process is centered at its aim value, many measurements will not be at aim, some higher, others lower. They will usually form a pattern characteristic of a normal distribution which permits the use of variable control charts to monitor performance. This random variability in a process determines its capability of meeting specification.

10.4.3 Control point management

A flow chart of the potato crisp process was pictured in Fig. 10.1. It shows the steps needed to produce the desired characteristics in the finished product. Each

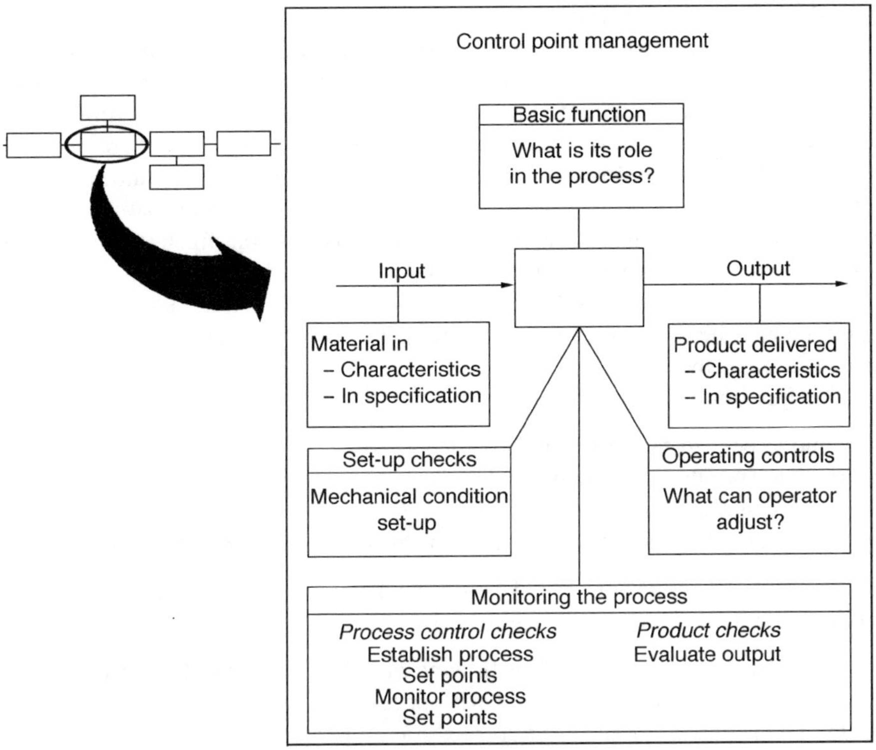

Fig. 10.2 Control point management model.

step, called a control point, is an independent process required to produce a specific characteristic in the finished product or to change the work in process so a subsequent control point can perform its function. A control point should be managed as a stand-alone process. Figure 10.2 pictures the elements that must be considered to effectively manage a control point.

Management of a control point requires a thorough understanding of its basic function in the process. This will define both its raw material needs and the characteristics of its output. The output of one control point is the input to the next. The equipment associated with operation of the control point must be in proper mechanical condition and set up according to the requirements of the product being produced. These conditions must be verified prior to the start of production. The key operating controls are adjusted to produce the control point's output at aim.

Process settings should be monitored periodically to make sure that they have not been changed inadvertently. Output of a control point should be evaluated to verify it is meeting its requirements. Variable control charts and attribute charts are very useful in making these determinations.

10.4.4 Potato preparation

The first two steps of the crisp process, peeling and the inspection/trim table, are to ready the potatoes for processing. Both steps make a major contribution to building quality and value into the finished crisp. In the first step, peeling, major processors use continuous peelers in their operations. However, there are still some batch peelers in operation. Peel is removed by tumbling potatoes against an abrasive for a given amount of time. In the continuous peeler the abrasive is attached to turning rolls to tumble the potatoes. During the fresh potato season some of these rolls may be replaced with brushes. In the batch peeler the abrasive is attached to the side wall of the peeling chamber and the potatoes are spun against the side wall to remove the peel. In both systems the role of the peeler is the same.

The function of the peeler is to

- remove peel from the potato
- remove sprouts from the potato
- remove surface dirt from the potato
- supply an adequate flow of potatoes to maintain the production rate.

The condition of a peeler should be checked prior to start-up and periodically during operation. Some of the items to be checked are as follows:

- the spray nozzles must be free flowing and directed at the bed of potatoes
- the abrasive must not be fouled with peel
- the level sensors must be set correctly and operating
- the safety devices must be set and operating

The amount of peel removed from a potato must be controlled. Insufficient peeling will give an undesirable appearance to the finished crisp, may cause an off-flavor tend to dull slicer blades and increase oil usage. Over-peeling will create excess waste. Inadequate sprout and dirt removal will increase slicer blade wear. Potato flow through the peeler is geared to downstream demand by level control switches in the slicer feed hopper. In normal operation a peeler should run about 90% of the time. Control settings must be established to hold the potatoes in contact with the abrasive long enough to peel them sufficiently but, at the same time not overpeel creating waste. In a continuous peeler the degree of peel is controlled by adjusting the discharge gate and the speed of the rollers. In a batch peeler the degree of peel is controlled by the amount of potatoes fed to the peeler and the time of peeling. Water sprays must have adequate force to wash away peel and other impurities while keeping the abrasive surface free of blinding. Actual settings are determined by examining potatoes at the peeler discharge for gouging and removal of dirt and sprouts and by periodic examination of crisps prior to finished inspection.

The function of the inspection/trim table is to

- cut oversize potatoes in half
- remove foreign material from feed stream

- remove deteriorated potatoes from feed stream
- remove undersize potatoes from feed stream
- trim surface blemishes from potatoes.

If potatoes are too large they are difficult to slice uniformily and may plug up the slicer. They are halved by screening them out of the feed stream with a sizing screw and passing them through a cutter or by cutting them by hand. Foreign material and deteriorated potatoes are inspected out of the feed stream and discarded. Undersize potatoes and potatoes with surface blemishes are hand inspected out of the feed stream and placed on the trim belt. The under-sized potatoes are removed by passing them under a sizing gate at the end of the belt and those with surface blemishes are trimmed and returned to the feed stream.

The condition of the inspection/trim table should be checked prior to start up and periodically during operation. Some of the items to be checked are as follows:

- the belt must track properly
- all rollers must rotate
- the potatoes must lie in a monolayer across the belt
- the sizer gap must retain potatoes larger than 77 mm in diameter
- the trim belt gap must pass potatoes less than 47 mm in diameter

Operating controls at the inspection/trim table are limited. The inspection station should be manned at all times. However the number of trimmers required will vary with potato quality and is determined by examination of the finished crisps at the end of the fryer. Results may be monitored by using attribute charts. Keep in mind that it is less costly to remove blemishes at the trim table than in finished inspection.

10.4.5 Slicing

Slicing is possibly the most important step in the potato crisp process. It impacts upon finished crisp appearance, texture, slice contour, bag fill, breakage, oil pickup and fryer production rate. Potatoes are dropped from the trim table into the slicer feed hopper and from there conveyed to the feed screws to deliver potatoes to the slicers one at a time. As a potato falls into a slicer, a rotating impeller forces it against the inside surface of the slicer head and drags it past the blades set in the wall of the head, cutting a new slice each time it passes a blade. The newly created slices drop down into the wash tank.

The function of the slicer is to

- cut slices to the correct thickness
- cut slices of uniform thickness
- contour slices
- control production rate.

Slice thickness determines the amount of curl, strength and texture of a crisp. Slices that are too thin may have a negative impact on color, tend to be oil soaked, easily broken and overfill the final package. Thick slices tend to have a higher moisture content, be less crisp and cause slack fill in the final package. Uniformity between slices is extremely important as it may create the undesirable characteristics of both thin and thick crisps.

The condition of the slicers should be checked prior to start-up and periodically during operation. Some of the items to be checked are as follows:

- the impeller must not wobble
- the hopper sensors must operate and control the hopper potato level at half to three-quarters full
- the incline belt must have no flights missing
- the locating pins must project 0.6–0.7 cm
- the feed screw must feed potatoes to the slicer one at a time
- the water flow must be adequate to remove scrap
- the safety guards must be in place

Slice thickness is controlled by the slicer blade gap. This setting will vary according to the specific gravity or solids content of the potato being processed. The higher the solids, the thinner the slice. For example, the slicer blade gap for a potato with a solids content in the range of 14% will be in the range of 1.40 mm and for a potato with a solids content of 18% in the range of 1.27 mm. In practice it is desirable to process successive potato lots with similar solids contents. The optimum setting is determined by examining finished crisps for foldovers, blisters and oil soaks. The presence of feathered edges indicates a damaged blade in the slicer. Upon placing a new set of blades on line, raw slice thickness should be measured for slice uniformity to make sure they are set at the correct gap and the thickness varies no more than 0.30 mm between slices. Periodic inspections of finished crisps should be made to determine if all blades are operating properly. Different crisp contours are obtained by using blades and holders designed for the desired contour. The production rate of the process line is controlled by the rate potatoes are fed to the slicers. For optimum quality and productivity a process should be operated at its design production rate at all times. Starts and stops should be avoided.

10.4.6 Slice washing

The slice washer transfers potatoes from the slicer to the fryer while rinsing starch from the slices and removing slicer scrap. The washers most commonly used in processing are the drum and the speed washer. The drum washer is a rotating perforated cylinder sitting in a water filled tank under the slicer discharge. It is fitted with spiral flights to tumble the slices and move them through the drum. The speed washer is a tank located under the slicer discharge equipped with water jets to agitate the slices, wash them and spread them out across the width of the drain conveyor.

The function of the slice washer is to

- remove starch from surface of slices
- remove slivers and scrap from feed stream
- deliver slices to fryer in monolayer
- remove excess water from surface of slices.

Water movement washes the surface of the slices to remove starch, potato slivers and slicer scrap. In the drum washer the slivers and scrap drop through the perforations in the rotating cylinder and are pumped to a fines box and strained out by passing the wash water through a mesh belt. Starch is carried out in the waste wash water. In the speed washer the slivers and scrap are also pumped to a fines removal box, strained and the starch-laden water dicarded. In both systems wash water should not be recycled. Fresh water must continually be added to replace the starch-laden water. It is not uncommon for the starch to cause excessive foaming during the wash cycle. This may be relieved by adding a food-grade anti-foam agent to the system. In some installations the wash water is heated to a temperature of about 60°C to blanch the slices in an attempt to improve color in the finished crisp. However, this practice has been found to have limited value by major crisp processors and is not commonly used.

Slices are carried from the washer to the fryer by the drain conveyor. As they are fed onto the conveyor they must be spread across the width of the conveyor in a monolayer so they will not be stuck together when dropped into the fryer. Slices that stick together, called clusters, do not fry throughout leaving an unfried portion in the center of the crisp having a high moisture which, when packaged, will be inedible. As the slices move up the drain conveyor, excess starch, slivers and moisture continue to be removed. In the drum system, fan-shaped water sprays wash the slices as the move up the conveyor. In the speed wash system, an air knife located above the conveyor flutters the slices as they pass under it to blow off moisture. A suction, located under the conveyor, draws away the excess moisture and helps to clear away scrap.

The wash system should be checked prior to start-up and periodically during operation. Some of the items to be checked are as follows:

- in the drum washer
  - the perforations must not be plugged
  - the water sprays must be directed at the monolayer of slices and cover the full width of the slice bed
  - the diverters must be placed to form a monolayer of slices on the feed conveyor
- in the speed washer
  - the water jets must create turbulence to form a monolayer on the feed conveyor
  - the air knife must be on and have sufficient force to flutter the slices
  - the suction must be on

- anti-foam must be added if excessive foaming persists
- the fines removal system must be operative with the screen not plugged

In the drum washer the rotation speed determines the amount of time slices stay in the washer and is set to wash the slices adequately, approximately 45 seconds. Water flow must be sufficient to carry away the starch and the small pieces of potato slices. The sprays over the conveyor must have enough flow for a final rinse and to wash away the remaining scrap. A word of caution: insufficient washing may result in clusters, excess oil usage and possible oil deterioration; on the other hand, excessive water usage is expensive.

The speed washer water jet flow should cause enough turbulence in the slice bed to separate slices and remove starch and scrap. The air knife pressure must be sufficient to flutter or bounce the slices but not move them enough to overlap. The suction must be strong enough to remove moisture and scrap from the conveyor. Crisps coming out should be checked periodically for soft centers and clusters.

10.4.7 Frying

The fryer makes the physical and chemical changes in the potato slice to convert it into a finished crisp. Hot cooking oil is pumped into the fryer as the slices enter the fryer and is discharged under the take-out belt. The fryer is equipped with flow wheels to retard the flow of slices through the fryer before they are fed under a submerger which controls the finished moisture content of the crisp.

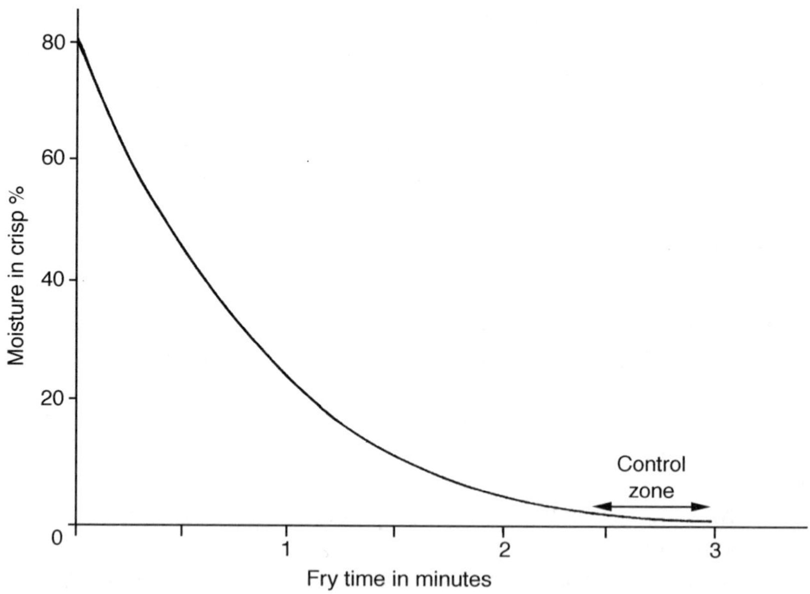

Fig. 10.3 Moisture content versus fry time.

After the slices have been held in the fryer the required amount of time, they are then removed by the take-out belt which allows the crisps to start cooling and for excess surface oil to drain from them. The function of the fryer is to

- remove moisture from the slices
- absorb oil into the slices
- develop color in the slice
- develop flavor
- develop texture.

Moisture is removed from the slices by holding them in the cooking oil for the required amount of time. When a slice is dropped into the fryer, the surface of the slice is sealed shortly after falling into the oil. This retards the release of moisture as the slices flow through the fryer and is a key factor in the development of texture. By the time a slice reaches the submerger, the slope of the moisture curve has flattened out making it possible, by adjusting the submerger speed, to fine tune the moisture content of the finished crisp. This area of the curve is identified as the control zone in Fig. 10.3. As moisture is removed from the slice it is replaced by oil amounting to as much as 35–40% of the finished crisp.

As moisture is released from a slice it will begin to curl, undergo some leavening and become rigid. It must be held in the oil long enough to cook throughout and develop its desired texture. As a slice nears 150ºC, potato starch begins to break down to release sugar and as it reaches around 155ºC it starts to develop color and the basic crisp flavor. As the temperature continues to rise the sugars in the crisp will begin to caramelize becoming darker and developing a burnt flavor. This effect is shown in Fig. 10.4.

A fryer should be checked prior to start-up and periodically during operation. Some of the items to be checked are as follows:

- flow wheels
  - must be timed with the submerger to give uniform flow of slices through the fryer
  - must be offset to control slice flow through the fryer
  - must have no broken or bent vanes

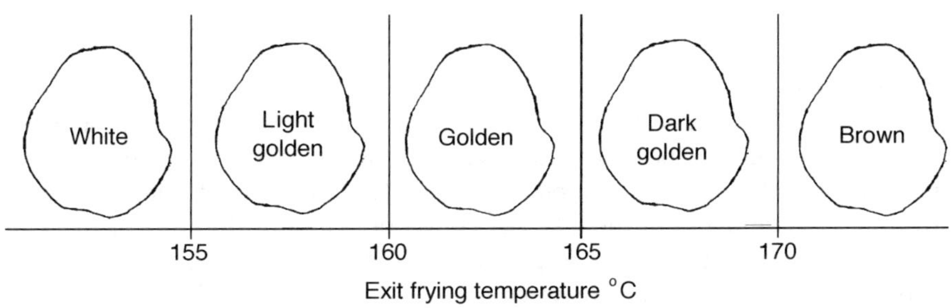

Fig. 10.4 Temperature effect on crisp color.

- the oil flow must be uniform from side to side through fryer
- the stack damper must be closed sufficiently to emit wisps of steam from the ends of the fryer
- the oil level must be set just above the submerger belt with the level control on automatic
- the drip back eliminator must be centered with the drain open
- the fines removal system conveyor and air jets must be on
- the safety devices must be operating

The oil level should be adjusted so that the bottom of the submerger lies just below the oil, thus ensuring that all the slices will be submerged in the oil during the final phase of the frying. The inlet oil temperature is adjusted to produce an exit oil temperature of approximately 157ºC and at no time exceed 163ºC. At higher finishing temperatures the residual sugar in the slices will begin to discolor. At standard production rates the inlet oil temperature will be in the range of 180–185ºC. The differential between the inlet and exit oil temperatures is achieved by speeding up or slowing down the potato feed rate to the slicer.

The flow wheels are used to hold the chips back in the oil to give time for texture to develop. The time under the submerger is controlled to fine tune the finished crisp moisture. The flow wheel speed should be adjusted to maintain a slight build up of slices at the entrance to the submerger. In many installations the flow wheel speed and the submerger speed are tied together so that any change in the submerger speed will automatically be reflected in the flow wheel speed. In cases where the flow wheels and the submerger are not tied together, the flow wheel speed must be adjusted whenever a change is made in the submerger speed. The total time in the oil for regular crisps should be about 2½ minutes and for thicker cuts 30–40 seconds longer. The take-out conveyor speed should be adjusted to maintain a crisp bed depth of approximately 10 cm on the take-out belt to allow the crisps to cool and drain off excess oil.

All of the controls should be monitored periodically to ensure that the process balance has not been disturbed. The crisps dropping off the end of the take-out belt should be periodically evaluated for moisture content, color, soft centers (clusters), flavor and texture. Results of moisture and color evaluations may be monitored by using variable control charts and the other characteristics by attribute charts.

10.4.8 Salting

Several types of salters are used in crisp processing, the pneumatic and the serrated roll salter. The pneumatic salter is located over a conveyor belt immediately following the take-out conveyor. The serrated roll salter is located over the discharge end of the take-out conveyor. The function of a salter is to apply a uniform coating of salt to crisps and to apply salt to crisps in a specified amount. Salting is a balance between product (crisp) flow under the salter and salt flow rate. To control salt application requires a steady flow of of crisps under the salter. In the pneumatic salter, salt is dropped onto a metering belt from the supply hopper and fed into a mixing chamber where it is blown through a distribution head onto the crisps. In the serrated roll salter, salt is dropped into the grooves of the salter roll and spilled onto the bed of crisps as the roller revolves.

Salting equipment should be checked prior to start-up and periodically during operation. Some of the items to be checked are as follows:

- pneumatic system
  - the hopper screen must be in place and clean
  - the hopper level probe must be operative and set to maintain the hopper half full
  - the low level warning light must be functional
  - the air pressure must be set at standard
  - all distribution header ports must be open
- serrated roller system
  - the hopper level probe must be operative and set to maintain the hopper half full
  - the roller grooves must be clean and undamaged
  - the weir plate gap must be uniform across the width of the salter
  - there should be no salt leaks
  - the roller rpm must be set at standard

In the pneumatic salter, the amount of salt applied to the crisps is controlled by the speed of the metering belt. In the serrated roll salter, the salt feed rate is controlled by the rpm of the roll. Salt application may be monitored by using a variable control chart.

10.4.9 Finished inspection

In the past finished crisp inspection was performed by hand. However, with the present throughput of fryers, hand inspection is not capable of making a significant improvement in crisp appearance. Nevertheless it is prudent to have at least one food safety inspector stationed on each side of the inspection conveyor to remove foreign material, soft centers and clusters. In recent years more and more processors have turned to optical sorters for finished crisp inspection.

A crisp sizer is frequently located in the process stream prior to transferring product to the packaging department. The sizer removes small crisps and scrap from the stream and directs them to the small package fill line. Sizing has several advantages: small crisps and scrap are packaged in small packages and usually eaten directly from a package where size is not a major appearance factor; large crisps are directed away from small bag fill lines where they have a tendency to plug the formers and disrupt production; reducing the amount of small crisps in large packages improves the appearance of the crisps when they are emptied into a bowl prior to eating; large crisps enhance bag fill.

The function of finished inspection in crisp processing is to remove foreign material, remove soft centers, remove clusters, remove blemishes, improve large bag crisp appearance and size crisps for packaging. In the case of the optical sorter, crisps are fed onto a high-speed belt in a mono-layer and passed under a high-speed video scanner. The scanner checks the color and size of blemish of each crisp on the transfer belt and, by means of a microprocessor, tracks its location on the belt to an air jet that rejects any crisp not meeting finished crisp color and blemish requirements.

The equipment should be checked prior to start up and inspected periodically during operation. Some of the items to be checked are as follows:

- the crisps must be monolayered on the sorter belt
- the blemish shade and shape reject levers must be set to meet the finished crisp appearance quality standards
- the jets must be set to reject no more than 25% good product
- the conveyor gap must be set to reject no more than 25% good product
- the crisp sizer must be adjusted to meet small package product demands

The scanner is adjusted to control the shade and size of defective crisps to be removed from the product stream. However, having the ability to remove defective crisps from the process line should not lessen the need to optimize processing conditions upstream. Rejection of crisps at this point in the process is costly. Finished crisp appearance may be monitored by use of attribute charts.

10.4.10 Seasoning application

Crisps are seasoned in several ways. In cases where the volume of a specified seasoning is large, a portion of the fryer output is diverted to a seasoning loop that consists of a seasoning tumbler, a seasoning auger and a separate conveyor to transfer the seasoned crisp to the packaging machines. The tumbler has a series of flights that tumble the crisps in order to season both sides. Seasoning is fed down a long tube by an auger placed over the center of the bed of crisps. A spinner located under the auger spreads the seasoning across the entire bed of crisps. A second way to season crisps is to pass them through a seasoning drum located on each individual packaging machine. The function of the seasoner is to apply a uniform coating of seasoning to both sides of a crisp and apply seasoning to the crisps in a specified amount.

In the loop system a low-level feeder receives the diverted flow from the process line and monitors the flow of crisps into the tumbler, turning the seasoning feed on and off to maintain a balance between product flow and seasoning flow. Seasoning is fed into the tumbler by an auger housed in a tube about 150 cm long with slots covering the last 60 cm. A sleeve fitted around the tube is adjusted to have seasoning flow out over the full length of these slots forming a curtain. It is adjusted so that very little seasoning falls from the end of the tube. Seasoning falls from the auger onto the spinner dispersing it over the width of the crisp bed. The amount of the seasoning delivered is controlled by the revolution speed of the auger.

Seasoning equipment should be checked prior to start-up and periodically during operation. Some of the items to be checked are as follows:

- loop system
  - the crisp feeder must be set to maintain uniform product flow through the tumbler
  - the auger must be centered over the product
  - the seasoning curtain must extend the length of the auger
  - the spinner must rotate
  - the tumbler pitch should be set at approximately 8 in.
  - the tumbler should rotate at standard rpm
- packaging machine drum system
  - the crisp feed must be uniform
  - the seasoning flow must be set at standard
  - the on/off controls must be operative

The amount of seasoning applied can be monitored by using a variable chart control and seasoning coverage by using an attribute chart.

10.5 Packaging

The impact of packaging on quality and productivity in crisp processing is frequently overlooked. Optimum results are obtained when a fryer is operated at its designed capacity and the finished crisps are packaged as quickly as possible after leaving the fryer. Many operations, however, do not have a sufficient number of packaging machines available to handle the designed fryer output when the packaging mix is skewed toward small bags. Two approaches are used to address this situation. Both are wrong.

One way is to slow down the fryer which not only affects quality but increases the cost of processing. When a fryer is slowed down it disrupts the temperature balance and dwell time required to develop color and texture in the crisp. It increases the labor cost per unit of production and provides less product to cover other operating costs.

The other way is to place an accumulator in the process line just prior to packaging. It is a walled bin fitted with a fill conveyor on top of it capable of delivering crisps over its full length and a moving belt along its bottom to feed crisps to packaging when called for. When packaging is unable to keep up with the fryer, crisps are held back in the accumulator building up a bed of crisps as high as 60 cm over much of its length. Once it is full, the fryer must be shut down. Crisps are often held in the accumulator thirty minutes or more prior to packaging. By exposing them to the processing floor environment at an elevated temperature they will not only pick up moisture reducing shelf life, but be subject to oil damage which may affect flavor. Handling crisps in this manner also will increase breakage.

While investment in additional packaging capacity may appear costly, a simple calculation by a qualified engineer will show a cost benefit to go along with the quality gains.

10.6 Future trends

10.6.1 The product

The basic potato crisp has changed little over the past forty years being produced as either a flat or ridged crisp and offering a variety of flavorings. Many attempts have been made to develop crisps to meet a special dietary niche requested by consumers but none have had a major impact in the market place. Crisps have been baked to eliminate oil, fried with a reduced oil content, seasoned with little or no salt, fried in oils without preservatives and, most recently, fried in a cooking oil substitute. To date, these products have had limited success probably because of their inability to reproduce the unique flavor of an oil-fried crisp. Nevertheless, it is expected that new products will continue to be introduced into the market to fill special demands of the consumer.

10.6.2 Process monitoring

In the last ten years considerable progress has been made to make continuous on-line measurements of a number of crisp characteristics. Continuous on-line moisture measurements have made it possible to use computerized controls to adjust fryer settings to control the moisture content of a finished crisp. This has led to more uniformity in the moisture content of a crisp and improved productivity by producing less out of specification product.

Optical scanners have been developed that can detect variations in the color of crisps as well as determine the size of any discoloration that is present in a crisp. It is placed at the end of the process line feeding information into a microprocessor to activate air jets to remove any undesirable crisp from the process stream.

Improvements in monitoring techniques are expected to continue. Since slicing is such an important step in the crisp process, efforts should be directed to developing on-line procedures to measure slice thickness and variation between slices.

10.6.3 Equipment

Until recently the only changes in equipment have been to increase the fryer production rate. Since processing labor is a relatively small part of the cost of producing crisps, this has produced only limited cost benefits. It is not anticipated that fryer production rates will increase significantly in the future.

The recent development of optical sorters, on the other hand, has resulted in improvements in both the quality of crisps and the cost of processing them. They deliver a more consistent crisp and provide a means to adjust crisp appearance to match or better competition. At the same time it has dramatically reduced the number of finished crisp inspectors required to perform adequate inspection.

A desirable improvement in the crisp process would be to be able to automatically adjust slice thickness while running. It is also desirable to develop slicer blades with a longer useful life. Methods to improve seasoning application to crisps should focus on uniformity of application on both sides of a crisp as well as uniformity crisp to crisp.

10.6.4 Raw materials

Research has made continued improvements in potato quality through the years. Varietal development specifically focused on crisp production has increased potato solids, improved size uniformity, improved storage ability and improved control of the sugar content of the potato. It is expected that progress will continue to be made in these areas.

Future efforts should address the storage and handling of potatoes from the time they are harvested until they are processed. It is felt that storage procedures and conditions can be developed to maintain the potato in harvest condition, eliminate storage losses and the need for reconditioning. New procedures in these areas can improve crisp quality and offer substantial savings in the cost of producing a crisp.

Oil research will continue to work on fat-free oil substitutes to match the flavor of a natural cooking oil in the finished crisp and not cause any digestive side effects.

10.7 References

1. BENNETT R M, *Potato Chip Process Control* June 1997.
2. HELLEVANG K J, 'Maintenance vital to keep potato storage functioning', *Valley Potato Grower*, 1996, August, 8.
3. SHETTY K K, 'Potato for disease control', *Valley Potato Grower*, 1996, Harvest, 18–19.
4. BELYEA S H, 'Storage Management', *Valley Potato Grower*, 1996, Harvest, 20–21.
5. AGUAGO RAFAEL, DR. DEMING, New York, Simon and Schuster, 1990.
6. BENNETT R M, *Process Control Basics* May 1997.
7. WALSH J, 'Obtaining optimum color for chipping and processing potatoes', *Valley Potato Grower*, 1996, Harvest, 22–23.
8. 3rd International Symposium on Deep-Fat Frying: March 20–21, 2000, Deutsche Gesellschaft für Fettwissenschaft e. V.

11

Effective process control in frying

G. B. Quaglia and F. M. Bucarelli, Istituto Nazionale della Nutrizione, Rome

11.1 Introduction

Frying is a cooking process using fat or oil as the heat transfer medium. Frying is a common treatment for many foods. The process includes:

- Deep-frying: cooking of food submersed in hot fat or oil;
- Shallow frying: cooking of food in shallow fat or oil in a pan;
- Roasting: cooking of food in an oven.

Of all the above methods, deep-frying is the one most often subject to technological study. This is because it is widely applied in industrial production and demands constant plant control. In deep frying, fat or oil is exposed for long periods to heat stress and alteration by food components.

This cooking method is used to give particular organoleptic characteristics (colour, aroma, consistency), to various food products of either vegetable or animal origin. Frying is also a means of preserving food because, in the process, heat destroys micro-organisms, renders enzymes inactive and reduces surface water activity. During the deep fat frying process, food and oil are subject to chemical and physical changes, including decomposition with a potential toxicological risk.

Many factors (the nature of fat used, process temperature, oil turnover, coating procedure, food ingredients, etc.) condition the control of the chemical changes taking place and related toxicological risks.

The most important requirement in terms of food quality is product safety, which can be guaranteed by the preventive analysis of risk and thorough monitoring of the entire process. This means adopting the Hazard Analysis Critical Point system.

This chapter examines methods of monitoring deep-frying, hereinafter referred to as frying. A short introduction to the HACCP (Section 11.2) system is followed by an examination of flow diagrams relating to the frying process (Section 11.3), and an evaluation of hazards and preventive controls (Section 11.4). Finally, frying techniques (Section 11.5) and future trends in frying technology control methods (Section 11.6) are reviewed.

11.2 The HACCP approach

Hazard Analysis Critical Control Point (HACCP) is a production control system for the food industry. It is designed to identify potential contamination areas (the critical control points or CCPs) which may then be managed and monitored in the interests of product safety. HACCP is, in fact, prevention rather than cure.

HACCP comprises the following seven principles:

1. Identification of the potential hazard(s) associated with food production at all stages – production, processing, manufacturing, distribution; assessment of hazard liability and identification of preventative measures;
2. Identification of the points/procedures/operational steps to monitor in hazard(s) prevention and minimise likelihood of occurrence – Critical Control Point (CCP) ('step' means any stage in food production and/or manufacturing including raw materials, their receipt and/or production, harvesting, transport, manufacturing, processing, storage, etc.;
3. Fixing targets and tolerance levels to ensure the CCP is under control;
4. The institution of a monitoring system to ensure CCP control by scheduled testing;
5. Deciding on corrective action in the event of indications that a particular CCP is not under control;
6. The institution of checks to include supplementary tests and monitoring of the HACCP's efficiency;
7. The initiation of a procedural record charting the application of these principles.

11.2.1 List all hazards associated with each procedure and consider hazard prevention measures (principle 1)

As its name implies, the HACCP hazard analysis is one of the most important measures. A badly conducted hazard analysis inevitably would lead to the development of an inadequate HACCP plan. Hazard analysis requires technical expertise and scientific grounding in various areas if all potential hazards are to be averted.

Hazard analysis consists in an evaluation of all procedures in the production, distribution and use of raw materials and food products; this, to identify potentially hazardous practices or sources of contamination, while at the same time assessing the risk and severity of hazards.

Hazards may vary from one establishment to another as the following factors vary:

- sources of ingredients
- treatment
- equipment and processing plant lay-out
- preparation/processing methods
- duration of process/storage
- experience/skill/attitude of personnel.

11.2.2 Identification of CCPs applying the HACCP decision tree to each step (principle 2)

The identification of critical control points (CCPs) is the second HACCP principle. A CCP is defined as any point, step or procedure at which control can be applied and a food safety hazard prevented, eliminated or reduced to an acceptable level.

CCPs are selected according to the following criteria:

- identified hazards and likely occurrence of unacceptable contamination;
- operations to which the product is subjected during processing and preparation;
- intended use of the product.

At each critical control point, critical limits are established.

11.2.3 Establishing critical limits for each CCP (principle 3)

At each critical control point, critical limits of unacceptability are established. These parameters, if kept within certain boundaries, will guarantee product safety. The critical limits will meet government regulations, corporate standards and scientific requirements. One or more critical limits will be set to control the identified hazard and may regard factors such as temperature, time (minimum exposure), physical characteristics, water activity, moisture level, etc.

11.2.4 Establishing a monitoring system for each CCP (principle 4)

Monitoring is the scheduled measurement or observation of a CCP according to its critical limits. The monitoring procedures must be able to detect loss of CCP control and should ideally provide this information in time for corrective action to be taken. Such action will be designed to regain control of the process before there is any need to reject the product. Data derived from monitoring must be evaluated by a designated person with the expertise and authority to carry out corrective action when indicated. Monitoring, if not continuous, must nevertheless be frequent enough to guarantee CCP control. Most monitoring procedures for CCPs will need to be rapid because they relate to on-line processes allowing no time for protracted analysis. Physical and chemical evaluations are often preferred to microbiological testing in monitoring CCPs because they may be carried out rapidly. The microbiological analyses are applied to verify the CCP control activity.

All CCP monitoring records and reports must be signed by the control director and by one or more corporate executives.

11.2.5 Establishing corrective action (principle 5)

Specific corrective action must be developed for each CCP in the HACCP system to address deviations when they occur. This action must ensure that the CCP will be brought under control and affected products re-treated. All deviations and treatment procedures must be reported in the HACCP records. Corrective action should also be taken when monitoring results indicate a loss of CCP control. It will be designed to bring the process back into control before deviation leads to a safety hazard proper.

11.2.6 Establishing checks (principle 6)

Establishing procedures to check that the HACCP system is working correctly entails: monitoring, controls and tests including random sampling and analysis. Checks should be made with a frequency such as to validate the HACCP system.

11.2.7 Establishing records (principle 7)

Efficient and accurate records are essential to the application of any HACCP system. HACCP procedures at all stages should be recorded in a manual.

11.3 Flow diagrams examination

Frying can be by the batch or continuous. It is used as a treatment for food on an industrial catering scale or in the home. The end product may have a long shelf life or take the form of frozen convenience food and fresh pan-to-table fried food. It is therefore impossible to develop a general prototype of frying flow diagrams because the process depends on the food and its production context.

However, the following are elements of the frying process:

Frying medium (fat or oil) selection and storage The quality of the frying medium is a critical element in the frying process. The medium's manufacturing processes (extraction, hydrogenation, additives, etc.) should all be accounted for in the HACCP analysis. The system must also incorporate acceptability testing of oil or fat and storage factors, and transport both from the refinery to the frying plant and from the storage tanks in the frying plant to the fryer itself. Segregation of the storage and transport of waste/reject oil from fresh fat must be ensured and fresh deliveries should not be loaded onto the top of old oil in partly filled tanks (see Chapter 7).

Food ingredient selection The HACCP system should examine the hazards associated with raw materials used in the process, and their relative acceptability. These aspects are not pertinent to this study, however, because they concern the nature of the specific fried food and not the peculiarities of the frying process.

Food preparation Preparation will depend on the food type and its own particular culinary process. It may include: washing, peeling, chopping, pre-cooking or bleaching, dressing, etc. There are many stages in the preparation of food, but all should be subjected to controls under the HACCP system according to the type of food. This study does not analyse risks inherent in some preparatory stages which do not directly concern the frying process. Food preparation treatments are reviewed where they involve heat or mass exchanges between the frying medium and the food, where they affect humidity or food surface structures.

Surface treatments Before frying, starchy food need not be coated with batter, flour or breadcrumbs to obtain the desired organoleptic property. Other food such as fish or meat are usually coated with batter and optionally breadcrumbs to inhibit migration of fat from the food into the frying oil.

The coating will act on the mass exchanges between the frying medium and food. The parameters of the coating (its composition, treatment, dehydration, etc.) are critical to fat absorption and the rate of oil degradation. These effects of coating will be examined in the HACCP study.

Proper deep-frying The deep-fried cooking of food in hot oil or fat is the core of the process. Frying destroys all vegetal bacteria though the food may absorb toxic compounds from the frying medium. The frying parameters (frying medium temperature, equipment, etc.) are critical and should be subject to HACCP controls.

Oil or fat management The proper handling and turnover of used oil or fat are critical to their quality and, consequently, to food safety. Different approaches may be adopted providing they comply with good practice criteria and the critical point monitoring procedure.

Fried food storage and heat preservation The diagram flows of the industrial fried snack or convenience fried foods such as crisps or frozen French fries should include their storage and retailing. In these steps, the auto-oxidation process should be evaluated taking into account the high content of thermal stressed fat present in the food. Other biological and mechanical risks associated with the nature of foods and their marketing should also be controlled independently of the cooking process. In catering, restaurants and in the home, fried food is generally kept hot until served. It may cool and be re-heated. Such cooling and post-frying procedures harbour the same risks of secondary microbiological contamination and proliferation that are inherent in other cooling and post-cooking treatments. So, equally strict control procedures must be adopted. The risks of fat breakdown during re-heating or heat conservation should be evaluated.

The risk inherent in the frying process is strictly chemical, compounded with the oil/fat degradation process and absorption by food. Monitoring frying operations according to the HACCP system will, however, take into account the entire food production cycle. Of particular relevance are the microbiological, mechanical or chemical hazards potential to the productive cycle, which will be analysed using the same criteria for non-fried foods. The flow diagram of frozen French fries production is shown by way of example (see Fig. 11.1). This chapter therefore examines only the specific chemical hazards of frying, identifying specific CCPs and relative control procedures.

11.4 Hazard evaluation and preventative measures

11.4.1 Thermal stress, neo-compound formation and toxicological hazard

There are numerous studies on the changes to the chemical composition of oil or fat during frying. Hydrolysis of triglycerides, oxidation and polymerisation occur rapidly in deep fat frying. The chemical compositions of oils and fats exposed to heat, oxygen, humidity and juices given off by food are notably altered. Free fatty acids, diglycerides, monoglycerides and glycerol formed in the hydrolytic process, produce polymerisation, isomerisation, pyrolisis and cyclisation reactions leading to the formation of a wide range of compounds which can be toxic or undesirable (acrolein, crotonaldehyde, alpha and beta unsaturated aldehydes, hydroperoxides, conjugates dienes, etc.). Some of these new chemicals are volatile and, at frying temperature, are partially emitted by the oil as hydrocarbons, aldehydes, ketones and carboxylic acids.

Other compounds are soluble or colloidally suspended in frying oil, such as cyclic and non-cyclic monomers, dimers and trimers, proteinaceus compounds, high MW compounds and gum (see Fig. 11.2). The rate of these reactions depends on frying conditions, chiefly temperature, duration and aeration. The quality of the oil or fat and the quality of the food produced are interdependent. Fat content, and especially polyunsaturated fatty acid content, salt, humidity and the batter used to coat food will determine the composition of the oil and its degradation process. In fact, the compounds transferred from the food to the oil (for example cholesterol from meat and polyunsaturated fatty acids from fish) may be involved in the oxidisation process or may catalyse degradation. The many reactions caused by components from the food may also release toxic substances.

Intermittent heating and cooling is more severely detrimental to oil or fat than continuous frying over a period of hours. This is because acyl peroxides increase as oil cools; upon re-heating, their decomposition causes further damage to the

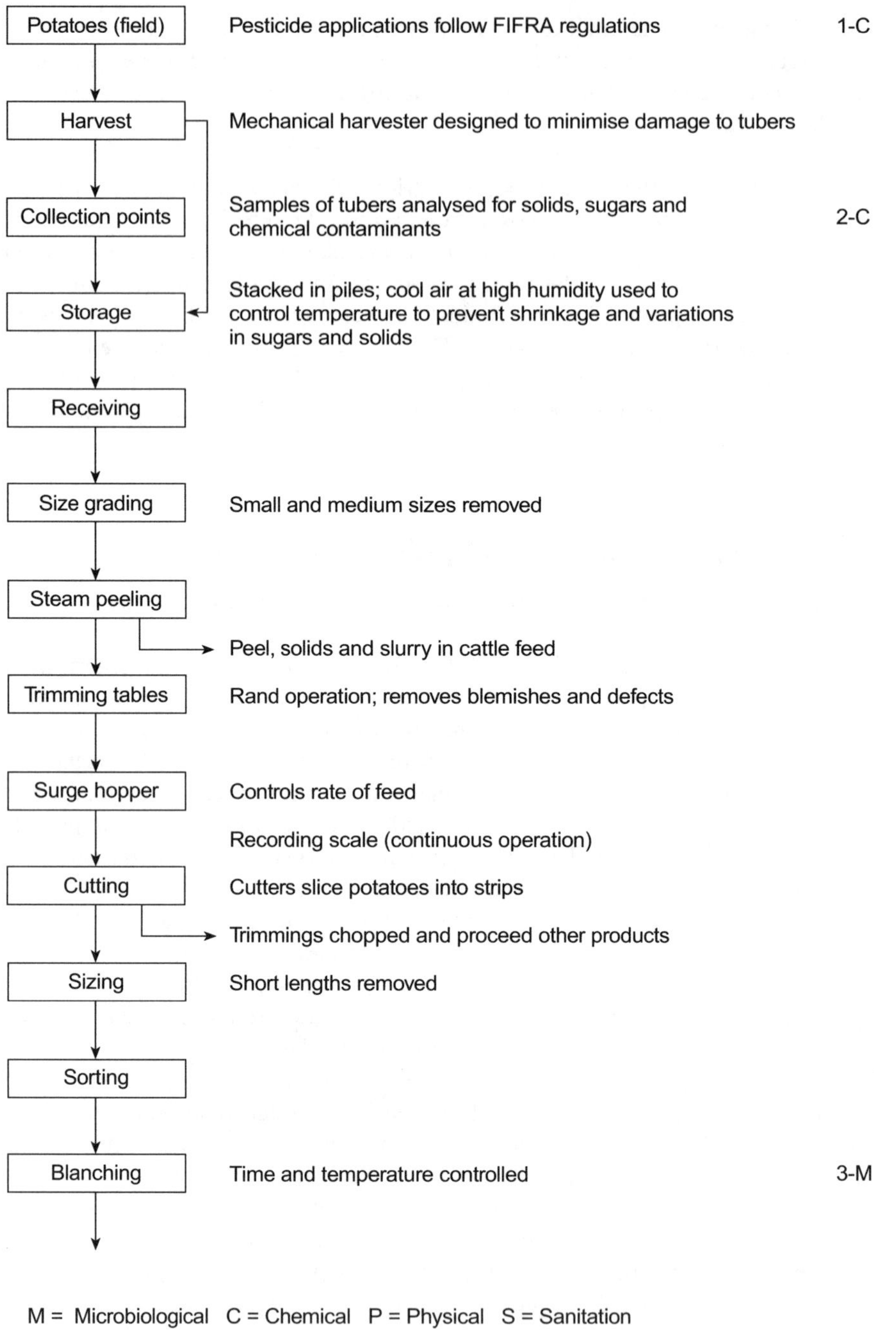

Fig. 11.1 French fries production flow diagram (Merle D. Piersons and Donald A. Corlett, Jr (eds) *HACCP – principles and applications*, Chapman and Hall, 1992).

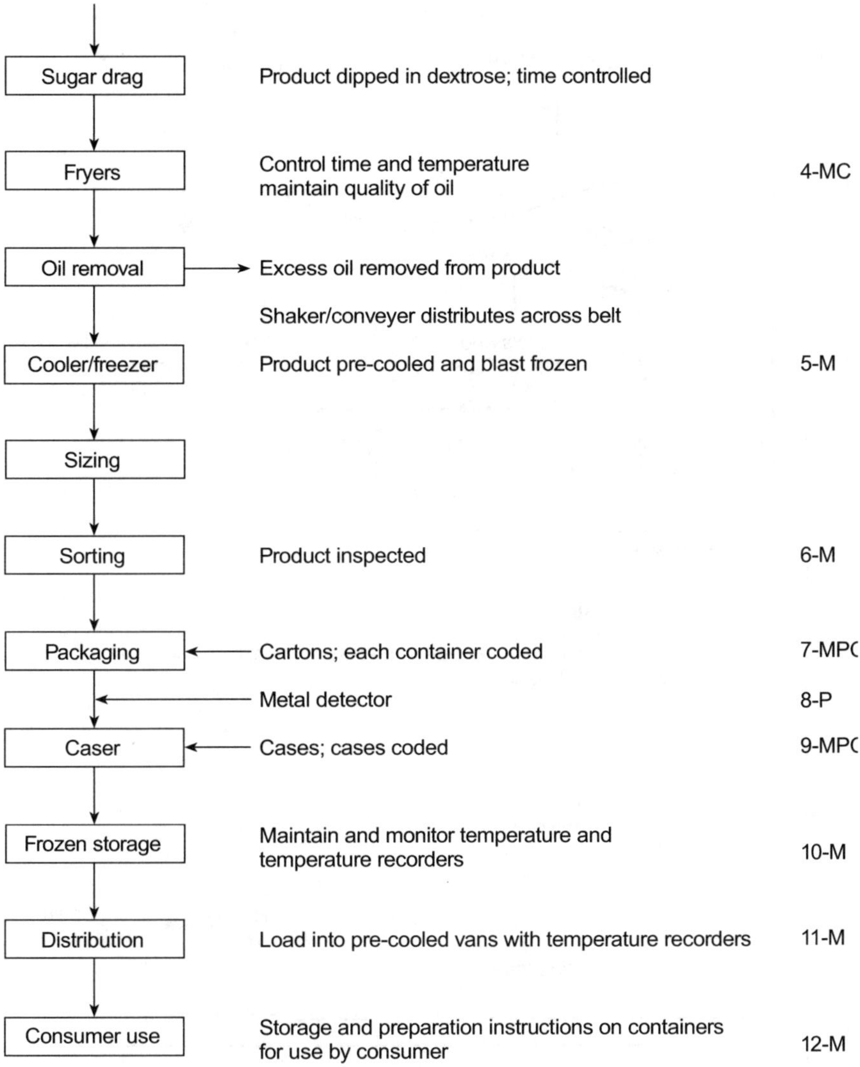

Fig. 11.1 (continued)

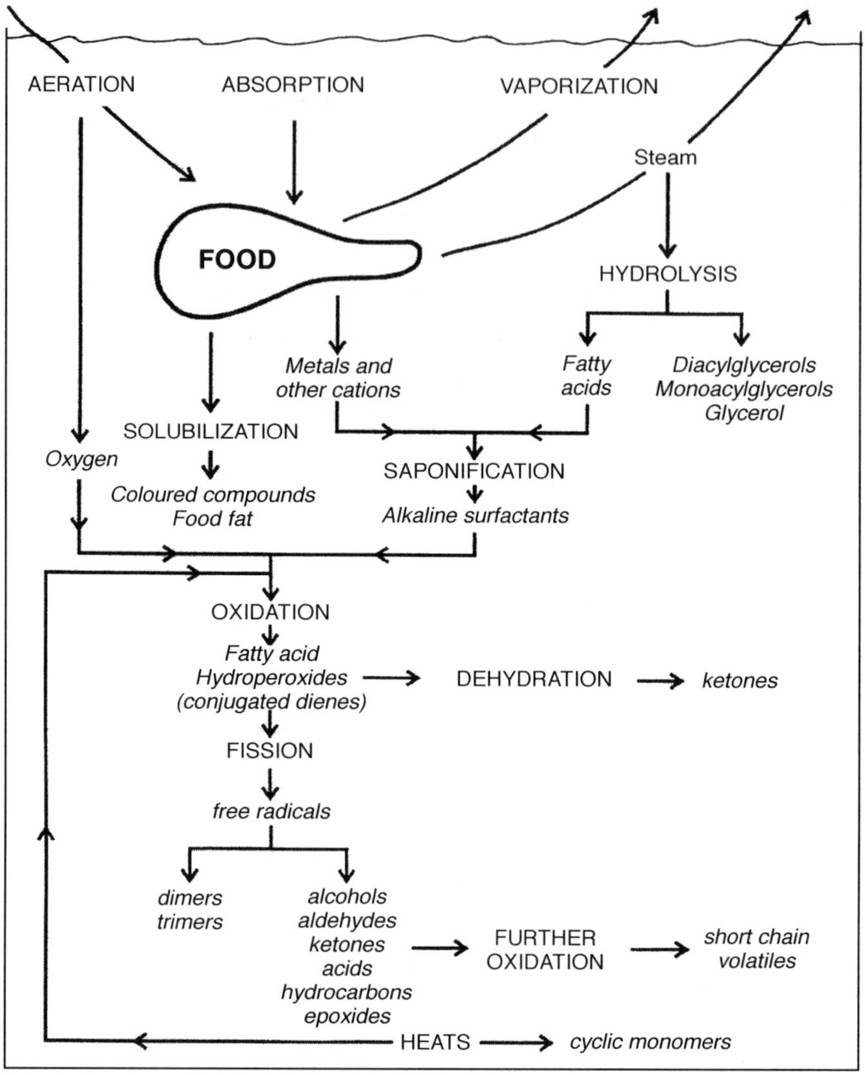

Fig. 11.2 Deep frying schematic picture showing mass transfer and chemical reaction during the frying process (C.W. Fritsch (ed.) *J.Am. Oil Chem Soc.* 58:272, 1981).

fat. Furthermore, moisture released as steam during frying distils breakdown products from the oil and also blankets the oil with a layer of the steam, restricting access of atmospheric oxygen. However, the hydrolytic processes of oil are not totally negative; in the beginning, this process improves technological properties – as a frying medium – without toxicological risks.

The oil degradation process in frying takes five stages (see Fig. 11.3): break-in oil (O-A), fresh oil (A-B), optimum oil (B-C), degrading oil (C-D) and runaway oil (D-E).

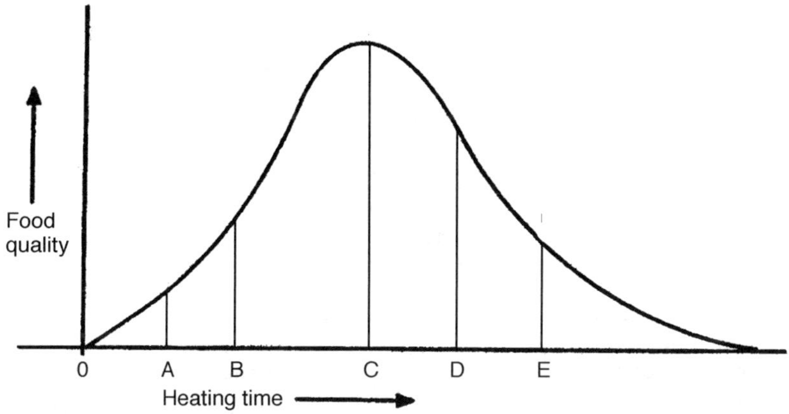

Fig. 11.3 Blumenthal curve (David D. Brooks (ed.) *Inform – Food Technology*, Vol 2, no. 12 1991).

For frying, heat must be transferred from the non-aqueous medium (oil) to the mostly aqueous product (food). During the process, the polar substances developed act as reducers of the surface tension between the materials being blended, increase the oil-food heat transfer rate. These surfactants are classified as:

- water-activated (soaps, sodium oleate, phospholipids, organic salts, FFA, etc.), forming an oily emulsion;
- lipid-activated (low polar compounds).

Low surfactant concentrations mean that food, both externally and internally, is under-cooked. Moderate concentrations allow for proper cooking, while high concentrations result in over-cooking on the outside and under-cooking internally.

At frying temperature, oxygen does not remain dissolved in fat but reacts with the oil to form thinly stretched films of bubbles. If the surfactant content is increased, the interfacial tension is reduced at the air interface and the rate of oxygenation (and the relative rate of oxidative degradation) rises. Surfactants also cause oil to rise up the side of a hot frying vessel. The oil carbonises at high temperature, producing cyclic acid and other undesirable substances (see Fig. 11.4). Thus, the amount of polar compounds is the principal quality control parameter for oil.

Several European countries have established limits for polar matter at between 25 and 27% in the worst case, i.e. at limit of acceptability. Supposing a content of 25 per cent more polar materials than non-polar degraded compounds, and supposing that 20% of the oil was absorbed by the food, the consumer eats more than five per cent of degraded fats by weight of fried food. The 20% oil absorption rate is an average value, but many products absorb much more (potato crisps absorb about 40% of oil by weight of fried food).

The polar compounds are preferentially absorbed in food with an aqueous surface, giving the possibility of high amounts of polar material in the oil

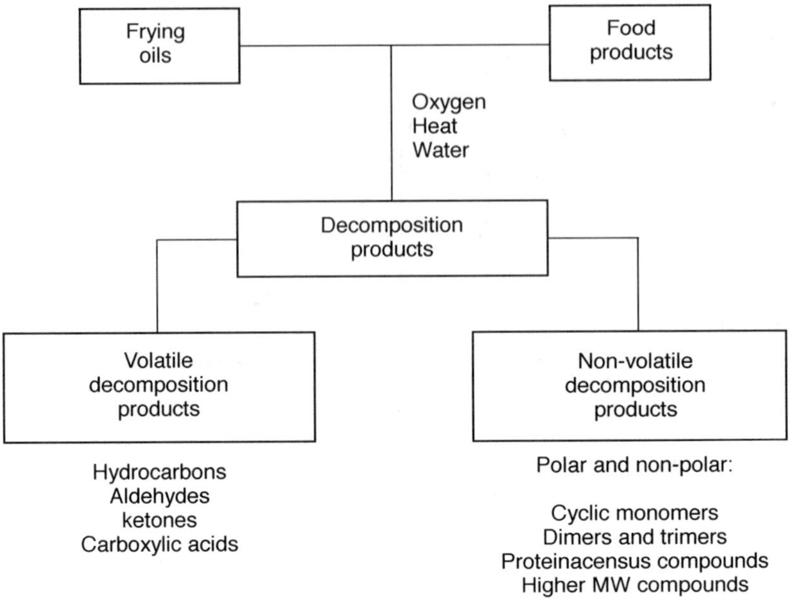

Fig. 11.4 Decomposition products in fat frying (G. Quaglia (ed.) *Grasa Y Aceites*, 49, 275–281, 1998).

absorbed on the food in comparison with the oil remaining in the frying bath. Scientific studies are lacking in data on the percentage that fried oil contributes to the total amount of fat in food. Not all toxicological studies agree in terms of the toxic hazard of fried food. Harmful effects on the membrane and enzyme activity and consequently mutagenicity and citotoxicity properties have been reported. Many authors do not believe that the notably harmful effects reflect the real condition because these effects are obtained in unrealistic and extreme frying processes. These aspects are discussed in great depth in Chapter 5.

Animals fed fried foods produced under routine frying conditions have not been found to have suffered from toxins. However, excessive frying does produce toxic compounds. This is unacceptable and should be rejected by the consumer. In the UK a lot of people maintain that the oil tastes very bad if it has 25% polar material. Quality critical limits of toxicological compound content or indicators of the degradation process are established by law but these limits vary from country to country (see Section 11.5).

11.4.2 Preventing the oil degradation process

The principal critical elements in controlling oil degradation are:

- the composition of the oil and its anti-oxidant property;
- the frying apparatus;
- good industrial practice.

Quality of oil: composition in fatty acids

Oil's resistance to degradation in the frying process depends principally on the ratio between saturated/monounsaturated/polyunsaturated fatty acids and antioxidant activity. The influence of the unsaturated fatty acids on the thermal degradation process of fried oil is well known; a high content of polyunsaturated fatty acids increases the rate of radical reaction. Belgian law has established a linolenic acid content for frying oil under 2%.

Soybean, rapeseed, sunflower, maize and generally all seed oils have a high iodine value and often require hydrogenation treatment to optimise the frying property. Hydrogenated oils have a good melt and smoke point, but the trans-fatty acid content of these oils should be monitored.

The use of frying oil rich in saturated fatty acids has a negative nutritional impact on the fat composition of fried foods; moreover, the use of animal fat also involves hazards because of oxidation of fat acid and cholesterol (animal fat has a low antioxidant activity). Australian law has established the saturated fat in frying oil at under 500 g/kg.

Oils rich in low MW fatty acid (i.e. lauric acid), such as tropical oil (e.g. coconut oil), have an excessively low smoke point. Palm oil has a higher smoke point than coconut oil but has a very high saturated fatty acid content. Palm oil browns early in the frying process, but its changes in colour are not such good indicators of its runaway property as the same process is for other oils.

Groundnut oil has an admirable fatty acid composition but is of limited use because of the widespread allergy to peanuts, Also the allergenic risk is principally related to cold pressed virgin oil (refined oil with no protein would not give allergenic reaction). In the production of groundnut oil, the hazard of aflatoxine contamination should be strictly controlled.

Every study has shown that olive oil is the premium oil in the frying process because of its high content of monounsaturated fatty acids and antioxidant properties, but it too is of limited use because of its high cost.

The guidelines for frying adopted in some European countries indicate heat-resistant oils and fats for frying, meaning oil or fat with controlled ratios of unsaturated fatty acids. However, high polyunsaturated oils, animal oils, tropical oils and heavily hydrogenated oils are commonly used in the frying process and their impact on human health remains an open question.

Quality of oil: antioxidant capacity

The oxidation reaction can be inhibited by antioxidants which are naturally present in oils. Research has highlighted the antioxidant activity and subsequent protection afforded by oils such as avenesterols, carotens, fucosterol, phitophytins, tocopherols and polyphenols. These compounds account for the overall antioxidant capacity of the oil. The ratio of each natural oxidant to total antioxidant capacity depends on the composition of the oil. However, the natural complex systems present in oils are believed to be the major protective factors. Olive oil and especially virgin olive oil are premium oils for antioxidant capacity.

The technological process of frying destroys the natural antioxidant systems of seed oils. They generally require the addition of natural tocopherol concentrates. Synthetic antioxidants (tertiary butylhydroquinone, butylated-hydroxyanisole, butylated-hydroxytoluene, gallates) can be used in the production of oils, but they are prohibited in most countries.

The prevention of oxidation can be obtained by forming a physical barrier between air and oil using dimetylpolysiloxane as anti-foam. Silicone also prevents oxidation at static conditions, and is suited to infrequent frying cycles. In many countries the use of anti-foam is prohibited.

Frying equipment: temperature control apparatus
Oil temperature is the principal control point in the frying process. The optimal frying temperature is between 160–180°C. To avoid localised surface over-heating, oil should be heated very slowly. The fryer should be also provided with a thermostat. Fryers for solid fats should have a melting cycle; in some, pulses of heat are followed by rest periods to allow solid fat to melt without over-heating. The oil can be heated by gas flames from the bottom of the vessel, by electric elements or by external heat exchangers: in the latter case, there is a toxicological risk in the event of leaks of heated transfer fluids into the oil. In some countries, the thermostat-equipped fryer is compulsory.

Frying equipment: surface metal contamination hazard
The construction material used in frying equipment can introduce contamination by metals (iron, copper, etc.) which catalyse the degradation process. All equipment components in contact with oil should be stainless steel. Copper containing alloys should be avoided. All pipes, especially valves, thermometer pockets, electric elements and other fittings should be checked to ensure they are not made of copper ore iron alloys (mild steel). Alkaline cleaning treatment will remove any protective films from these alloys.

Frying equipment: filter systems and eliminating polar contaminants
One of the most important aspects of any frying system is the filtering apparatus. It needs to remove charred batter and breading, commonly referred to as crackling. This may damage equipment activating the oil degradation process. Industrial fryers should be equipped with a continuous filtering apparatus and automatic discharge, or with two batch-type filter units working in unison.

It is useful to arrange for a cool zone at the bottom of the fryer – below the heating elements, for example – so that scraps may fall into it without damaging the oil. The gas-heated fryer has no such possibility.

Various types of screens, cartridges and paper filters may be used to remove crackling with or without filter aids. Paper and woven cloth filters can remove particles as small as about 5 mm and reduce the development of pyrolisis products. Non-woven filters will remove particles as small as about 1 mm, including micro-emulsified water. This will slow down the oil deterioration process. Cleaning filters regularly will ensure that foreign matter is removed from the oil. The bed of debris may feature materials rich in iron, copper or other chemicals, which catalyse the degradation process. Particularly, breadcrumbs fortified with iron can hasten oil oxidation, so fortified cereals should be avoided.

The traditional filter system will remove particles but not soluble chemicals, such as surface tension reducers which play a major role in accelerating oil degradation. Removal of polar contaminants by active filtration is possible using an absorbent product. Active filtration will keep the oil in an optimum condition for longer. Attapulgites, zeolites, active carbon, kaolin, active silica, diatomaceous substances, or other miscellaneous substances activated with citric acid or other chemicals are the active filtration absorbents used. Some absorbents, such as activated carbon, will indiscriminately also remove good components, such as tocopherol.

Other active absorbents on the market are more specifically designed for surface tension-reducers, and will slow down deterioration caused by the process. However, these products will be chosen according to the specific filter system in use and the results of active filtration remain controversial.

Frying equipment: fume extraction apparatus and fryer lid

Fume extraction systems condition contact between oil and air. An excessive draught of fresh air cools the oil and promotes oxidation. It is also important that fumes condensing in the extraction equipment do not drip or run back into the fryer. Cooling and then covering the vessel between batches and at the termination of the frying operation will avoid light-catalysed reactions and reduce air contact. If the lid is positioned above the surface of the oil, condensation will drip back into it, causing deterioration. For this reason, a floating lid is best and it is also more effective in excluding air.

Good industrial practice: storage of oil

Oil oxidation is autocatalytic; oxidised oil can therefore promote and accelerate oxidation of fresh oil. Care should therefore be taken not to load deliveries of fresh oil into a tank on top of old stock.

The oil should have no contact with air and, if possible, the oil in the tank should be blanketed with nitrogen. The oil should be loaded into the tank through a bottom outlet, not poured from the top down; this, to reduce aeration. Oils with melting points above room temperature should be kept warm in the tank, but not over-heated.

Good industrial practice: turnover of oil

The most critical element in the frying process is the quality of fried oil and its turnover. It is not possible to schedule turnover because the degradation process will depend on many factors (oils, fried foods, fryer, filter system, etc.). The last DGF recommendations indicated total polar (<24%) and polymeric matter (<12%) as the principal test for classifying oil as abused. Quick quality tests are useful instruments for establishing critical limits for oil turnover (see Section

13.5). However, sensory parameters such as darkened colour, viscosity, odour and tendency to smoke can be used as an oil quality index in checking critical limits.

Since fried food absorbs oil, the level of oil in the fryer will fall and is usually topped up with fresh oil, especially in continuous frying. The oil turnover schedule is calculated as the rate between the weight of oil in the fryer and the average weight of the fresh oil added per hour. In the continuous fryer the turnover is between five and 12 hours. Fryer idling will put more stress on the oil. Oil quality cannot be maintained if turnover is set at over 20 hours. The critical limit is best set at 12 hours. For fatty foods (fish, meat, etc.), turnover limits should be reduced.

For discontinuous cycles typical of restaurants, fried oil should be discarded at least once every three or four days. The Hungarian National Institute of Food Hygiene and Nutrition fixed the turnover schedule as follows: 8–10 hours for sunflower oil, 10–13 hours for corn oil, 18–20 hours for lard.

Many authors disagree with reconditioning oil (the addition of fresh oil to fried oil). Italian Ministry of Health regulations explicitly prohibit this because the fresh oil is rapidly altered on contact with used oil. Other studies indicate that blending fresh and fried oil optimises the oil-breaking process. The partial discarding of oil is said to prevent problems with fresh oil, reducing its bad qualities and absorption. The minutes of the DGF Symposium 2000 confirm that used but not abused frying oils may be topped up or diluted with fresh oil with no adverse effects on quality.

The use of active filtering aids may reduce the slow degradation process of fresh oil catalysing the surface tension reducers contained in used oil. However, this treatment of oil demands continuous quality control.

Good industrial practice: avoiding water and other polar contaminants from foods

Wet food should not be fried as the additional moisture will cool the oil and cause additional hydrolysis. However, the moisture normally present in food has a beneficial effect on the oil because steam generated during the frying process deodorises the oil, as it is distilled from the volatile oil compounds such as aldehydes and ketones.

The polar contaminants of food juices accelerate the oil degradation process; food should not be salted before frying as this will lead to sodium chloride contamination of the oil along with other impurities present in salt, which are also prooxidant.

Good industrial practice: loading of fryer

Over-loading the fryer will reduce the oil temperature. The cooling and heating of oil increases the oxidisation process. In an over-loaded fryer, food will take longer to cook increasing its general absorption of oil. The overload critical limit is a 1/6 ratio between food weight/oil. Restaurants without adequate equipment are recommended to pre-fry foods, and so reduce cooking time and over-loading at peak hours. Also, an over-sized deep fryer will require a great quantity of oil, which has to be kept hot and leads to longer turnover time, which is bad for the oil.

Good industrial practice: surface treatment and draining of fried food
The surface of food influences the mass transfer between oil and food. The mass transfer from food to oil includes water, fats, polar compounds and the crackling particles; crackling promotes the oil degradation process. The mass transfer from oil to the food takes place in the absorption of oil.

A critical element in reducing crackling is to shake or blow loose breadcrumbs and batter from coated food prior to frying. The optimisation of the air flow to the blower on the coating unit can, therefore, improve oil quality. Draining excess oil from the food surface by straining is another critical point. The food should be drained immediately after frying to avoid general oil absorption.

Good industrial practice: correct cleaning of fryer
It is vital to remove traces of polymerised oil, which will hasten degradation. This may be achieved by using a hot alkaline detergent solution. This cleaning treatment should be carried out once a week for fryers subjected to much use with unsaturated oils.

Fryers should be well rinsed and dried to remove traces of detergent before refilling with oil. Failure to remove alkaline cleaning compounds is said to be a primary cause of oil degradation. However, the gum deposits and carbon formation will be minimal in a proper frying process incorporating controls of the polar contaminant content.

11.4.3 Controlling oil absorption

The possibility of hazards in the frying process depends on the quality of fried oil and the food's absorption tendency. The following is a study of the principal factors reducing oil absorption and its relative control point.

Quality of oil
The surface substances formed in the degradation process reduce surface tension and increase oil absorption; the more degraded the oil, the more absorbent it is.

Temperature and duration of process
The duration of the frying process and the amount of oil absorbed are correlated. Less will be absorbed at temperatures of between 155 and 200ºC. Lower temperatures will extend cooking time and increase oil absorption.

The physical shape of food
There is a linear relation between the surface area of food and the fat absorption rate. Reducing the ratio between surface and weight lessens the food's oil uptake.

Moisture content
It is well established that oil absorption occurs as moisture and is removed from the food during frying. Low initial moisture content would lessen the amount of oil uptake, but it is not a general rule. Some additives, such as alginates and cellulose, increase water retention and reduce oil absorption.

Composition of food
In some products, such as dough, higher fat content increases the mass exchange, causing high fat uptake. However, this does not hold for every food matrix. High ionic content, especially sodium and calcium, generally reduces oil uptake, but the dynamic of the reaction has not been determined.

Pre-frying treatment
Reducing the moisture content of food prior to frying reduces oil absorption, while pre-drying generally reduces fat uptake and freeze-dying causes the product to absorb more oil. Blanching reduces oil absorption. Steam treatment can increase or reduce oil uptake. Wetting the food surface with emulsifier or treating it with hot water vapour will reduce its oil absorption capacity. Generally, freezing prior to frying means less fat uptake in French fries.

Surface treatment
The process whereby a surface crust develops giving the fried product the desired organoleptic characteristic, is a critical element in controlling fat uptake. The formation of a surface crust is helped by the presence of starch, which is a constituent of the product itself (for example in potatoes), or is added to the food in the form of bread-crumbs, batter or flour.

The crust generally should be a barrier to the mass transfer of oil to fried food (water discharge and fat uptake), but many other factors (increased food surface, development of heat transfer and relative cooking time, etc.) can modify the crust's fat absorption capacity. However, the results of research on surface treatment have often been inconsistent.

Breading does not generally influence oil absorption. Battering reduces oil uptake and water loss. The incorporation of hydrocolloid or long fibre cellulose additives within the battered food lessens oil uptake. Information on the use of these additives in fried food coatings are generally of a proprietary nature. Batter coatings have been shown selectively to absorb oxidised fractions of oil.

11.5 Monitoring critical control points in the frying process

11.5.1 Used oil quality control

The changes in fat or oil during frying involve four major degradation processes:

- Fat hydrolysis;
- Fat oxidation and subsequent polymeration or cyclation;

- Fat polymeration and the formation of new carbon-carbon linkages in the absence of oxygen;
- Mass transfer from food to oil or fat or polar contaminant.

Gauging heat abuse of oils should be based on these changes. There are many methods for measuring the degradation process by analysing the formation of fat degradation components or correlated indicators. The degraded product may be classified as a volatile or non-volatile compound.

Volatile substances of decomposition

Many of the volatile substances of decomposition are distilled during frying, so care must taken to capture these compounds for an accurate measure of heat abuse. A portion of volatile substances of decomposition remains in the oil and is absorbed on the surface of fried food; this substance may also be inhaled by the operator and dispersed in the environment. The effect of these substances for human health should be investigated.

The volatile fraction in used oil is very complex and depends on frying conditions and the nature of fried foods. More than 220 volatile degradation compounds were identified in fried oils by gas chromatographic analysis; many of these compounds are known to be toxic, while others contribute to the flavour of fried foods.

Measuring volatile substances of decomposition in fried oil is very time-consuming and laborious. Little additional work has been done on volatile decomposition compounds since the 1970s because of the complexity of the task and the difficulty in quantifying the volatile fraction. The chromatographic gas inhalation technique is applied to study aromagrams of fried food. These studies were designed to analyse the organoleptic property of frying, but not to check the quality of the frying process or the fried food.

The development of air-space sensors (electronic noses) has renewed interest in accurate controls of the frying process by volatile oil fraction analysis (see Section 11.6).

Non-volatile substances of decomposition

The non-volatile substances of decomposition remain in the oil too and will promote degradation. They are absorbed by fried food and are thus eaten by the consumer. The accumulation of non-volatile substances of decomposition is also responsible for the physical changes evident in frying fat, such as the increase in viscosity, darkness of colour and foaming. Most methods for assessing deterioration of fat are therefore based on these chemical or physical changes.

Polar components

The total polar component content is the most usual parameter of control adopted for oil or fat. The polar component method of analysis is an IUPAC- and AOAC-approved method. It is accurate, simple and reproducible. It was one of the methods of analysis recommended by the last international DGF symposium on Deep Fat Frying and Used Oil Quality Control. The polar component approach is adopted in many countries as a legislative parameter for monitoring the quality of used oil.

The critical limits accepted for polar components in used oil swing from 24% to 30% depending on the country and researcher. The length of time (about three hours) needed to evaluate one sample is this system's major drawback. The polar component content modifies the dielectric constant of the frying medium and it is measured accordingly. The apparatus measuring the dielectric constant of frying oil was a failure because it was too sensitive to moisture in the oil.

Fatty acid analysis

The ratio between saturated/polyunsaturated fatty acids and polyunsaturated fatty acid degradation rises during frying.

Particularly, the 18:2/16:0 ratio is used as a parameter of control for the frying process. Obviously, the critical limits should be defined by calibrating the system according to the quality of fresh oil. The measurement of fatty acids is considered non-reproducible by the DGF.

Organoleptic property

In runaway oil the taste, odour and colour become unpleasant or unacceptable because of the chemical changes which have occurred. A clean flavour and the absence of any objectionable odour in used oil is considered in some countries to be the critical limit. An acceptable appearance (colour and foaming) is adopted by other legislative systems as the official parameter. Some countries prohibit the use of anti-foam agents in that they mask the natural foaming in deteriorated oil. The electronic nose approach could improve this subjective evaluation of odour and appearance.

Smoke point

The smoke point is a traditional indicator for both fresh and used oil; the smoke point falls during frying because of fat hydrolysis. In many countries, the critical smoke point has been set at between 170 and 180°C. The minutes of the last international DGF symposium establish that the smoke point method is not reproducible.

Viscosity

Viscosity increases during frying as a consequence of polymerising. Thus, viscosity may be adopted as a critical parameter for checking the degradation of oil. In Belgium the critical limits for fat frying is fixed at a rate of 37 mPa-sec at 50°C, and for oil at a rate of 27 mPa-sec at 50°C.

Free fatty acids (acid value)

The free fatty acid (FFA) content is an indicator of the hydrolytic degradation process; it is a common control parameter for fresh and used oil. In many countries, the critical limit for oil is between about 2.5% and 4.5%. The international DGF symposium found that FFA content is not a reproducible method to control the quality of frying oil; in fact FFA content is not well correlated with quality of oil because FFA are involved in the complex kinetic of the degradation process.

Conjugated dienes and trienes

When polyunsaturated fatty acids are oxidised, a shift of the double bonds occurs so that the conjugate diene (and triene) content increases. This degradation process can be measured by absorption in an ultraviolet field at 283 nm (the official AOAC (1983) and IUPAC (1987) methods are available). This test is useful in measuring heat abuse of polyunsaturated oils, but it is less applicable to fats containing few unsaturated fatty acids. The critical limit should be defined by calibrating the methods applied according to the specific process under examination.

Dimer and polymer triglycerides (DPTG)

Condensation occurs between the double bonds of oxidised fatty acids of triglycerides during the frying process. Polymeric compound content is a direct indicator of the degradation process in abused oil. To determine polymeric compounds chromatography gel permeation methods are generally used. However, the most reproducible results are obtained using GL chromatography. The last international DGF symposium recommendations indicate polymeric matter, as well as the total polar matter as the critical parameter for controlling used oil. The DPG fixes the critical limit for polymeric matter at 12%. In other countries, the critical limits swing from ten per cent to 16%.

Near IR reflection, the spectroscopy technique, can also be usefully adopted for the fast detection of the DPTG.

Moisture

Most of the water released by food is steamed out by the oil; however, some remains in the oil as inverse drops. This water increases the hydrolytic reaction rate, triggering the degradation process. In some countries, the regulation amount of water in fried oil is set at below three per cent.

Oxidised fatty acids

The content of oxidised fatty acids in used oils is a direct parameter for the control of damage by oxidation. In some European countries, the limit for fatty acids insoluble in petroleum is fixed at between 0.75 per cent and one per cent.

Peroxide value

Peroxide value (PV) is a parameter, an alternative to oxidised fatty acids, for measuring stress from oxidation in both fresh and used oil. The critical limits for peroxide value are generally fixed at 2 meq/kg.

PV is conventionally analysed by the iodometry. Near IR spectroscopy scanning at 1100–2500 nm can be applied as a fast method for the determination of PV. PV is not a reliable measure with frying oils as the peroxides decompose at frying temperature.

Iodine value
Iodine value is a measure of the unsaturated bonds in oil. The iodine value decreases during the frying process as a consequence of the oxidation of the double bonds. The Iodine value decrease (compared to that of the unused oil) is a control parameter of quality of oil. In Finland the Δ-Iodine Value limit is fixed at 16 for used frying oils.

Cyclic monomers
Cyclic monomers are specific products of the degradation process during frying and are potentially toxic; they could be specific indicators of the hazards in using abused oils. Cyclic monomers are determined by means of liquid gas chromatography once the sample has been hydrogenated. The method's reproducibility is low owing to its complexity, to the large number of cyclic monomers (some still chemically unknown) and to their low concentration in the sample. However, high contents of cyclic monomers have been found in abused oil with high polyunsaturated fatty acid content. Critical limits on the content of cyclic monomers have never been defined.

Antioxidant capacity
Frying oils have a complex natural antioxidant system and/or blend of natural and synthetic antioxidant additives. The total content of antioxidants is an indicator of the oil's resistance to oxidative stress during frying. This measure is useful for the selection of fresh oil. However, antioxidant capacity could also be adopted to detect the residual antioxidant capacity in used oils; this, to optimise the frying process.

Total antioxidant capacity is generally detected by the crocine bleaching inhibition method, measured by UV spectrophotometer at 443 nm. Critical limits for the antioxidant capacity have not been defined.

11.5.2 Fresh oil quality control

Hazards in the frying process depend on the oil's capacity to resist heat stress, on environmental and technological contaminants (micotoxins, solvents, trans-fatty acids, additives, etc.), on nutritional value and on the presence of allergens. The selection of frying oil and subsequent quality control methods should be designed to control these risks.

Resistance to heat stress depends on the unsaturation of fat, on the rancidity of the product and on its antioxidant properties. These characteristics may be detected on the basis of iodine and peroxide values, total antioxidant capacity, free fatty acids (acid value) and fatty acid analysis, as explained above for the control of used oils. In many countries, it is compulsory to control the polyunsaturated fatty acid content of oil and, for linolenic acid in particular, the critical limit is generally fixed at two per cent.

Peanuts have a high risk of contamination by micotoxins (especially aflatoxins). Cereals, especially maize, are also subject to contamination, chiefly by Ocratoxin A. The EU has fixed the legal limits of aflatoxins in peanuts at 8 ppm for aflatoxin B1, at 15 ppm for the sum of B1 B2 G1 G2 and for aflatoxin in cereal products at 3 ppm. Erucic acid is a natural component of traditional colza oil with a toxicity property. In the EU, the legal limit of erucic acid content in oils is 5%. Oil should not contain mineral oils. There is no set critical limit for trans-fatty acids in oil, although many studies have pointed to the health hazard inherent in a rich trans-fatty acid diet. There are European legal regulations for baby food, which limit the trans-fatty acid content to 4% of total fat.

Oil additive content regulations vary from country to country. Synthetic antioxidant and silicone are prohibited in many countries and limited in others. Silicone additives, where admitted, are generally limited to(3 mg/kg in Belgium and France, 10mg/kg UK, USA and Australia). Gallates and other synthetic antioxidants are generally limited to below 0.1%.

11.5.3 Fried food quality parameters

Chemical risk

The principal critical element for fried food is correlated to:

- the quality of the frying medium;
- the amount of fat absorbed;
- the fat degradation process during storage (especially for pre-fried frozen food and fried snacks).

The qualitative analysis of fat in fried food involves complex analytical procedures to assess alteration. However, the qualitative parameter adopted for oil can also generally be used for fried foods; to identify the specific frying degradation process, qualitative analysis of fat in fried food should be made by the chromatographic technique. The matrix of food interferes in these measures and the quantitative results are not very reproducible in general. Legal critical limits have not been fixed for the quality of fat in fried food.

There are no critical limits for the amount of fat in fried food because it depends on the nature of the food itself. However, in some countries the amount of fat has to be labelled both for snacks as well as in large-scale catering so that the consumer is aware of the fat content of his food.

The biological risks of bacterial proliferation

During frying the core of food is heated at about 100ºC and the crust will be in contact with oil frying at 170ºC. Thus, bacteria are destroyed and a large proportion of their spores rendered inactive. But fried food and especially pre-fried, oil-blanched food used in catering and in the home, has high water activity. These products are perishable and require the same controlled storage temperatures as other food products.

Recent research has highlighted pathogenic bacteria (staphylococcus aureus, bacillus cereus, etc.) growth on home-made French fries, which will eventually produce specific enterotoxins if kept at the wrong storage temperature. In the storage of fried food, the catering industry should use the same safe procedure they adopt for other hot foods.

11.5.4 Quick frying process control tests

In general, the HACCP monitoring systems should be sufficient indication. All the better if deviations are detected immediately. Several commercially available procedures have been developed to provide a rapid means of testing for frying-oil abuse. However, this field of research is rapidly developing. The most common quick tests are reviewed below. Measuring change in the dielectric constant has been used by the Northern States Instrument Corporation in the development of an instrument, the Food Oil Sensor (FOS), responsive to total polar components. A value of 4.0 with the FOS reading should be the limit for used oil. Many external factors, such as water or fat extracted from fried food and the nature of fresh oil influence the measure value. Nonetheless, the method is useful in certain circumstances. The instrument is calibrated according to the freshness of the oil and the range of measurement is from 0 to 6. The Merk has set up two quick-tests for oil: the Oxifrit-test and Fritest. The Oxifrit-test is a colorimetric method, which contains a redox indicator that reacts with the total quantity of oxidised compounds in the oil sample. The colour evolving from contact between the sample and reagents is compared with a colour scale featuring four qualitative indicators: 1 good, 2 still good, 3 intermediate quality, 4 poor. The Fritest provides a colorimetric measurement of the carbonyl compounds. The mixture of the sample and reagents is compared with a three-colour scale: 1 good quality, 2 intermediate quality, 3 poor. The Spot Test is a quick colorimetric pH-monitoring test. To test the oil a drop should be placed on a silica gel-covered slide which has bromocresol green incorporated into the gel as pH indicator. The test monitors the free fatty acid content as an indicator of hydrolytic rancidity. Veri-test, commercialised by Libra Laboratory Inc., also features redox indicators changing from blue to green as the polar compounds increase. Alkaline Contaminant Material colorimetric quick tests are available. The procedure measures soaps at levels of 0 to 200 ppm.

Near infrared spectroscopy (NIR) is applied to control the quality of oil. However, unlike the colorimetric test, the NIR technique needs considerable instrumentation and specialised personnel, but it is more reliable. Dimer and polymer triglycerides and acid value in used oils were determined by Near IR reflection spectroscopy. Scanning of NIR at 1100-25500 nm is applied for peroxide value in oils, although PV is not a good yardstick, as explained earlier.

Headspace sensors (artificial noses) are highly likely to be applied for the continuous checking of oil quality in the interests of automatic oil turnover control. Processing the information from aspecific sensors by a neural network, the artificial nose provides an immediate answer on the state of the oil. The sensors used are not specific for assessing the degradation of volatile substances, such as enals, dienes and carbonyl compounds. The Electronic Nose apparatus is cheap and easy to use. However, at the moment, there are not enough instruments on the market suitable for attachment to the fryer.

11.6 Future trends

Increased knowledge of the biological effects of enals and other degrading substances on human health will drive the development of frying process control technology. The technological approach to deep frying should be extended to shallow frying and other techniques. Commercial chemical sensor development (such as the artificial nose) will be a major way of enhancing fryer equipment engineering, especially for continuous processes. The use of NMR, MRI Analysis is improving knowledge of surface transformation during frying. Coating technology will be optimised to reduce oil uptake while preserving the characteristics typical of fried food.

However, the main problems in frying remain: heat stability; the use of vegetable oil blends rich in oleic acid, and making the most of the natural antioxidant system to improve the stability of polyunsaturated oils. All these elements will be the object of study. The active filtering procedures developed thus far should be subsidiary to the study of oil blends to reduce discard.

11.7 Bibliography

AGUILERA JM, STANLEY DW (1990) *Microstructural Principles of Food Processing & Engineering*. Elsevier Science Pub., UK, pp. 161–92.

ANNAPURE US, SINGHAL, RS, KULKARNI PR (1998) Studies on deep-fat fried snacks from some cereals and legumes. *J Sci Food Agric*, 76(3): 377–82.

AW B, DORNIER M, DUBOIS C, GUILLAUMONT A, AYMARD C, REYNES M (1998) Optimisation des conditions de confisage friture de tranches d'ananas par la methodologie des plans d'experiences. *Science des Aliments*, 18: 313–22.

BACETTI LB, FALCONE M (1995) Production of green fried banana using the Musa cavendishii Lamb variety: a study. *Rev. Instituto Adolfo Lutz*, 55(1): 1–6.

BARTLETT PN, ELLIOTT JM, GARDNER JW (1997) Electronic noses and their application in the food industry. *Food Technol*, 51: 44–8.

BERDEAUX O, MARQUEZ-RUIZ G, DOBARGANES MC (1999) Characterization, quantitation and evolution of monoepoxy compounds formed in model systems of fatty acid methyl esters and monoacid triglycerides heated at high temperature. *Grasas-y-Aceites*, 50(1): 53–9.

BERRY SK, SEHGAL RC, KALRA CL (1999) Comparative oil uptake by potato chips during frying under different conditions. *J Food Science and Technology, India*, 36(6): 519–21.

BERSCIA L, MOREIRA R G (1997) Modelling and control of a continuous frying process: simulation study. II. Control development. *Food Bioproducts Processing*, 75(C1): 12–16.

BLAHOVEC J, VACEK J, PATOCKA K (1999) Texture of fried potato tissue as affected by pre-blanching in some salt solutions. *J Texture Studies*, 30(5): 493–507.

BOSKOU D (1998) Frying temperatures and minor constituents of oils and fats. *Grasas-y-Aceites*, 49(3/4): 326–30.

BREWER M S, VEGA J D, PERKINS E G (1999) Volatile compounds and sensory characteristics of frying fats. *J Food Lipids*, 6(1): 47–61.

CHARPENTIER J-C, TRARNBOUZE P (1998) Process engineering and problems encountered by chemical and related industries in the near future. Revolution or continuity? *Chem Eng Processing*, 37: 559–65.

CHE MAN Y B, IRWANDI J (2000) Effect of rosemary and sage extracts on frying performance of refined, bleached and deodorized (RBD) palm olein during deep-fat frying. *Food Chemistry*, 69(3): 301–7.

CHE MAN Y B, LIU J L, ABDUL-RAHMAN R, JAMILAH B (1999) Shelf-life of fried potato chips using RBD palm olein, soybean oil and their blends. *J Food Lipids*, 6(4): 287–98.

CHIU Y J (1993) US Patent #5 263 406.

CHRISTENSEN C M, VICKERS Z M (1981) Relationship of chewing sounds to judgment of food crispness. *J Food Sci*, 46: 574.

CLARK L, SERBIA G W (1991) Safety Aspects of frying fats and oils. *Food Technology*, February: 84–9.

D'AMICO A, DI NATALE C, MACAGNANO A, DAVIDE F, MANTINI A, TARIZZO E, PAOLESSE R, BOSCHI T (1998) Technology and tools for mimicking olfaction: status of the Rome Tor Vergata Electronic Nose. *Biosensors and Bioelectronics*, 13: 711–21.

DE MARISCAL P, BELL D A (1996) Fiber based fat mimics methylcellulose gums. In: Roller S, Jones S A, (Eds), *Handbook of Fat Replacers*, p. 153. CRC, Boca Raton, FL.

DGF (2000). Eight recommendations for frying oils were discussed and adopted by the delegates. [Announcement]. Hagen (Germany): German Society for Fat Science (Deutsche Gesellschaft für Fettwissenschaft). Retrieved 13 September, 2000, from the World Wide Web: www.gdch.de/dgf/recomm.htm (2000).

DIAZ A, LISSE I, RAOULT WACK A L, BRICAS N (1998) Consumption trends of plantain chips and consumer appreciation of their quality in Cali (Colombia). *Tropical Sci*, 38(3): 171–8.

DIAZ A, TOTTE A, GIROUX F, RAOULT-WACK A L (1996) Deep-fat Frying of Plantain (Musa paradisiaca L.) : I. Characterization of Control Parameters. *Lebensm.-Wiss. u. Technol.*, 29: 489–97.

DI NATALE C, PAOLESSE R, MACAGNANO A, MANTINI A, GOLETTI C, TARIZZO E, D'AMICO A (1998) Characterisation and design of porphyrins-based broad selectivity chemical sensors for electronic nose application. *Sensors and Actuators B*, 52: 162–8.

DI NATALE C, PAOLESSE R, MACAGNANO A, MANTINI A, D'AMICO A, LEGIN A, LVOVA L, RUDNITSKAYA A, VLASOV Y (2000) Electronic nose and electronic tongue integration for improved classification of clinical and food samples, *Sensors and Actuators B*, 64: 15–21.

DOBARGANES MC, MARQUEZ-RUIZ G (1998) Regulation of used frying fats and validity of quick tests for discarding the fats. *Grasas-y-Aceites*, 49(3/4): 331–5.

DOMAGALA A, JANUS P (1999) The effect of the temperature of fat on heat energy consumption and the quality of French fries during the process of frying. *Nahrung*, 43(5): 347–8.

EISENBRAND G, TANG W (1993) Food-borne heterocyclic amines. Chemistry, formation, occurrence and biological activities. A literature review. *Toxicology*, 84: 1–82.

EL BAYOUMY K, JOHNSON BE, ROY AK, UPADHYAYA P, PARTIAN SJ (1994) Biomonitoring of nitropolynuclear aromatic hydrocarbons via protein and DNA adducts. *Res Rep Health Eff Inst*, 64: 1–27.

FARKAS BE, SINGH RP, J MM (1992) Measurement of oil/water interface in foods during frying. In: Singh RP and Wirakartakusumah, MA (Eds.), *International Drying Symposium '92*. Boca Raton, Ann Arbor, London, Tokyo: CRC Press.

FILLION L, HENRY CJK (1998) Nutrient losses and gains during frying: a review. *Int J Food Sci Nutrition*, 49(2): 157–68.

FIRESTONE D (1993) Worldwide regulation of frying fats and oils. *Inform–Food Technology*, 4(12): 1366–71.

FUNAMI T, FUNAMI M, TAWADA T, NAKAO Y (1999) Decreasing oil uptake of doughnuts during deep-fat frying using curdlan. *J Food Science*, 64(5): 883–8.

GAMBLE, MH, RICE P (1987) Effect of pre-fry drying on oil uptake and distribution in potato crisp manufacture. *International J of Food Science*, 22: 535–48.

GARDNER JW, BARTLETT PN (Eds.) (1992) *Sensors and Sensory Systems*. Kluwer, Dordrecht, NL.

GREWAL SP (1996) UK Patent Application GB 2 286 108 A.

GURAYA HS, TOLEDO RT (1996). Microstructural characteristics and compression resistance as indices of sensory texture in a crunchy snack product. *J Texture Studies*, 387–701.

HAGEMAN G, HERMANS R, TEN HOOR F, KLEINJANS J (1990) Mutagenicity of deep-frying fat, and evaluation of urine mutagenicity after consumption of fried potatoes. *Fd Chem Toxic*, 28: 75–80.

HAGEMAN, G, VERHAGEN H, SCHUTTE B, KLEINJANS J (1991) Biological effects of short-term feeding to rats of repeatedly used deep-frying fats in relation to fat mutagen content. *Fd Chem Toxic*, 29: 689–98.

HATCH FT, KNIZE MG, FELTON JS (1991) Quantitative structure-activity relationships of heterocyclic amine mutagens formed during the cooking of food. *Environmental Molecular Mutagenesis*, 17(1): 4–19.

HAYWOOD R M, CLAXSOM A W, HAWKES G E, RICHARDSON D P, NAUGHTON D P, COUMBARIDES G, HAWKES J, LYNCH E J, GROOTVELD M C (1995) Detection of aldehydes and their conjugated hydroperoxydiene precursors in thermally-stressed culinary oils and fats: investigations using high resolution proton NMR spectroscopy. *Free Radic Res*, 22: 441–82.

HOUNHOUIGAN J, ROUZIÈRE A, NOËL J M, BRICAS N, MAROUZÉ C, RAOULT-WACK A L (1998) Relance de la production d'huile de coco par la technique de séchage friture. *Plantations, Recherche, Développement*, 3: 111–18.

JACKSON J C, BOURNE M C, BARNARD J (1996) Optimization of blanching for crispness of banana chips using response surface methodology. *J Food Sci*, 61(1): 165–6.

JARCZYK A, FEDEJKO E, RYBAK A, SZUMIELEWICZ P (1996) Technology for making fried fruit chips. *Przemysl Fermentacyjny i Owocowo Warzywny*, 40(5): 21–23.

JASWIR I, CHE MAN Y B (1999) Use optimization of natural antioxidants in refined, bleached, and deodorized palm olein during repeated deep-fat frying using response surface methodology. *J American Oil Chemists Society*, 76(3): 341–8.

JASWIR I, CHE MAN Y B, KITTS D D (2000) Use of natural antioxidants in refined palm olein during repeated deep fat frying. *J Food Research International*, 33: 501–8.

JEONG HEE KIM, SU JEONG JANG, YONG U KWON (2000) Formation of cholesterol oxides in saury (Cololabis seira, kongchi) during pan frying, deep fat frying, and microwave cooking. *Food Science and Biotechnology* 9(1): 48–51.

JINTIAN F, SINGH P R, PINTHUS E J (1997) Physicochemical changes in starch during deep-fat frying of a molded corn starch patty. *J Food Processing Preservation*, 21(6): 443–60.

KILIBWA M (1999) Polydextrose as a fat absorption inhibitor in fried foods. United States Patent.

KRESS-ROGERS E (1996) *Handbook of Biosensors and Electronic Nose*, CRC Press, Boca Raton, FL.

KROKIDA M K, OREOPOLOU V, MAROULIS Z B (2000) Water loss and oil uptake as a function of frying time. *J Food Engineering*, 44(1): 39–46.

KROKIDA M K, OREOPOLOU V, MAROULIS Z B (2000) Effect of frying conditions on shrinkage and porosity of fried potatoes. *J Food Engineering*, 43(3): 147–54.

KUMAR-SHARMA H, SINGHAL R S, KULKARNI P R, GHOLAP A S (1999) Carboxy-methyl cellulose (CMC) as an additive for oil reduction in deep-fat fried paneer. *International J Dairy Technology*, 52(3): 92–4.

LEMAIRE H, REYNES M, MGALANI J A, TCHANGO T, GUILLAUMONT A (1997) The suitability of plantain and cooking bananas for frying. *Fruits*, 52(4): 273–82.

LIN S, AKOH C C, REYNOLDS A E (1999) Determination of optimal conditions for selected adsorbent combinations to recover used frying oils. *J American Oil Chemists Society*, 76(6): 739–44.

LINNEMANN AR, MEERDINK G, MEULENBERG MTG, JONGEN WMF (1999) Consumer-oriented technology development. *Trends Food Sci Technol*, 9: 409–14.

LISSE I, RAOULT WACK AL (1998) Drying of meat materials (lean and fat) by deep-fat frying in animal fat. *Sciences des Aliments*, 18(4): 423–35.

LOLOS M, OREOPOULOU V, TZIA C (1999) Oxidative stability of potato chips: effect of frying oil type, temperature and antioxidants. *J Science of Food and Agriculture*, 79(11): 1524–28.

McDONOUGH C, GOMEZ MH, LEE JK, WANISKA RD, ROONEY LW (1993) Environmental scanning electron microscopy evaluation of tortilla chip manufacture during deep-fat drying. *J Food Science*, 58: 199–203.

MARTIN YG, PEREZ-PAVON JL, CORDERO BM, PINTO CG (1999) Classification of vegetable oils by linear discriminant analysis of electronic nose data. *Analytica Chimica Acta*, 384: 83–94.

MEYERS MY (1990) Functionality of hydrocolloids in coating systems. In: Kulp K and Loewe R (Eds.), *Batters and Breadings in Food Processing*, pp. 117–41, AACC St. Paul, MN.

MOREIRA RG, BARRUFET MA (1995) Spatial distribution of oil after deep-fat frying of tortilla chips from a stochastic model. *J Food Eng*, 27: 279–90.

MOREIRA RG, CASTELL PEREZ ME, BARRUFET MA (1999) *Deep-fat frying: Fundamentals and Applications*. Aspen Publishers Inc.

MOREIRA RG, SUN X, CHEN Y (1997) Factors affecting oil uptake in tortilla chips in deep-fat frying. *J Food Eng*, 31: 485–98.

NI H, DATTA AK (1999) Moisture, oil and energy transport during deep-fat frying of food materials. *Food Bioproducts Processing*, 77(C3): 194–204.

NORMEN E, ROVEDO CO, SINGH RP (1998) Mechanical properties of an immersion fried potato starch-gluten gel during post-frying period. *J Texture Studies*, 29(6): 681–97.

NUNEZ E (1990) US Patent Application 4 935 254.

PFLUG IJ (1997) Evaluating a ground-beef patty cooking process using the general method of process calculation. *J Food Protection*, 60(10): 1215–23.

PINTHUS EJ, SINGH RP, SAGUY IS, FAN J (1998). Formation of resistant starch during deep-fat frying and its role in modifying mechanical properties of fried patties containing corn, rice, wheat, or potato starch and water. *J Food Processing Preservation*, 22: 283–301.

POKORNY J, REBLOVA Z (1999) Effect of food components on changes in frying oil. *Food Technology and Biotechnology*, 37(2): 139–43.

QUAGLIA GB, COMENDADOR J, FINOTTI E (1998) Optimization of frying process in food safety. *Grasa y Aceites*, 49(3–4): 275–81.

QUAGLIA G, FINOTTI E, PAOLETTI F, BERTONE A, GALASSI P (1998) Antioxidant capacity determination of extra virgin olive oils unsaponifiable fraction by crocine bleaching inhibition method *Nahrung*, 42(5): 324–5.

QUAGLIA GB, PIZZOFERRATO I, MINSTRETTA R, LOMBARDI M, SINESIO F, VECCIA-SCAVALLI D (1985) Trattamenti termici ed oli alimetari. Monografie dei quaderni della nutrizione, nuova serie n.3, Istituto Nazionale della Nutrizione, Roma.

QUAGLIA GB, DI NATALE C, MACAGNANO A, MANTINI A, TARIZZO E, D'AMICO A, BOSCHI T, SINESIO F, BUCARELLI FM, MONETA E (1998) Electronic Nose and Sensorial Analysis: Comparison of Performances in Selected Cases. *Sensors and Actuators B*, 50: 246.

RAYNER M, CIOLFI V, MAVES B, STEDMAN P, MITTAL GS (2000) Development and application of soy-protein films to reduce fat intake in deep-fried foods. *J Science Food & Agriculture*, 80(6): 777–82.

REBLOVA Z, KUDRNOVA J, TROJAKOVA L, POKORNY J (1999) Effect of rosemary extracts on the stabilization of frying oil during deep fat frying. *J Food Lipids*, 6(1): 13–23.

REDDY GV, DAS H (1993) Kinetics of deep-fat-frying of potato and optimization of process variables. *J Food Sci*, 30(2): 105–8.

REEVE RM, NEEL EM (1960) Microscopic structure of potato chips. *American Potato J*, 37: 45–52.

SAGUY IS, GREMAUD E, GLORIA H, TURESKY RJ (1997) Distribution and quantification of oil uptake in french fries utilizing radiolabeled ^{14}C palmitic acid. *J Agric Food Chem*, 45: 4286–9.

SCHWEIZER M, VAHINGER S, GÖPEL W (1994) Characterisation of fish freshness with sensor array. *Sens & Act. B*, 18–19: 282–90.

SHYI LIANG S, LUNG BIN H, SUN HWANG L (1998) Effect of vacuum frying on the oxidative stability of oils. *J Am Oil Chem Soc*, 75(10): 1393–8.

SHYI LIANG S, SUN HWANG L (1999) Changes of chemical components of apple slices during vacuum frying. *Food Science Taiwan*, 26(5): 507–16.

SIJBRING PH, VELDE JVD (1969) Principles of vacuum frying and the results of vacuum frying of chips in practice. [Potatoes]. *Food Trade Rev*, 39(6): 39–42.

SINESIO F, DI NATALE C, QUAGLIA G, BUCARELLI F, MONETA E, MACAGNANO A, PAOLESSE R, D'AMICO A (2000) Use of electronic nose and trained sensory panel in the evaluation of tomato quality. *J Science of Food and Agriculture*, 80: 63.

SINGH PR (1995) Heat and mass transfer in foods during deep fat frying. *J Food Technology*, 4: 134–7.

SINGH RP (2000) Moving Boundaries in Food Engineering. *Food Technol*, 54(2): 44–48, 53.

SINGH RP, VIJAYAN J (1998) Predictive modeling in food process design. *Food Sci Technol Int*, 4(5): 303–10.

STELLA R, BARISCI J, SERRA G, WALLACE GG, DE ROSSI D (2000) Characterisation of olive oil by an electronic nose based on conducting polymer sensors. *Sensors and Actuators B*, 63: 1–9.

STIER RF, BLUMENTHAL MM (1991) Frying and Health. *Baking Snack*, 13(9), 27–30.

SWACKHAMER R (1995) Responding to Customer requirements for improved frying system performance. *Food Technology*, April: 151–2.

SZCZESNIAK A S (1988) The meaning of crispness as a textural characteristic. *J Textural Studies*, 19: 51–9.

TRYSTRAM G, TRÉLÉA I C, RAOULT-WACK A L, DIAZ A, COURTOIS F (1999) Indirect measurement and control of moisture content during dehydration performed by frying. *Drying Technol*, 17(7/8): 1627–37.

UFHEIL G, ESCHER F (1996) Dynamics of Oil Uptake during Deep-fat Frying of Potato Slices. *Lebensm. Wiss. u. Technol.*, 29: 640–4.

VITRAC O, DUFOUR D, TRYSTRAM G, RAOULT-WACK A L (2000) Deep-fat-frying of cassava: influence of raw material properties on chips quality. *J Sci Food Agric*, in press.

VITRAC O, TRYSTRAM G, RAOULT-WACK A L (2000) A simulation tool for the on line assessment of heat and mass transfer during deep-fat frying. In: Thiel D (ed.), *Foodsim' 2000*. Society for Computer Simulation, Delft (Netherlands), pp. 101–5.

VITRAC O, TRYSTRAM G, RAOULT-WACK A L (2000) Deep-fat frying of food: heat and mass transfers, transformations and reactions inside the frying material. *Eur J Lipid Sci Technol*, in press.

WHITE P J (1991) Methods for Measuring Changes in deep-fat frying oils. *Food Technology*, February: 75–80.

XIN QING XU, VIET HUNG TRAN, PALMER M, WHITE K, SALISBURY P (1999) Chemical and physical analyses and sensory evaluation of six deep-frying oils. *J Am Oil Chemists Society*, 76(9): 1091–9.

XIN QING XU (1999) A modified VERI-FRY Registered quick test for measuring total polar compounds in deep-frying oils. *J Am Oil Chemists Society*, 76(9): 1087–9.

12

Flavour and aroma development in frying and fried food

P. Gillatt

12.1 Introduction

The objective of this chapter is to review the scientific literature on the development of flavour and aroma in fried food, particularly fried potatoes. Five areas are addressed:

1. an introduction covering the flavour components within the raw potato tuber, oil oxidation and subsequent flavour release, and the chemical reactions that take place to produce some of the organoleptic properties for which fried food is so desired (sections 12.2–12.4);
2. the development of fried food flavour in the production of potato chips (crisps), French fries and other foods (sections 12.5–12.7);
3. the factors affecting oil uptake in frying and related organoleptic issues (sections 12.8–10);
4. sensory testing of fried food (section 12.11);
5. the application of flavours in fried food manufacture (section 12.12).

Within the constraints of this chapter it is not possible to deal with these issues in depth, however, the review of the literature contained below is reasonably well referenced to enable the reader to access the primary literature and texts that the author has used in the production of this article. The first part of the chapter considers:

1. the odour and flavour components within the raw potato (section 12.2);
2. primary and secondary lipid oxidation which can occur within oils at frying temperatures (section 12.2)
3. the Maillard reaction and Strecker Degradation (section 12.3)

All of these are key influences on flavour development.

12.2 Flavour of raw potatoes

The major chemical components in potato tubers are shown in Table 12.1. Raw potatoes have an odour and flavour and although it has proved difficult to collect the volatile fractions it has been achieved by a number of authors. For example, Meigh *et al.* (1973), Buttery and Ling (1973) and Murray and Whitfield (1975) collected and identified the following compounds from raw potatoes: heptanol, octanol, octenol, 1-octen-3-ol, non-3-enol, 1-methylnaphthalene, trimethylnaphthalene, 1,4-dimethylnaphthalene, 1,6-dimethylnaphthalene, 2,6-naphthalene, and the following pyrazines: 2-butyl-3-methoxypyrazine, 2-isobutyl-3-methoxypyrazine, 2-isopropyl-3-methoxypyrazine. The major classes of volatile compounds released by raw potatoes are: acids (e.g. fatty acids), aldehydes (e.g. hexanal, 2-hexenal, octanal, 2-nonenal, 2,4-decadienal, 2,4-heptadienal, 2,4-nonadienal), alcohols, amines, esters, ethers, furans, hydrocarbons, ketones, pyrazines, pyridines and thiazoles. It has also been observed that raw potatoes contain the following volatile compounds: acetaldehyde, propionaldehyde, buytraldehyde, methanol, ethanol, 1-propanol, 1-butanol, 2-propanol, pentanol, octanol, acetone, butanone, heptanone, nonanone, fatty acid esters, acetoin and others.

It is generally regarded that the bulk of flavour-producing compounds are volatile. However, Maga (1994), in an excellent review of potato flavour suggested that potato flavour might be influenced by non-volatile compounds such as amino acids and sugars. It was postulated that there might be precursors to volatile compounds, which is supported by studies described later. Kiryukhin and Gurov (1980) reported that on heating, lipids in the raw potato gave rise to pleasant flavours although it was noted that much of the lipid fraction was lost during the process: this loss was associated with the formation of flavour

Table 12.1 Typical composition of potato tuber (abridged from Lisinska, 1989)

Substance	Mean content (%)
Dry matter	23.7
Starch	17.5
Reducing sugars	0.3
Total sugars	0.5
Crude fibre	0.71
Crude protein	2.00
Lipids	0.12
Ash	1.1
Organic acids	0.6
Ascorbic acid and dehydroascorbic acid	10–25 mg/100g
Glycoalkaloids	3–10 mg/100 g

components. It was concluded that raw potatoes with a high fat content tended to give rise to products with preferential organoleptic characteristics.

The potato plant and its tubers are protected through the presence of glycoalkaloids; these compounds have antimicrobial and antipest properties. The major glycoalkaloids are: α-solanine, α-chaconine, β_2-chaconine, solandine and caffeine and, at elevated concentrations, can give rise to bitter off-flavours or burning sensations (Maga, 1994).

12.3 Degradation reactions occurring in edible oils and fats during frying

12.3.1 Oxidation – free radical route

The role of lipid oxidation in the production of flavour components is crucially important as discussed by Josephson and Lindsey (1987) and Fischer and Muller (1991) and is discussed below. When frying oils are heated to elevated temperatures (e.g. 180°C) in the presence of oxygen, they may undergo oxidation and form hydroperoxides via a free radical pathway. These compounds are unstable and at frying temperatures will decompose to form an array of secondary oxidation products, many of which are volatile giving rise to pleasant and sometimes offensive odours and flavours. The reaction pathway involves the production of free radicals $R^{\cdot}$ from lipid molecules RH following reaction with oxygen. The oxidation reaction is frequently initiated by a catalyst such as heat, light or metal ions, such as iron and copper.

Initiation

$$RH + O_2 \xrightarrow{\text{catalyst}} R\cdot + \cdot OOH$$

$$RH \xrightarrow{\text{catalyst}} R\cdot + \cdot H$$

Propagation

$$R\cdot + O_2 \longrightarrow RO_2\cdot$$

$$RO_2\cdot + RH \longrightarrow RO_2H + R\cdot$$

Termination

$$R\cdot + R\cdot \longrightarrow R{-}R$$

$$RO_2\cdot + R \text{ ——— } RO_2R$$

where RH = unsaturated lipid; $R\cdot$ = lipid radical; and $RO_2\cdot$ = lipid peroxy radical.

Fig. 12.1 Auto-oxidation of lipids.

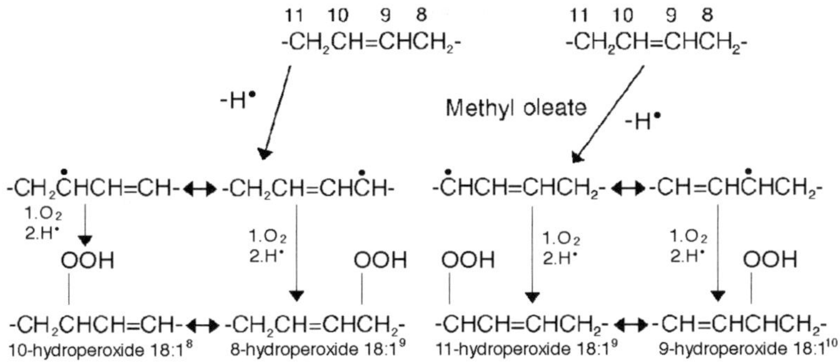

Fig. 12.2 The auto-oxidation of oleic acid (Hamilton, 1983).

The free radical R· produced in the initiation process can then react to form a lipid peroxy radical ROO· which reacts further to give a hydroperoxide ROOH. The second stage involves propagation steps which again involve free radicals (R·) and the process is, therefore, self-propagating. Consequently, once initiated by a catalyst (e.g. copper) the reaction described above will proceed rapidly leading to hydroperoxides which ultimately breakdown to give rise to rancid flavours and odours. The self-propagating reactions can be terminated when two radicals combine to give products which do not feed the propagating reactions (Fig. 12.1).

When this type of mechanism is applied to the auto-oxidation of oleic acid, hydrogen abstraction from the C-8 and C-11 results in two allylic radicals, each of which has two canonical forms. This explains why it is observed that not only the 8-hydroperoxide but also the 10–hydroperoxide from one allylic radical and the 9- and 11-hydroperoxides from the other allylic radical are formed. It is also the reason why it should not be expected to see hydrogen abstraction at any other position in the fatty acid chain. The double bond position is randomised to some extent because there are other hydroperoxides present in addition to the n-9 hydroperoxide and the configuration may be changed from the usual *cis*- to *trans*- (Fig. 12.2).

12.3.2 Lipid secondary oxidation products

Unsaturated lipids are prone to oxidation, particularly in the presence of pro-oxidant catalysts. The resulting hydroperoxides are unstable and decompose as shown in Fig. 12.3 to yield a series of secondary oxidation products. These are generally volatile and their presence at elevated temperatures can lead to the characteristic off-flavours and odours associated with rancid oils.

12.3.3 Hydrolytic rancidity

In addition to secondary oxidation products, methyl ketones, lactones and esters may be formed primarily by hydrolysis of frying oil. Hydrolysis involves the

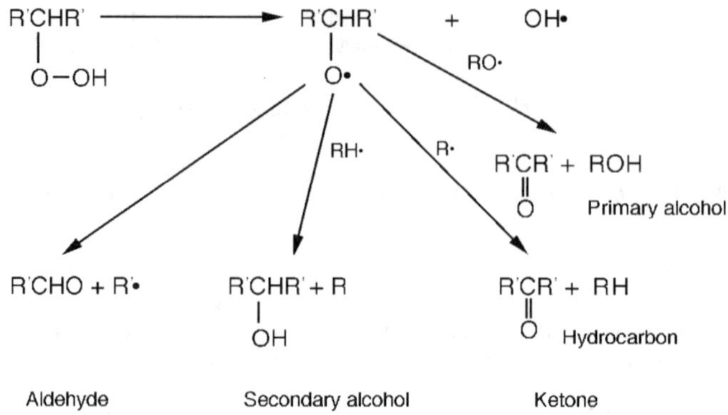

Fig. 12.3 The decomposition of lipid hydroperoxides.

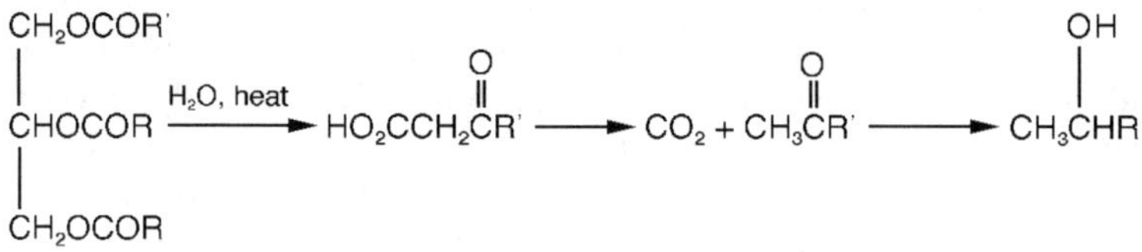

Fig. 12.4 Production and decomposition of keto acids following hydrolysis of lipids (Hamilton, 1983).

reaction of water (acting as a weak nucleophile) with the ester linkage in the triacylglycerol molecule to produce a diglyceride and free fatty acid. It is also possible for the triacylglycerol molecule, under the action of heat and moisture, to break down to keto acids, which lose carbon dioxide readily (Fig. 12.4). The release of hydroxy fatty acids can provide the precursors for γ- or δ-lactones

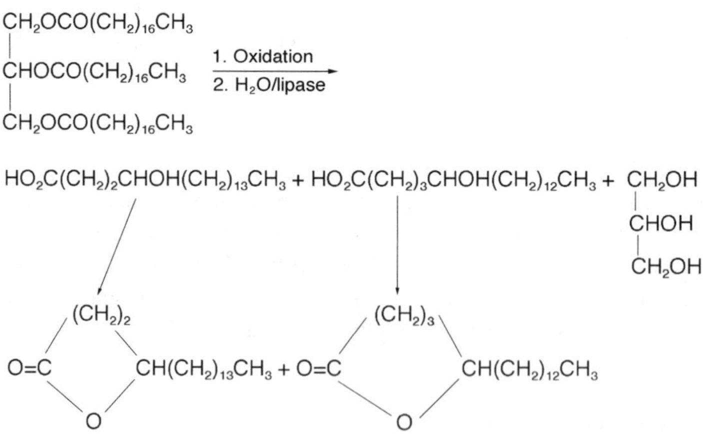

Fig. 12.5 The production of lactones (cyclic esters) following fat hydrolysis (Hamilton 1983).

No. of Carbon atoms	Homologous series		
	Saturated	2-Enals	2,4-Dienals
C_2	fresh, pungent		
C_3	fresh, milky		
C_4	↓	sweet, pungent	
C_5	↓	sweet, green	
C_6	fresh, green	↓	
C_7	↓	↓	sweet, oily
C_8	fresh, green	↓	↓
C_9	↓	sweet, fatty, green	↓
C_{10}	↓	↓	↓
C_{11}	fatty	sweet, fatty	↓
C_{12}	↓	↓	↓

(See also Appendix on page 335.)

Characters and threshold levels of selected aliphatic aldehydes

Alcohol/aldehyde	Threshold level (ppm) in paraffin oil	Character
1-Penten-3-ol	4.2	
1-Octen-3-ol	0.0075	
Hexanal	0.6	
Heptanal	0.04	
Octanal	0.6	
Nonanal	0.32	
Decanal	1.00	
cis-3-Hexenal	0.09	'Green bean'
trans-2-Hexenal	0.60	'Green'
cis-4-Heptenal	0.0005	Cream to tallow
trans-6-Nonenal	0.00035	'Hydrogenation'
trans-2-*cis*-6-Nonadienal	0.0015	Beany
trans-2-*trans*-Nonadienal	0.02	Cucumber
trans-2-*trans*-Decadienal	0.10	Stale frying oil

Fig. 12.6 Flavour characteristics of homologous series of aldehydes and various threshold levels of aldehydes and alcohols (expanded from Hamilton, 1983).

(Fig. 12.5). As with lipid secondary oxidation products, the compounds from fat hydrolysis (e.g. free fatty acids and lactones) give rise to flavours and odours.

In summary, oils exposed to elevated temperatures (ca. 160–190°C) in the presence of pro-oxidants and water can yield: volatile acids, aldehydes, esters, alcohols, epoxides, dimeric fatty acids, mono- and bicyclic compounds, aromatic compounds, glycerol, free fatty acids (FFA), and mono- and diglycerides.

Unsaturated fatty acids are associated with oxidation. However, at high temperatures, it has been found that saturated fatty acids react similarly producing alcohols, lactones, alkanes, acids, aldehydes and methyl ketones (Belitz and Grosch, 1999). Figure 12.6 shows some of the flavours and aromas resulting from the presence of aldehydes produced from lipid hydroperoxide decomposition.

In the Appendix a list of edible oil breakdown products is given together with the flavour notes generally associated with these components.

12.4 The Maillard and Strecker degradation reactions

The flavour and odour of foods is generally associated with volatile organic compounds. In the preceding text these have been discussed in terms of lipid oxidation. However, in foods that have been cooked in hot oil additional reactions can take place. It is these reactions that lead to the characteristic fried foods flavour. For instance, the Maillard reaction between reducing sugars, lipid oxidation products and amino acids can take place producing a variety of heterocyclic compounds which give rise to food-related flavours and odours. It also produces non-enzymic browning. The Strecker degradation, an important aspect of the Maillard reaction, may also occur and consequently both reactions are discussed below.

The Maillard reaction is complex and involves a number of steps (important functional groups are shown in parentheses):

1. a primary amino acid group of an amino acid or protein reacts with the carbonyl group (C=O) of an open chain reducing sugar;

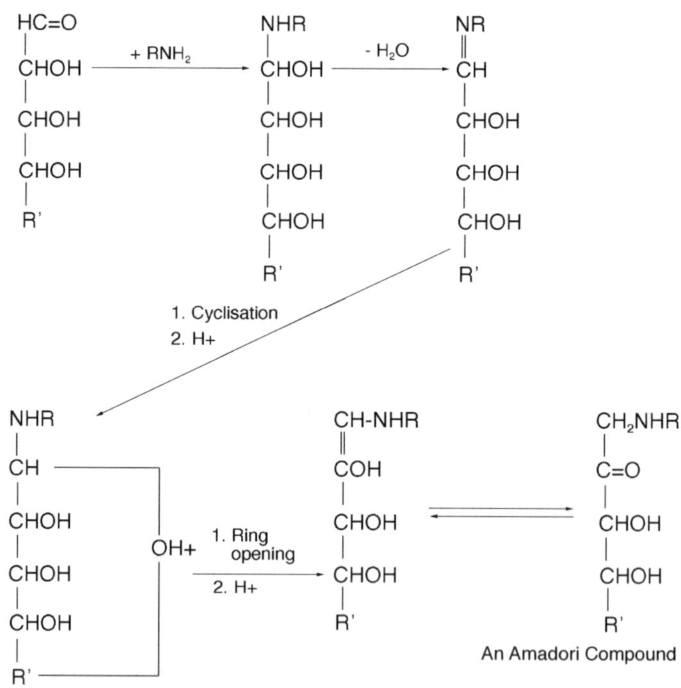

Fig. 12.7 An outline of the Maillard Reaction (abridged from Whitfield, 1992).

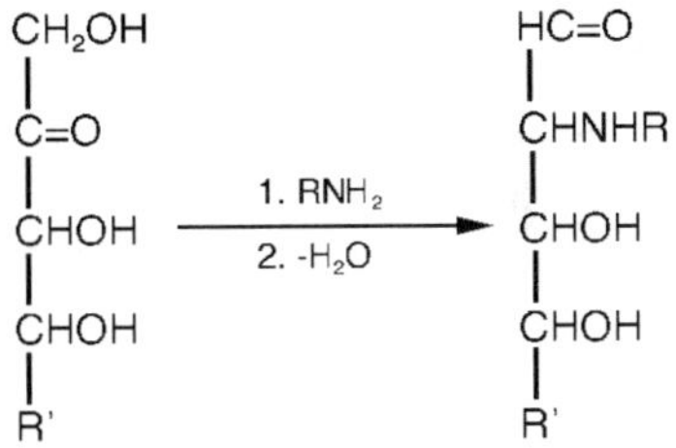

Fig. 12.8 Heynes product formation.

2. a Schiffs base is formed (-C=N-) following a condensation reaction to remove water;
3. the Schiffs base undergoes a cyclisation reaction to form a N-substituted aldosylamine;
4. which undergoes the Amadori rearrangement to give a 1-amino-1-deoxy-2-ketose (an Amadori product) (Fig. 12.7). If a ketose, rather than aldose, sugar is involved, a ketosylamine is formed which then undergoes a Heynes rearrangement to form a 2–amino-2–deoxyaldose (a Heynes product) (Fig. 12.8).

In an excellent review of the Maillard and associated reactions, Whitfield (1992) suggests that the products of this reaction do not contribute directly to taste but that they are important contributory precursors. As a result of their thermal instability they undergo dehydration, deamination and various rearrangement reactions (Figs 12.9 and 12.10).

It is possible that the Amadori and Heynes products will undergo decomposition via a series of retroaldolisation reactions which can lead to α-dicarbonyl compounds such as 2-oxopropanal, 2,3-butanedione, and 1,2-ethandial, and hydroxycarbonyl compounds such as 1-hydroxy-2-propanone, 2-hydroxyethanal, and 2,3-dihydroxypropanal.

As will be seen from the structure of the compounds above, they have functional groups capable of further reaction and the later stages of the Maillard reaction involve interaction of furfurals, furanones, and α-dicarbonyls, with other reactive products and degradation products from the Strecker reaction.

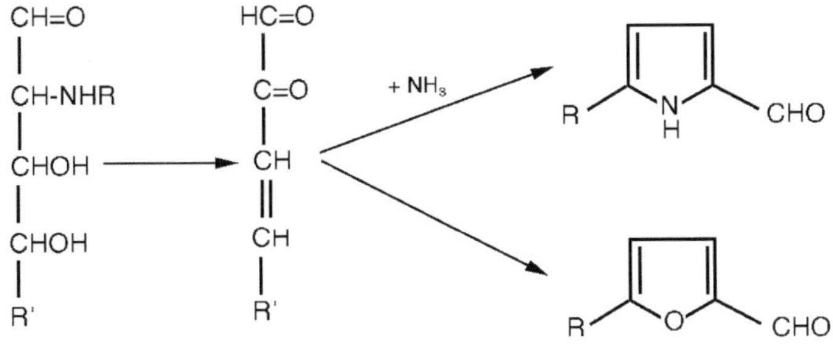

Fig. 12.9 Decomposition of Heynes product in the Maillard Reaction (abridged from Whitfield, 1992).

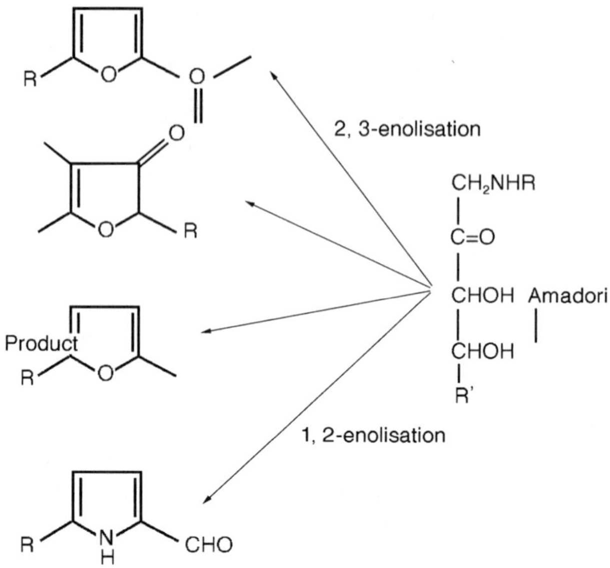

Fig. 12.10 Decomposition of Amadori Compound during the Maillard Reaction (adapted from Whitfield, 1992).

This involves the oxidative deamination and decarboxylation of an α-amino acid in the presence of a dicarbonyl compound, and tends to form an aldehyde and an α-aminoketone (Fig. 12.11).

The interaction of products from the Maillard and the Strecker reactions is very important as these lead to the formation of flavour and odorous heterocyclic compounds including: pyrazines, oxazoles, thiophenes, and heterocyclic compounds that contain more than one sulphur atom. It is also possible that lipid and lipid oxidation products will become involved in the above reactions with the formation of volatile flavour-inducing compounds. The following is a synopsis of reactions between:

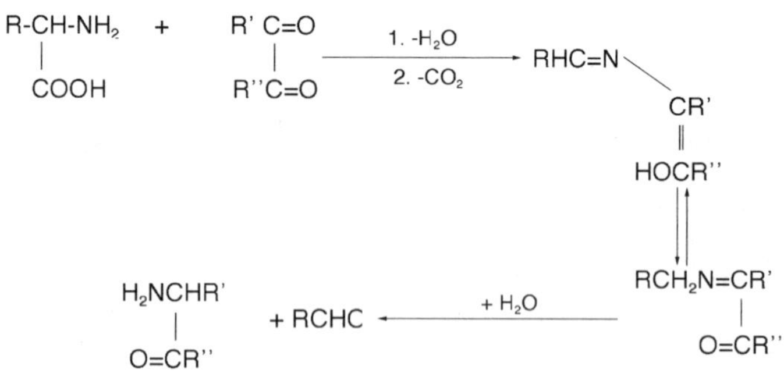

Fig. 12.11 Strecker Degradation as applied to α-amino acids (abridged from Whitfield, 1992).

1. amino acid degradation products and lipid secondary oxidation products;
2. amino acids and lipid secondary oxidation products;
3. fatty acids and amino acids;
4. sugar breakdown products and lipid oxidation products;
5. sugars and lipid secondary oxidation products (Whitfield, 1992).

1. *Amino acid degradation products and lipid secondary oxidation products.* Long-chain alkyl substituted pyrazines have been identified in a number of cooked foods such as fried potatoes (Carlin *et al.*, 1986). Using model systems, aldehydes (pentanal and hexanal – well known secondary lipid oxidation products) were reacted with the carbonyl compound 1-hydroxy-2-propanone (a potential Strecker degradation product), while the amino acid degradation group was supplied by ammonium acetate. It was found that the reaction of the two aldehydes with above mixture yielded twenty pyrazines including: dialkypyrazines; trialkylpyrazines; tetra-alkylpyrazines; and alkenyl dialykylpyrazines (Chiu *et al.*, 1990).

2. *Amino acids and lipid secondary oxidation products*, It was shown (Suyama and Adachi, 1980) that when a mixture of glycine, propanal, and 2-butenal was heated at 180ºC. the following were formed: 2,5-dimethylpyridine and 3,4-dimethylpyridine. Further studies using model systems were conducted to establish the identity of volatile compounds that developed during food frying. This involved heating mixtures of 2,4-decadienal (a well known lipid secondary oxidation product) with either cysteine or glutathione to 180ºC at pH 7.5. (Ho *et al.*, 1989; Zhang and Ho, 1989). Almost fifty compounds were identified from the reaction of cysteine and the dienal, the most abundant being 3,5-dihydro-2,4,6-trimethyl-4*H*-1,3,5-dithiazone. The analogous reaction involving glutathione led to the production of 42 compounds, the most abundant being 2–pentylpyridine.

3. *Fatty acids and amino acids.* When tributyrin was heated at 200–220ºC with cysteine the following were identified: 2-propylthiazoline, 2-propylthiazole, 2-methylthiazolidine, 2-methylthiazoline, N-ethylbutanamide, butanamide, 3,6-dimethyl-2,5-piperazinedione, and N,N-(2,2'-dithiobisethyl) dibutanamide (Severin and Ledl, 1972).

When linoleic acid (C18:2n-6c) was heated at 250ºC for 5 hours with valine, the most abundant components formed were: 2–pentylypyridine, N-(2–methylpropyl)octanamide, and N-(2–methylpropyl)linoleamide. In addition, alkylpyrroles and alkylpyridines were also seen. It was suggested (Henderson and Newar, 1981) the formation of the alkylpyrroles might be formed via two routes: (i) the reaction of ammonia with aldehydes through a Schiffs base; and (ii) the interaction of 1,3–butadiene with ammonia to produce pyrrole. It should, of course, be noted that the above was conducted under harsh conditions that would not be seen normally in frying operations.

Ohnishi and Shibamoto (1984) indicated that, unlike a sugar-amino acid mixture, a beef fat/glycine system did not produce a pleasant cooked flavour.

Earlier studies (Yamoto *et al.*, 1970) suggested that in order to obtain a pleasant meat-like flavour and smell, fat should be included in a sugar-amino acid mixture. Whitfield (1992) took this to mean that sugar-amino acid reactions produce compounds that then react with lipids or the oxidation products and in doing give rise to smells and flavours associated with cooked/fried meat.

4. *Sugar breakdown products and lipid oxidation products.* Thermal degradation of sugars has been shown to yield: alkyl- and alkenylfurans, 2-furfurals, 2-furylalkanones, 2-furylcarboxylic esters and furanones. Whitfield (1992) suggested that secondary lipid oxidation carbonyl groups would react with these to produce volatile compounds although he had not found evidence to this effect in the literature; neither has this author. However, it is conceivable that breakdown products from lipids and sugars could react, forming compounds that might react further with amino acid decomposition products producing increasing flavour and aroma materials, as is shown by the reaction of hydrogen sulphide with furfural and furanones, which lead to a variety of components with cooked food aromas.

5. *Sugars and lipid secondary oxidation products.* Lipid oxidation leads to the formation of secondary breakdown products (e.g. aldehydes and ketones), and such compounds react with sugars to form acetals and ketals. Although there is little information in the literature to support this claim, it is possible that the latter could interact with amino acid breakdown products (NH_3 and H_2S) to produce flavour compounds.

Chun and Ho (1997) studied the development of volatile nitrogen-containing compounds generated from the Maillard reaction under simulated frying conditions. This involved treating cotton balls with solutions of amino acid/glucose solutions, and 'frying' these in corn (maize) oil at 180–183°C for one minute in four instalments at intervals of 3 minutes. Four amino acids/glucose solutions were used: L-glutamine, L-glutamic acid, asparagine and aspartic acid. A total of twenty-nine volatile nitrogen-containing compounds were formed, with alkylpyrazines being the most important flavour compound generated. It was found that glutamine, which released free ammonia under the experimental conditions, gave the highest yield of alkylpyrazines. The profiles of pyrazines varied being dependent on the amino acid. The compounds of major concentration formed in the glutamine system were: 2-(2-furyl)pyrazine and 2-(2-furyl)-6-methylpyrazine.

Although many of the above reaction schemes have been developed in model systems, some of which have been subjected to conditions outside those normally associated with food production methodology, the above demonstrates that fried food flavours can be produced from the oxidative decomposition of cooking fats and oils, and the interaction of the oxidation breakdown products with Maillard and Strecker reaction products. The production and identification of aroma and flavour compounds in French-fried potatoes and potato chips (crisps) will be discussed in the following section.

12.5 Flavour development in foods

As has been described in the introduction, the flavour of fried food arises through a series of complex reactions, involving the interaction of amino acids, lipids, carbohydrates and their oxidation and breakdown products. This part of the chapter will deal with:

- the formation of flavour components in potato chips and French-fried potatoes and other snack-type products from the above reactions;
- the importance of fatty acids on flavour production in the frying process, in general.

12.5.1 Potato chip flavour development

Potato chips are usually known as crisps within the UK and the terms shall be used interchangeably in the text that follows. The production of potato chips is essentially a straightforward process involving the immersion of sliced potatoes, often blanched, in hot oil for a few minutes. The resultant product has found great favour with the consumer but its appeal is often taken for granted. The textural characteristics and colour of fried foods are discussed in Chapter 13. The production of flavour components and their identification is discussed briefly below; greater detail can be found elsewhere (Maga, 1994).

In the early to mid-1960s, volatile compounds formed during the potato chip cooking process were identified by Dornseifer and Powers (1963, 1965), and included common fatty acids, as well as aldehydes, ketones and mercaptans (-SH). The later work indicated that on storage under varying conditions, 9 different compounds were produced and the concentrations of these changed with time. For example, concentrations of ethanal decreased while 2-propenal increased. A summary of the compounds identified is shown in Table 12.2.

Table 12.2 Identification of volatile compounds formed during the production of potato chips (Dornseifer and Powers, 1963, 1965)

Compounds formed during frying: (* indicates observed during storage)
Acids acetic acid; propionic acid; isobutyric acid; butyric acid; valeric acid; caproic acid; enanthic acid (heptanoic acid); caprylic acid; capric acid; lauric acid.
Aldehydes Acetaldehyde*; 2-Propenal*; Propanal; Butanal*; 2-Butenal; Pentanal; 2-Pentenal*; Hexanal; 2-Hexenal*; Heptanal*; 2-Heptenal*
Ketones acetone*; 2-butanone; 2,3-butanedione*; 2-pentanone*; 2-hexanone*; 2-heptanone; 2-octanone; 2-nonanone
Sulphur compounds methanethiol; ethanethiol; 1-propanethiol; hydrogen sulphide

Mookherjee *et al.* (1965) studied the formation of carbonyl compounds in fresh and stored stale potato chips. It was interesting to note that in fresh and stale chips the same 18 compounds were observed, the difference being that in stale chips 2-hexenal was also seen. It was found that during storage, the concentration of lipid secondary oxidation products increased (e.g. hexanal, 2-pentanone, 2-octenal); it was noted, however, that the concentration of 2,4-decadienal decreased. There is no doubt that the formation of compounds from the degradation of lipids is of major importance in flavour formation during frying. However, the identification of the formation of another group of compounds, the pyrazines, was a significant step forward in the understanding of fried food flavour.

Deck and Chang (1965) identified 2,5-dimethylpyrazine in potato chips and also indicated that its flavour could be detected when present at 2 mg/kg in oil and 1 mg/kg in water. This is probably a result of pyrazines' differing solubility in oil and water. It was found that the majority of Deck and Chang's taste panel members described the flavour as earthy or raw potato when 2,5–dimethylpyrazine was present at 10 mg/kg in oil. This work was built on by Buttery *et al.* (1971) in which a series of pyrazines and pyridines were identified. Sensory evaluation indicated that a major component of this fraction, 2-ethyl-3,6-dimethylpyrazine, was likely to be a significant contributor to the potato chip aroma (Koehler *et al.*, 1971).

Buttery and Ling (1972) identified 46 compounds in the nonbasic steam volatile fraction of potato chips. It was found that methional, 3-methylbutanal, phenylacetaldehyde, and 2,4-decadienal were important in determining flavour profiles. It was suggested that the compounds formed were a result of lipid-carbohydrate interactions involving thermal degradation and various rearrangement reactions. Further work in this area by Buttery established the structure of a number of aldehydes that were formed by aldol condensation of aldehydes (see Table 12.3).

Guadagni *et al.* (1972) examined the steam volatiles from oil used to fry potato chips and determined the odour thresholds of these in both water and oil. The authors used this information in combination with the relative quantities found in potato chips to develop a relative odour intensity index for each compound. They concluded that the most important compound was methional, followed by 2,4-decadienal; 2-octenal; 2-ethyl-3,6-dimethylpyrazine; 2,6-diethylpyrazine; and 2-phenylacetaldehyde. Deck *et al.* (1973) identified more than 50 compounds from fried potato chips and indicated that the most important flavour components were the pyrazines. This was because the panel that evaluated the profile of these compounds described the odour as 'strong potato, baked potato, and earthy potato' (Maga, 1994).

In a comprehensive piece of research, Maga and Sizer (1978) investigated the formation of pyrazines using sensory and analytical techniques, and related this to the frying time and temperature. Two experiments were conducted, one in which a 30-strong taste panel evaluated the flavour profile of potato chips that had been prepared by frying at 150°C for a known time between 3–7 minutes. In

Table 12.3 Synopsis of compounds identified as contributing to potato chip flavour by Buttery and colleagues (abridged from Buttery and Ling, 1972; Buttery *et al.*, 1971; Buttery, 1973; Maga, 1994 and Koehler *et al.*, 1971)

Compound class	Identification
Alcohols:	2-butanol; 3-methyl-1-butanol; 1-pentanol; 2-furfuryl alcohol; α-terpineol.
Aldehydes:	2-methylpropanal; 2-methylbutanal; 3-methylbutanal; 2-isopropyl-2-butenal; 4–methyl-2-pentenal; *trans*-2-hexenal; 4–methyl-2-hexenal; *trans*-2, *trans*-4–octadienal; *trans*-2-nonenal; *trans*-2, *trans*-4–nonadienal; *trans*-2, *cis*-4–nonadienal; *trans*-2, *cis*-4–decadienal; *trans*-2, *trans*-4–decadienal; benzylaldehyde; 2-phenyl-2-butenal; 4–methyl-2-phenyl-2-pentenal; 5-methyl-2-phenylhexanal
Furans:	2-butylfuran; 2-pentylfuran; 2-hexylfuran; furfural; 5-methylfurfural; 2-methyldihydro-3(2*H*)-furanone; 2-acetylfuran; furfuryl alcohol;
Hydrocarbons:	1-decyne
Ketones:	2-butanone; 2,3-butanedione; *trans*-3-penten-2-one; 2,3-pentanedione; 5-methyl-2,3-hexanedione; *trans*-2-nonene-4–one; 2-decanone; acetophenone.
Pyrazines:	2-methylpyrazine; 2,3-dimethylpyrazine; 2,5-dimethylpyrazine; 2,6-dimethylpyrazine; 2-ethylpyrazine; 2-ethyl-3-methylpyrazine; 2-ethyl-5-methylpyrazine; 2-ethyl-6-methylpyrazine; 2,3,5-trimethylpyrazine; 2-methyl-5-vinylpyrazine; 2-ethyl-3,6-dimethylpyrazine; 2-ethyl-3,5-dimethylpyrazine; 2,6-diethylpyrazine; 2-isobutyl-3-methylpyrazine; 2,3-diethyl-5-methylpyrazine; 2-isobutyl-3,6-dimethylpyrazine; 2-methyl-6-vinylpyrazine; 2,5-dimethyl-3-vinylpyrazine; 2,5-dimethyl-6-isopropylpyrazine; 2-isoamyl-5-methylpyrazine; methylethylisobutylpyrazine; 2-isobutenyl-3-methylpyrazine; 2-isoamyl-3,6-dimethylpyrazine; isobutenyldimethylpyrazine.
Pyridines and pyrroles:	pyridine; 2-acetylpyridine; 2-acetylpyrrole.
Sulphur compounds:	(methylthio)acetaldehyde; methional.

a second experiment, the frying temperature was varied (T = 120, 135, 150, 160 and 180°C) for five minutes. It was found that the total pyrazine content increased from 1.1 mg/kg after frying for 3 minutes, to 15 mg/kg after 7 minutes frying. In terms of temperature, as this increased from 120°C to 180°C the total pyrazine concentration increased from 0.64 mg/kg to 24 mg/kg. The pyrazine data was highly correlated with sensory panel results.

The role of nonenzymatic browning arising from the Maillard reaction has been considered in terms of flavour production. In model systems potato slices were presoaked in various combinations and at varying concentrations of sugars and amino acids. The slices were fried at 210°C for 7 minutes (which are conditions at the extremes for frying potatoes slices) and then analysed for colour. It was found that the addition of L-aspartic acid or L-glutamic acid decreased the extent of browning in lysine-glucose and lysine-fructose reaction mixtures. (Nafisi and Markakis, 1983). Along similar lines, Talley and Eppley (1985) applied various amino acid-sugar solutions to filter paper to mimic potato slices. These were heated in fat for varying lengths of time at 103°C. The results indicated that fructose reacted more quickly than glucose; glycine reacted more quickly than any other amino acid. It was also noted that, when added in combination, amino acids could act in synergy.

It is well known that potatoes that have high concentrations of reducing sugars (e.g. glucose and fructose) are more likely to produce dark-coloured chips in comparison with those that have low levels. However, it may be argued that although the colour of the product may be undesirable, it is conceivable that the flavour is greater and/or preferable in comparison with that derived from a lighter coloured product. Maga (1973) studied this phenomenon by using a blindfolded sensory panel to monitor the flavour and odour profile of dark chips in comparison with conventional product whilst being stored at room temperature over four weeks. It was established that as the storage time increased, the sensory panel had a preference for both the odour and the flavour of the dark chips over the lighter product; commenting that the darker chips had more flavour or a more characteristic potato chip flavour.

12.5.2 French-fried potato flavour

There is a significant amount of data on the compounds responsible for the flavour formation in French-fried potatoes (also known as fries and, in the UK, as chips), although this appears to have been produced by relatively few authors. It has been shown that the components in the frying medium can be transferred to the food in the cooking process. For example, French fries cooked in beef tallow took on a beef-like taste which, it was suggested, came from the movement of butanoic (butyric), 2-methylbutanoic, 3-methylbutanoic, heptanoic, 4-methylheptanoic, and nonanoic acids from the tallow to the food (Ha and Lindsay, 1991). All these acids are volatile and will impart flavour characteristics to the food.

Extensive work conducted to identify the volatile compounds in French fries has been conducted by Carlin *et al.* (1986). The experiments involved heating the potato slices in fat at 150–190°C for 1.5–3.5 minutes. A summary of their findings is shown in Table 12.4. In excess of four hundred compounds have been identified in French fried potatoes and may be grouped as 24 acids; 37 alcohols; 30 aldehydes; 10 esters; 20 furans; 105 hydrocarbons; 36 ketones; 5 lactones; 25 oxazoles; 33 pyrazines; 13 pyridines; 1 pyrrole; 50 thiazoles; 19 thiophenes; 16 miscellaneous sulphur compounds; 16 miscellaneous nitrogen compounds.

It was noted by Whitfield (1992) that 59 of these compounds might have been formed by Maillard/lipid reactions and include pyrazines, pyridines, thiophenes, thiazoles, benzothiazoles, and oxazoles. It was noted by Whitfield, and can be

Table 12.4 A synopsis of volatile compounds formed in French fried potatoes (Carlin, 1986; Ha and Lindsay, 1991; Maga, 1994)

Compound class	Identification
Acids:	Acetic, hexanoic, 2-methylbutanoic, 3-methylbutanoic, nonanoic, 2-methylpropanoic, octanoic, pentanoic, benzoic, 4-methylbenzoic, heptanoic, *cis*-4-heptenoic, 6-methylheptanoic, 4-methyloctanoic, octenoic, 7-methyloctanoic, 8-methylnonanoic, decanoic, *trans*-2-nonenoic.
Aldehydes:	3-methylbutanal, 2-methylpropanal, hexanal, heptanal, octanal, nonanal, pentanal, 2,4-decadienal, decanal, benzaldehyde, 3-hydroxybenzaldehyde, 2-phenyl-2-butenal, phenylacetaldehyde, 4-methyl-2-phenyl-2-pentenal.
Alcohols:	homologous series of 1-butanol to 1-decanol, 2-hexanol, 3-heptanol, 4-octanol, *cis*-2-penten-1-ol, 2-methyl-1-propanol, 3-methylphenol, 2-methyl-5-isopropylphenol.
Ketones:	2,3-butanedione, 2-hexanone, 3-octanone, cyclopentanone, 4-methyl-3-pentene-2-one, acetophenone, 1-phenyl-1-pentanone, phenylacetone, 2-tridecanone, 2-ethyl-2-cyclopenten-1-one, 4-hydroxy-5-methyl-1,2-cyclopentanedione.
Ester:	ethyl acetate, phenyl acetate, di-isobutyl adipate, 2,2,4-trimethyl penta-1,3-diol di-isobutyrate, diethyl phthalate, 3-methylbenzoate.
Furans:	2-pentylfuran, furfural, dihydro-2-methyl-3(2*H*)-furanone, 2-propionylfuran, 2-allylfuran, 2,6-dimethylbenzofuran.
Hydrocarbons:	hexadecane, 2,6,10,14-tetramethylhexadecane, 2-methylpropylcyclohexane, 2-methyl-4-nonene, 4-propyl-3-heptene, 1,3-dimethyl-*trans*-cyclopentene, 4-decyne, 1,2,4-trimethylbenzene, 5-phenylundecane, 2-methyl-3-phenyl-1-propene.
Lactones:	4-hydroxyundecanoic acid lactone; *cis*-4-hydroxy-3-methyldecanoic acid lactone; 4-hydroxyoctanoic acid lactone.
Nitrogen-containing compounds:	2-formyl-3,4-dihydro-2*H*-pyran; cyclohexylamine; indole; 2-butylpyrrolidine; benzonitrile; hexanamide; 5-pentylpyran-2-one.
Oxazoles:	trimethyloxazole; 4-methyl-5-ethyl-2-isopropyloxazole; 4,5-dimethyl-2-isobutyloxazole; 5-methyl-4-propyl-2-butyloxazole.
Pyrazines:	2,6-dimethylpyrazine; pyrazine; 2-methyl-5-vinylpyrazine; 3,5-diethyl-2-methylpyrazine; 2-butyl-3,5,6-trimethylpyrazine; 2-methyl-6,7-dihydro-5*H*-cyclopentapyrazine.
Sulphur-containing compounds:	dimethyl sulphide; 2-thiaheptane; 4,5-dimethyl-2-butyl-2-thiazoline; 3-(methylthio)heptanal.
Thiazoles:	4,5-dimethyl-2-butylthiazole; 2-heptylbenzothiazole; 2-nonylbenzothiazole; 4-butylthiazole; 2-isopropylthiazole; 2-octylbenzothiazole.
Thiophenes:	2-heptylthiophene; 2-acetylthiophene; 2-phenylthiophene.

seen from the summary data in Table 12.4, that many of these could contain long-chain alkyl substituents from C4 to C7. In fact, some of the thiazoles contained longer alkyl groups. While it is possible that many of the compounds might be produced by a variety of mechanisms, for the Maillard reaction to yield thiazoles and oxazoles it might require the involvement of aldehydes (Vernin and Parkanyi, 1982). It has been suggested aldehydes derived from lipid oxidation might produce the thiazoles and oxazoles found in potato chips. (Whitfield, 1992). This might relate to the poor flavour formed when potatoes are fried in fresh fat in which aldehydes have not yet formed. Of the compounds identified, it has been suggested that 4,5-dimethyl-2-hexyl-thiazole; 5-methyl-4-ethyl-2-heptylthiazole; 4-methyl-5-ethyl-2-octylthiazole; and 4-methyl-5-ethyl-2-pentylthiazole are important contributors to French-fried potato flavour. More recent work by Carlin *et al.* (1990) showed that 3-(methylthio) butanol and 3-(methylthio) heptanol were produced in French fries and gave rise to characteristic aroma.

The above is supported by the results reported by Wagner and Grosch (1997). They heated potatoes (cv. Agria) bought from retail sale with a composition of starch (14.4%), glucose (0.26%), and fructose (0.21%). Palm oil was used as the frying oil, which contained the following unsaturated fatty acids: oleic acid (38.6%), linoleic acid (10.5%), and linolenic acid (0.1%). From the fresh French fries produced, 48 aroma compounds were identified. These include: methional; 2-ethyl-3,5-dimethylpyrazine; 2,3-diethyl-5-methylpyrazine; *t,t,*-2,4-decadienal; 4-hydroxy-2,5-dimethyl-3(2H)-furanone; methanethiol; dimethyl trisulphide; 3-methylbutanal; and 2,3-butanedione. The authors reported that a deep-fried note predominated when the fried products were nasally evaluated and that this was caused by *t,t*-2,4-decadienal, whereas deep-fried and boiled potato smells tended to prevail when analysed by a retronasal test; these were due to the presence of methional.

With regard to the production of compounds that give rise to fried food aromas and flavours, the Maillard reaction involving interactions with lipids and their oxidative breakdown products appear to play an important part. This is supported by studies with model systems (Whitfield, 1992). A significant proportion of compounds associated with fried food flavour are heterocyclic (i.e. cyclic systems containing one or more nitrogen, oxygen or sulphur atoms) and are substituted with alkyl groups containing 4 to 9 carbon atoms.

It has been noted, again in model systems, that there is a reduction in the production of Maillard reaction products, particularly if phospholipids are present in high concentrations. It was suggested that the reduction in the concentration of the alkylpyrazines, alkylpyridines, and alkylthiazoles in the reactions containing egg lecithin (i.e. phospholipid) might be due to preferential reaction of free ammonia or the amino acid nitrogen with compounds produced from lipid oxidation, a supposition that is supported by the knowledge these compounds are known to react with ammonia and amino acids to yield pigments. However, while it is known that pH can affect the Maillard and subsequent reactions, little work has been conducted in this area.

Not all research on model systems has, however, led to the identification of pyrazines, pyridines and thiazoles as reaction products. Mandin *et al.* (1999), for example, heated aqueous mixtures of methionine, glucose, linoleic acid and starch at 100–105°C for 2 hours in a modified Nickerson-Likens apparatus. No pyrazines, pyridines or thiazoles were observed. It was found that the majority of volatile compounds resulted from the breakdown of linoleic acid and included: hexanal, 2,4-decadienal, and 2-pentylfuran. All systems that contained methionine (a sulphur-containing amino acid) were found to contain dimethyl disulphide and dimethyl trisulphide. Furthermore, when methionine was heated with glucose and linoleic acid, it was observed that 3-(methylthio)propanal (methional); 2–methyl-5-(methylthio)furan, 2-hexylthiophene, and methyl methane-thiosulphonate ($CH_3SO_2SCH_3$) were present. When starch was added to the above mixture of substrates and heated, 2,4,5-trithiahexane and dimethyl tetrasulphide were produced. Volatile compounds did not appear to bind to the starch. In contrast, it seemed that the starch acted as a source of reactive carbohydrate. It should be noted that although pyrazine and related heterocyclic compounds were not present, the mixtures were heated to a comparatively low temperature and in aqueous media.

May *et al.* (1978) synthesised a series of lactones. When γ-lactones unsaturated at the 2 or 3 positions were added to cottonseed oil at 2.5 mg/kg (ppm), sensory panel evaluation indicated that they gave rise to a deep-fried flavour. When evaluating fried food flavour formation, it is important to consider the frying conditions. For example, further studies by May *et al.* (1983) involved heating triolein at 185°C. with periodic injections of steam for 75 hours. The compounds identified following exposure to these conditions included saturated and unsaturated acids, aldehydes, hydrocarbons; dibasic acids, ketones, esters, keto acids, alcohols and lactones. The authors noted, however, that there was a lack of dienals, particularly 2,4-decadienal, which it was suggested might explain the lack of a strong, deep-fat fried aroma under the experimental conditions.

It may be concluded that the complete composition and understanding of fried potato flavour has not been fully established. This is largely a result of the differing composition of the main ingredients used in the production of fried potato products, and variety of processing conditions, all of which will affect the production of flavour compounds. This was compounded by the observation that the volatiles produced are often formed in quantities at the limits of analytical detection. Nonetheless, it appears that desirable fried potato flavour arises from the production of alkyl pyrazines, alkyl thiophenes and alkyl thiazoles. Many of these compounds and other aroma inducing compounds are likely to be produced through the reaction of amino acids, sugars and lipids and their breakdown products (oxidative and thermal). However, it is also necessary to consider off-flavour production which may arise through lipid oxidation. In addition, if present at high concentrations, naturally occurring glycoalkaloids tend to produce a bitter taste to potatoes; this can mean that potato waste disposal can result in objectionable odours.

12.5.3 Flavour components in fried and roasted products other than potatoes

Pfnuer *et al.* (1999) established that 5-methyl-(E)-2-hepten-4-one (filbertone) was a flavour component of hazelnut and its oil. It was, however, present in much greater quantities in oil from roasted than unroasted hazelnuts (approx. 316 and 6 μg/kg, respectively). It was observed that the concentration of the flavourant could be increased by factors of about 600–800 times by pan frying raw hazelnuts for 9–15 minutes or boiling the crushed nut material for one hour in water. The precursor to the flavour is unknown. Lee *et al.* (1981b) and Ho *et al.* (1983) have identified the flavour and aroma compounds from roasted peanuts, as shown in Table 12.5.

Shimoda *et al.* (1997) investigated the effect of the extent of roasting on the formation of volatile compounds formed in sesame seeds; in each case the oil was analysed. The recoveries of volatiles from deep-roasted and light-roasted oils were approximately 9.7 and 2.0 mg/kg, respectively. The flavour compounds identified included: mono-, di- and trialkylpyrazines; mono- and dialkylthiazoles; a variety of pyrroles, furans and pyridines; aliphatic aldehydes, alcohols, ketones and acids; aromatic compounds.

Lee *et al.* (1981a) and Hartman *et al.* (1983) identified methyl-3,4-dimethyl-5,6-dihydro-α-pyran-6-carboxylate and a total of 44 nitrogen-containing heterocyclic compounds associated with the flavour of roasted beef. These included: 15 thiazoles, 1 thiazoline, 6 oxazoles, 11 pyrazines, 6 pyrroles, 2 piperidines, and 3 pyridines. Work conducted by Cerny and Grosch (1993) involved roasting beef for 7 minutes in either glass or stainless steel pans, which gave rise to roasty-sweet and a roasty-harsh flavours, respectively, the latter

Table 12.5 Synopsis of flavour components in fried and roasted peanuts (abridged from Lee *et al.*, 1981b and Ho *et al.*, 1983).

Compound type	Identification
Oxazoles:	4,5-dimethyloxazole; 2-acetyloxazole; 2,4,5-trimethyloxazole; 5-butyloxazole; 2-pentyloxazole; 2-ethyl-5-butyloxazole; 2,4-diethyl-5-propyloxazole; 2–methyl-4-butyloxazole; 2,4-dimethyl-5-propyloxazole; 2-methyl-4-butyloxazole; 2-isopropyl-4,5-dimethyloxazole; 2,4-diethyl-5-propyloxazole.
Thiazoles:	2-methylthiazole; 4-methylthiazole; 5-methylthiazole; 5-butylthiazole; 2-isopropyl-4,5-dimethylthiazole; 4-butyl-2,5-dimethylthiazole; 2-propyl-4,5-diethylthiazole; 2,4-dimethyl-5-ethylthiazole; 2-methyl-4-ethyl-5-propylthiazole; 2-isopropyl-4-ethyl-5-methylthiazole; 2-isopropyl-4-propylthiazole; 2,4-diethyl-5-propylthiazole; 2,5-diethyl-4-propylthiazole; 2,5-dipropyl-4-methylthiazole; 2-butyl-4-methyl-5-ethylthiazole; 4-methyl-5-(2-hydroxyethyl)thiazole.
Oxazolines:	2-methyl-3-oxazoline; 2,4-dimethyl-3-oxazoline; 2,4,5-trimethyl-3-oxazoline.
Others:	2-(2-aminoethyl)piperidine.

reaching a higher temperature than the glass pan. The following flavour compounds were identified: 4-hydroxy-2,5-dimethyl-3(2H)-furanone (A), 2-acetyl-2-thiazoline (B), 2-ethyl-3,5-dimethylpyrazine (C), 2,3-diethyl-5-methyl-pyrazine (D), guaiacol (E), and methional (F). The differences in flavour between the two frying implements were attributed to variations in the concentrations of compounds A, B, D, and F.

Flavour volatiles from fried chicken were studied by Tang *et al.*(1983), who found a total of 130 compounds which included acids, alcohols, aldehydes, esters, furans, hydrocarbons, ketones, oxazoles, oxazolines, pyrazines, pyridines, pyrroles, thialdine, thiazoles, thiazolines, thiophene, trithiolane and trithiane.

12.6 The importance of fatty acid composition on the flavour production in the frying process

Fried foods, particularly those high in fat such as potato chips, are inclined to undergo oxidation and produce components that can affect the flavour of the product. This issue is considered in the following text.

It was found, for example, that a taste panel preferred potato chips that had been fried in sunflower oil in comparison with those that had been produced in a blend of cottonseed oil (70%) and corn (maize) (30%) oil (Evans and Shaw, 1968). However, Robertson *et al.* (1972) found that when a taste panel evaluated potato chips that had been fried in partially hydrogenated sunflower seed oil and a blend of cottonseed and maize oils there was no statistical difference in the flavour characteristics of the food. Later studies by Robertson *et al.* (1978) examined the flavour of potato chips fried in sunflower, cottonseed, and palm oils and then stored for 10 weeks at 31°C. The sensory panel could not correlate potato chip flavour with frying oil type. It was concluded that the oils were compatible for producing potato chips without affecting product flavour. However, many chip producers claim that the frying oil is very important as this can inpact flavour.

Oil in chips can oxidise and deteriorate giving rise to off-flavours and odours characteristic of rancid products. A frequently proposed technique of monitoring potato chip deterioration involves observing the production of hydrocarbons from lipid breakdown (Horvat *et al.*, 1964; Evans, *et al.*, 1969; Selke, *et al.* 1970; and Jarvi, *et al.*, 1971). Warner *et al.* (1974) measured the formation of pentane in potato chips produced from maize oil, and found that the observations of a sensory panel mimicked those of analytical measurements. For example, the entire sensory panel of 18 members indicated that the product was rancid when the pentane concentrations reached 0.08 ppm. In a similar vein, Jeon and Bassette (1984) used headspace analysis to measure the production of pentanal and hexanal in potato chips that had been: (i) exposed to fluorescent light; (ii) stored at 45°C or 120°C; or (iii) stored in the dark at 25°C. The hexanal concentrations fluctuated initially but tended to increase as storage time increased, while the pentanal concentrations in poor quality chips were approximately four times those in control chips. Similar studies by Chan *et al.* (1978) indicated that illuminating potato chips followed by storage in the dark resulted in products that had sensory and oxidative properties that were different to those of potato chips stored only in the light. This may have practical implications to the snack food industry.

For several decades now it has been well known that soyabean and rapeseed oils do not perform as effectively as other frying oils. This is generally attributed to their comparatively high content of linolenic acid (C18:3n-3c) which undergoes oxidation more rapidly than most other commonly occurring unsaturated fatty acids (e.g. oleic and linoleic). Foods fried in these oils tend to take on a fishy taint, and when stored for lengthy periods these foods exhibit the characteristics of lipid oxidation producing rancid flavours. It has been shown that hydrogenating oil can increase its frying performance, a direct result of the oil having a lower degree of unsaturation. However, hydrogenation can impart an odour and flavour that is not overly desired and, furthermore, the process introduces *trans* double bonds (*trans* fatty acids) which have received less than favourable press in recent years.

By way of example, Przybylski and Hougen (1989) measured the formation of the volatile carbonyl content (related to lipid oxidation) in freshly made potato chips that had been fried in either rapeseed or cottonseed oils (linolenic acid approximately 9.0 and 0.2%, respectively), and stored under both light and dark conditions. Before frying started both oils had low concentrations of carbonyl compounds. However, chips made from rapeseed oil whether stored or fresh had higher levels than cottonseed oil. No sensory comparisons were reported and, consequently, it is not possible to link carbonyl value measurement with rancidity evaluation. Nonetheless, the above tends to support the contention that frying with oils high in linolenic acid will increase the risk of producing products with a greater propensity to rancid flavour formation.

Warner and Mounts (1993) studied the efficacy of frying Idaho russet potato slices with a series of soyabean and canola (rapeseed) oils. The linolenic acid (C18:3n-3c) content of some oils had been altered through breeding and/or hydrogenation. The soyabean oils had linolenic acid concentrations of 6.2% (standard soyabean oil), 3.7% (low-linolenic acid soyabean oil) and 0.4% for hydrogenated soyabean oil. The rapeseed oils had linolenic acid contents of 10.1% (standard canola oil), 1.7% (canola modified by breeding), and 0.8 and 0.6% for oils modified by breeding and hydrogenation.

It was found by sensory testing that the modified oils (low C18:3n-3c) had lower odour intensity after initial heating tests at 190°C than the standard oils. The sensory panels also indicated that the standard oils had higher intensities of fishy, burnt, rubbery, smokey and acrid characteristics than the low-linolenic acid oils. Furthermore, the low-linolenic acid oils had lower FFA, total polar and foam height than conventional oils after 5 hours frying; again indicating that the former were performing more ably. It was also noted that chips fried in the modified, low-linolenic acid oils, had better flavour quality than conventional grades.

Warner *et al.* (1994) produced potato chips from six canola (low-erucic rapeseed oils) oils. The compositions of the oils are described briefly in Table 12.6. The chips were fried for 130 seconds at between 192°C (inlet) and 187°C

Table 12.6 Fatty acid compositions of canola oils (abridged from Warner *et al.*, 1994)

Fatty acid	Canola	A	B	C	D (blend of A&C)	Hardened canola
C16:0	4.2	4.0	4.0	3.5	3.8	4.8
C18:0	2.0	2.3	2.4	2.3	2.3	5.5
C18:1-*cis*	61.9	64.2	66.8	78.3	68.4	57.4
C18:2-*cis*	20.6	23.6	21.3	8.5	19.7	11.3
C18:3	7.7	2.8	2.9	4.2	3.1	0.8
Iodine Value (IV)	108.8	103.1	101.6	93.0	100.8	86.0

Note: The hardened (hydrogenated canola) contained approximately 17.1% *trans*-fatty acids.

(outlet). At the completion of nine hours frying on day 1, the oil was cooled and pumped to a holding tank overnight. The oil was pumped back into the fryer on day 2, and the frying process repeated. Fresh oil was added periodically. Complete turnover of oil was attained on day 2 after 12–15 hours frying.

Oil samples were collected from the fryers at 3, 6 and 9 hours on day 1, and 12, 15 and 18 hours on day 2. Potato chips were taken for analysis eight times during the frying process. With regard to the sensory characteristics of chips, it was found that those produced from oil exposed to 3 or 6 hours frying, had better flavour quality than chips produced from fresh oil. It was also found that chips cooked at the beginning of day 2, had lower flavour scores than those produced at the end of day 1. It was observed that total volatile component concentration in the frying oil and the fresh and aged potato chips increased overnight between days 1 and 2. This resulted in the sensory evaluation of the chips being lower at the beginning of day 2 in comparison with chips produced at the end of day 1. It was suggested that the process of cooling and holding oil overnight might require alteration to reduce the extent of lipid oxidation. This observation might be explained as follows. The steam produced when food is being cooked tends to steam – distil off – flavours from the oil. However, at the end of the shift, as the frying medium cools, no steam is generated but the oil still deteriorates. The off-flavours created are not steam distilled from the oil and are, therefore, present at the start of the next day. Furthermore, during frying, oil is absorbed by the food, the loss being made good by topping up with fresh oil. This is not the case as the oil cools or warms, exacerbating the problem. After start-up the next day, steam distillation and top-up oil restore the oil to its previous condition but this can take 30 minutes or so. The first few batches of food after start-up might, therefore, be worse than those at close down the previous day, and worse than those later on in the same day.

It was found that the conventional oil produced chips had a lower flavour score than those of the modified oils from the end of frying on day 1 to the completion of the trial on day 2. The highest flavour quality ratings were given to chips produced from the blended oil D, oil B and the hardened canola oil. Oil

C experienced the lowest fluctuations in flavour, had the lowest total polar components but did not have the best flavour quality scores. Following storage of chips at 25ºC for four months, the quality of the aged product mimicked that for fresh chips. Again those fried in the conventional oil were of lowest quality, whereas those produced from oils C and D were best quality.

When all the overall quality scores were pooled for all oil samples, it was shown that the chips improved in quality on going from fresh oil to oil 3 hours old and then a decrease in quality at the start of day 2. Importantly, it was seen that the characteristic potato chip flavour had the greatest impact on product quality. Other indicators such as stale, rancid, waxy, fishy, and hydrogenated had less effect on perceived chip quality.

As might be expected, the initial fatty acid composition of the oil had an influence on the flavour quality of the chips. It was noted that, with one exception, as the concentration of linolenic acid increased the quality score decreased in a linear manner. The relationship between linoleic acid and chip quality was not found to be linear, and the content of linolenic acid in the oil had a greater influence on chip quality than did linoleic. Interestingly, it was found that a plot of quality scores against oleic acid gave a parabola. It was suggested by Warner that this may indicate that there is an optimum concentration of oleic that is required to produce a desirable chip and that too much or too little causes a reduction in quality.

The authors plotted fatty acid composition against three flavour descriptors in an attempt to explain changes in flavour quality. It was found that as the linolenic acid concentration rose above 3%, the fishy flavour intensity also increased. The parameter potato chip flavour (the most important flavour characteristic) tended to increase with linoleic acid. Importantly, it was established that, to an extent, the off-flavours described as 'hydrogenated' (from hydrogenated oil) and 'fishy' (from conventional canola oil) masked the potato chip flavour. The waxy note increased with oleic acid concentration.

It was suggested that the above may indicate that there are optimum concentrations for major fatty acids in frying oils to maximise flavour perception. The authors indicated that an oil with linolenic acid of 3% would not give rise to an excessive fishy taint and that the oil with approximately 19% linoleic acid had the best overall potato chip flavour. It was noted that although an increasing oleic acid content (at the expense of the more unsaturated linoleic and linolenic acids) increased oil stability, there was a need to ensure the optimum was not exceeded as this might impart a waxy flavour.

In conclusion to this study, the authors suggested that chips produced in fresh oil had a lower flavour quality than those cooked in oil which had been used for frying for 3–6 hours. The higher the linolenic acid the lower the chip quality. The canola oil with the highest oleic acid concentration (C) was the most stable to oxidation as determined by total polar materials and total volatile compounds. However, chips produced using this oil did not have the best observed quality, although they did have the best flavour stability having been produced in oil exposed to 18 hours frying and stored for four months at 25ºC.

Table 12.7 Simplified fatty acid compositions of canola, modified canola and high-oleic sunflowerseed oils used in above frying trials (Xu *et al.*, 1999a)

Fatty acid	Oil A	Oil B	Oil C	Oil D	Oil E	Oil F
C16:0	3.9	3.9	3.9	4.5	4.9	3.4
C18:0	1.8	2.2	2.2	2.3	9.5	3.8
C18:1	67.4	69.7	70.3	59.9	73.6	86.2
C18:2	21.2	15.8	15.6	20.7	2.9	4.3
C18:3	2.6	5.3	5.2	8.4	0.5	0.1

Key: oil A = high oleic-low linolenic acid canola oil; oil B + C = high oleic canola oils (monola); D = standard canola oil; E= partially hydrogenated canola oil; F = high oleic sunflowerseed oil.

It was noted that the blended oil (D) which had the following fatty acid composition: oleic acid 68%; linoleic acid (20%); and linolenic acid (3%), produced chips that tended to have a higher flavour quality and stability more often than any of the other oils. This level of linoleic acid corresponds roughly with the 19% optimum concentration suggested above. It should be noted that while the fatty acid composition will, as described in the above summary of Warner's paper, affect the flavour quality of the chips, so too will other aspects, particularly the tocopherol content. This issue does not appear to have been addressed in this work.

These results are supported by work conducted by Xu *et al.* (1999a). In this study five different canola oils and a high oleic sunflower seed oil were analysed. Chips were fried in the oils at 190°C in two experiments each lasting 80 hours; the products were sensory evaluated. It was found that high oleic canola oil with low linolenic acid concentration (oil A, Table 12.7) had a high flavour and oxidative stability. Those canola oils with higher linolenic acid contents (approx. 5%, oils B and C) performed well but were not as effective as oil A. It was stated that these oils (A, B and C) were more effective than high oleic sunflowerseed oil (oil F) and significantly more so than standard canola (oil D) and hydrogenated oils. (With regard to oil deterioration, it was suggested that changes in FFA, total polar compounds, dielectric constant, colour index, and iodine value were significantly related to extent of frying and could be used to monitor the oil. The TPM VERI-FRY® Quick Test for total polar compounds and measurement of dielectric constant by the Foodoil Sensor were the most

Table 12.8 Simplified fatty acid composition of oils (Xu, 1999b)

Fatty Acid	LLC	MLC	HLC	PHC	SO	PO
C16:0	4.1	3.9	4.1	5.1	5.3	40.1
C18:0	1.9	1.6	2.2	10.0	4.8	4.5
C18:1	68.7	71.0	69.3	71.1	55.4	41.1
C18:2	20.1	16.5	13.8	2.8	32.4	10.5
C18:3	2.5	4.4	6.8	0.5	0	0.1

Table 12.9 Simplified fatty acid composition of most effective frying oils (Warner *et al.*, 1994; Xu, 1999a; Xu, 1999b)

Fatty acid	Warner *et al.*, 1994	Xu, 1999a	Xu, 1999b
C16:0	3.8	3.9	4.1
C18:0	2.3	1.8	1.9
C18:1	68.4	67.4	68.7
C18:2	19.7	21.2	20.1
C18:3	3.1	2.6	2.5

convenient and reliable ways to monitor individual oil quality during deep fat frying.)

Slightly more recent studies by Xu *et al.* (1999b) examined the performance of three high-oleic canola oils all with varying linolenic acid concentrations (low = LLC; medium = MLC; and high HLC), a medium high-oleic sunflowerseed oil (SO); a commercial palm olein (PO); and a commercial partially hydrogenated canola (PHC). The fatty acid composition of the oils is given in Table 12.8. Sensory analysis was also conducted. As might be expected, the concentration of linolenic acid in an oil was inversely proportional to oxidative stability of that oil (as measured by colour, FFA and total polar compounds) and the sensory quality of the food fried in it. In terms of sensory evaluation, the highest quality was obtained from chips that were fried in LLC and medium-high oleic acid sunflowerseed oil, although LLC was more stable oxidatively. MLC was equivalent to palm olein in sensory terms but was more oxidatively stable. The partially hydrogenated canola oil performed least well of the six tested in sensory evaluation. Of those examined, it was concluded that the high-oleic, low-linolenic acid (2.5%) was the oil best suited to frying.

It is interesting to compare the fatty acid compositions of the oils that were shown by the above three studies to perform most effectively (Table 12.9). The data in Table 12.9 tends to indicate that good frying results may be achieved through the use of an oil that has a fatty acid composition of approximately: C16:0 (4.%); C18:0 (2%); C18:1 (68%); C18:2 (20%); and C18:3 (3%).

Hawrysh *et al.* (1995) conducted chemical analysis on oils (canola, cottonseed, partially hydrogenated canola, and soyabean) and sensory analysis on chips produced from these products. It was found that after very limited frying, those chips produced from canola and cottonseed oils had higher characteristic potato chip odour and flavour and lower off-flavour than chips produced from other oils. It was noted that after storage at 23°C for 18 weeks, chips from canola oil had the highest potato chip odour and flavour and the lowest off flavour and odour. While chip odour and flavour was affected by the frying oil used, the colour and texture were not generally affected. This is an interesting finding and, on the basis of fatty acid composition, one might have expected the cottonseed oil to have performed most effectively as it has a low linolenic acid content (0.8%) in comparison with canola oil (9.3%). However, in this instance, it might be that the high concentration of linoleic acid in

Table 12.10 Simplified fatty acid composition of oils used by Hawrysh *et al.* (1996) and the mean from Table 12.12.

Fatty acid	Canola oil	Soyabean oil	Cottonseed oil	PHCO*	Mean from 12.12
C16:0	3.7	9.9	21.0	3.7	3.9
C18:0	1.8	3.8	2.7	3.3	2.0
C18:1	60.3	25.1	20.0	75.8	68.2
C18:2	20.4	51.0	53.7	10.2	20.3
C18:3	9.3	7.9	0.8	1.7	2.7

* Partially hydrogenated canola oil.

cottonseed oil acts to reduce the flavour characteristics. (see Table 12.10). The tocopherol content of the oils might have influenced some of the above findings but discussion of this matter was limited.

Similar studies by Hawrysh *et al.* (1996) used canola (C18:3 = 9.9%), corn (C18:3 = 1.4%), hydrogenated soyabean (C18:3 = 1.5%), partially hydrogenated canola (C18:3 = 1.2%) and low linolenic canola (C18:3 = 3.4%) oils to produce tortilla chips. The products were subjected to sensory and analytical testing after being exposed to a Schaal oven at 60°C for either 8 or 16 days, and under practical storage conditions (i.e. 16 and 24 weeks at 23°C in the dark). After 16 weeks at 23°C chips produced from partially hydrogenated canola oil and low-linolenic acid canola had marginally higher off flavour and odour scores than other chips. After storage for 16 and 24 weeks at 23°C all stored tortilla chips had lower off flavour and odour scores than the reference chips. It was noted, however, that some of the chips stored for 16 and 24 weeks at 23°C had rancid, painty, buttery odours and flavours.

Przybyslski *et al.* (1993) examined the oxidative stability of a conventionally produced canola oil (Westar) and a genetically modified canola, with linolenic acid contents of 11.5% and 3.1% respectively. The modified canola was more stable to the onset of rancidity under the conditions of the Schaal Oven Test as shown by both analytical and sensory tests. Studies conducted by the same group of researchers (Malcolmson *et al.*, 1994) compared the stabilities of canola, sunflowerseed, soyabean and cottonseed oils stored under variations of the Schaal Oven test. It was found that canola and sunflowerseed oils had similar induction periods, while that of soyabean oil was longer indicating a greater stability. Interestingly, canola seemed to be more stable to oxidation under fluorescent light than cottonseed and soyabean oils but less so than sunflowerseed oil. Products that had been fried in canola and soyabean oils had similar storage stabilities. It is worth noting that the authors suggested that there should be a standardisation of the Schaal Oven testing procedure and the rating scales used by sensory panels.

Two margarines, one based on rapeseed oil and one on sunflowerseed oil, were prepared and used to fry and bake various foods (Gustafsson *et al.*, 1993). Margarines, pancakes and minced beef were heated to 165, 175 and 185°C for

Table 12.11 Composition of rapeseed and sunflowerseed oil margarines (abridged from Gustafsson *et al.* 1993)

Component	Rapeseed oil	Sunflowerseed oil
Fat (g)	80	80
Water (g)	18	18
Protein	0.2	0.2
Carbohydrate (g)	0.4	0.4
Vitamin D (μg)	7.5	7.5
α-tocopherol (mg)	18	46
β-tocopherol (mg)	0	2
γ-tocopherol (mg)	34	5
Vitamin E	22	48

3–5 minutes. Sweet bread and cakes were baked. The sensory characteristics of the products were determined. The fried rapeseed (canola) oil margarine was considered to be less rancid than that made from sunflowerseed oil. A small difference was noted between the margarines when they were added to pancake batter but none when used to fry minced beef. The use of rapeseed margarine was preferred for the production of sweet white bread and cake and it was concluded that this margarine was more resistant to heating than the sunflowerseed oil margarine. The fatty acid and margarine compositions are shown in Tables 12.11 and 12.12.

This is an interesting study and the results are in some ways slightly surprising when it is considered that the rapeseed oil has: (i) a higher linolenate concentration; and (ii) a lower tocopherol level which would tend to indicate that it might have a lower oxidative stability than sunflowerseed oil. However, no mention is made of the tocotrienol concentration nor of the observation that the sunflowerseed oil contained 1.0% linolenic acid which is higher than authenticity studies have suggested is present in pure sunflowerseed oil (Rossell, 1998). As seen from the above, much work has been conducted on the suitability of canola (rapeseed oil) as a frying medium. However, as shown below, a significant amount has also been conducted on the use of soyabean oil.

Mounts *et al.* (1988) investigated the efficacy of using soyabean oils (SBO) produced by hybridisation breeding to yield oils with altered fatty acid

Table 12.12 Simplified fatty acid composition of rapeseed and sunflowerseed oils and margarines (abridged from Gustafsson *et al.*, 1993)

Fatty Acid	Rapeseed oil	Rapeseed margarine	Sunflowerseed oil	Sunflowerseed margarine
C18:0	1.7	2.6	4.6	5.0
C18:1 n-9c	60.1	59.2	22.9	22.7
C18:2 n-6c	21.5	21.5	63.9	63.0
C18:3 n-3c	11.4	10.7	1.0	1.1

compositions with low linolenic acid contents. In the three oils tested these were 3.3%, 4.2% and 4.8% (test oils). The stability of these oils was compared with those of a conventional SBO and a commercial winterised-hydrogenated soyabean oil (linolenic acid contents 7.7% and 3.0%, respectively.) The oils were tested by a sensory panel after accelerated storage test at 60ºC and during use at 190ºC in room tests. It was established that there was no significant difference in flavour stability between the control and test oils. It was noted, however, that the test oils had improved overall room odour intensity scores and did not give rise: (i) to fishy taints associated with the control SBO; (ii) nor the hydrogenated odours of the commercial oil, suggesting that the low-linolenic acid oils were more stable.

Mounts *et al.* (1994a) examined the performance of six soyabean oils (SBO), three that had low linolenic acid contents (5.5%, 2.9% and 1.9%) following hybridisation breeding, a conventional SBO (linolenic acid 6.2%) and two hydrogenated SBOs (linolenic acid 4.1% and 0.2%). The oils with reduced linolenic acid contents gave rise to lower odour intensity than the commercial frying oils. For example, the modified oil containing 1.9% C18:3n-3c, did not produce a fishy odour after it had been used to cook batches of potatoes at 190ºC for 10 hours, and lower burnt and acrid smells after 20 hours when compared with the commercial oils. Furthermore, potato chips cooked in SBO with 1.9% linolenic acid at 190ºC for 10 or 20 hours, had a better flavour quality than those produced from unhydrogenated SBO and hydrogenated oils.

It was found that the flavour quality of potato chips fried in modified oil (linolenic acid = 2.9%) was equivalent to those produced from hardened SBO (C18:3n-3c = 4.1%). It was suggested by Mounts *et al.* (1994a) that fishy flavours associated with chips produced from low linolenic oils were lower than those cooked in the control oil (C18:3n-3c = 6.2%); furthermore, the product lacked the hydrogenated flavours associated with chip production from hardened oils. It was concluded that SBO with lower than conventional linolenic acid contents, produced as a result of hybridisation breeding, could act as potential alternatives to hydrogenated frying oils.

Liu and White (1992) intermittently heated one canola and six soyabean oils, two being commercial and four experimental varieties (i.e. A6, A16, A17 and A87). The fatty acid compositions of these are shown in Table 12.13.

Each oil was heated to 180º±5ºC in a fryer. Bread cubes were fried, half being used for analysis the remainder for storage at 60ºC for seven days. The oils were also exposed to a regime whereby they were heated for 20 hours, cooled for ten hours and reheated for a further 20 hours. Following this temperature cycling, bread cubes were fried in the oil. The analytical data (e.g. PV, and conjugated dienoic acid levels) of the oils demonstrated that A17, A16, A87 and A6 were more stable than the commercial soyabean oils and canola oil to oxidation. This was supported by the results of sensory panels. It was found that the initial linolenic acid concentrations tended to predict the oils' flavour and oxidative stabilities.

Table 12.13 Relative fatty acid composition of oils before and after heating at 180°C (Lui and White, 1992)

Oil	C18:0	C18:1	C18:2	**C18:3**	C18:2+C18:3
Hardin					
Start	3.6	25.2	54.8	**5.9**	60.7
End	6.8	37.5	36.1	**1.2**	37.3
BSR101					
Start	3.8	22.6	56.7	**6.8**	63.5
End	7.0	34.0	39.4	**1.6**	41.0
A17					
Start	5.1	29.3	49.4	**1.5**	50.9
End	9.3	39.3	25.3	**0.3**	25.6
A16					
Start	5.6	31.8	50.7	**1.9**	52.6
End	9.6	42.8	29.0	**0.3**	29.3
A87					
Start	4.5	29.1	54.7	**1.8**	56.5
End	7.3	40.7	33.9	**0.5**	34.4
A6					
Start	27.7	21.5	40.4	**4.1**	44.5
End	38.9	26.1	21.8	**1.2**	23.0
Canola					
Start	1.4	63.0	21.3	**10.3**	31.6
End	2.1	74.4	13.1	**3.8**	16.9
Control Oil 1	4.1	24.8	53.5	**6.7**	60.2
Control Oil 2	1.9	65.0	20.4	**8.6**	29.0

Note: Start and end refer to the oils' fatty acid composition before and after heating and frying bread cubes.

Shen *et al.* (1997) examined the frying stability of low-linoleate oil and those that were devoid of lipoxygenases, by frying bread cubes at 180°C. It was found that those oils with low linoleate contents were more stable than those high in this fatty acid moiety. However, sensory panels could not distinguish products fried in oils that had high or low contents of linolenic acid. Lipoxygenase activity did not appear to affect the PV of oil extracted from the bread cubes.

Warner *et al.* (1997) investigated the effect of fatty acid composition of frying oils on the fried food flavour and off-flavours in potato chips. This was achieved by mixing commercially processed cottonseed oil (CSO) with high oleic sunflowerseed oil (HOSFO) to produce blends containing 12–55% linoleic acid and 16–78% oleic acid. Food fried in CSO (oleic acid = 16%; linoleic acid = 55%) had the highest fried food flavour. This positive characteristic decreased as the linoleic acid concentration was reduced. Although HOSFO was more stable than other oil mixes, potato chips produced from this oil had the lowest intensities of fried food flavour. It was concluded that oil compositions containing 16–42% oleic acid and 37–55% linoleic acid produced fresh fried chips with moderate fried food flavour intensity, good overall flavour quality and chips that contained low concentrations of total polar compounds. In food that had been cooked in used oil, compositions of 42–63% oleic and 23–37% linoleic acid gave best flavour stability. This corresponds roughly with the optimum linoleic acid level of 19–20% suggested in earlier work (Warner, 1994) – see page 288.

Brewer *et al.* (1999) examined the relationship between lipid-oxidation derived volatiles in French fried potatoes and sensory characteristics. The oils used included: low-linolenic acid soyabean oil (IV = 118.4), creamy partially hydrogenated soyabean oil (IV = 83.7), liquid hardened soyabean oil (IV = 95.5), liquid partially hydrogenated soyabean oil (IV = 80.6). In all frying trials, the hexanal concentration increased in fried products as oil-use time increased. Hexanal concentration was negatively correlated with overall odour quality of the fried product and positively with grassy, rancid and painty odours. Positive odours (e.g. buttery and French fried) decreased and painty and rancid odours increased as oil-fry time increased.

Goburdhun and Jhurree (1995) investigated the frying performance of soyabean oil (SBO), a blend of soyabean and palm kernel olein (i.e. the liquid fraction of palm kernel oil) and SBO containing a blend of the following antioxidants: butylated hydroxyanisole, butylated hydroxytoluene, propyl gallate and citric acid. Potato slices were fried in the oils at 180°C on 15 occasions for 337 minutes. The oils were analysed and the products subjected to sensory evaluation. It was found that during frying: (i) any reduction in the concentration of unsaturated fatty acids was minimal; (ii) FFA and PV increased in the three oils during frying with the lowest increase seen in the blended oil and the highest in SBO; and (iii) SBO with antioxidants had the lightest colour at the conclusion to the trial. Sensory evaluation indicated that panelists could not link the fried products to the oils used in their manufacture. The blended oil produced the highest sensory ratings and chips fried in SBO the lowest.

The frying performance of RBD palm olein and RBD coconut oil were compared (Man and Hussin, 1998). Potato chips were fried intermittently for 5 hours per day for five days. Extensive analysis was conducted on the frying oils. The results indicated that palm olein was a better frying oil than coconut oil, in particular, with regard to FFA, smoke point and foam production. This is not surprising since the coconut oil contains significant amounts of short chain fatty acids which have higher volatilities and are more susceptible to hydrolysis which will lead directly to higher FFA smoking and foaming. It was, however, noted that coconut oil was more resistant to oxidation than palm olein producing lower concentrations of polar and polymeric material (probably because of coconut oil's lower IV). Sensory evaluation indicated that chips fried in palm olein were preferred to those cooked in coconut oil.

French fries and potato crisps were produced from sunflowerseed oil and high oleic sunflowerseed oil, with or without the antifoaming agent polydimethylsiloxane (PDMS). Control samples were prepared from palm olein and a hydrogenated rapeseed oil/palm oil mixture. The potato chips were stored at ambient for six months and the French fries at −20°C for 12 months. It was found that crisps prepared in sunflowerseed oil had a lower sensory quality than those in palm oil; crisps produced in high oleic sunflowerseed oil were comparable with those from palm oil. Differences were noted in the texture and mouthfeel of the products but were not thought to be caused by the use of different oils. The addition of PDMS did not have a positive effect on the storage quality of the products produced in sunflowerseed or high oleic sunflowerseed oils; however, it was noted that French fries produced from these oils gave rise to a sweet fruit odour and flavour, which decreased intensity as the storage period increased. (van Gemert, 1996).

It has been shown in this section that some oxidation is necessary for the development of fried food flavour, and it appears from experiments carried out by Warner (1994), and others, that linoleic acid is involved in some way in the production of this flavour. However, a high concentration of linoleic acid gives rise to fishy and painty taints. The optimum concentration, therefore, appears to lie in the range 19–37%.

It should not be forgotten that although lipid oxidation and the associated secondary products are important in determining the flavour characteristics of fried food, the presence of FFA in the oil as a result of hydrolysis can cause flavour deterioration. Ledahudec and Pokorny (1991) added fatty acids to oils and established that free lauric acid (a major component of palm kernel and coconut oils) imparted a soapy taste to edible oil and fried food when present at less than 1 g/kg. This is not unexpected as lauric acid is known to have a soapy taste (Rossell, 1994). It was also reported that the presence of free linolenic acid at 1 g/kg gave rise to a bitter and rancid taste. In contrast, palmitic, stearic and oleic acids were not detected by sensory panels at 20 mg/kg.

12.7 The effect of antioxidants in frying oils

Antioxidants are often classified according to mode of production and, in particular, may be regarded: (i) as being naturally-occurring compounds (e.g. the tocols); or (ii) synthetic (e.g. BHA and BHT). This fairly brief section is added

Table 12.14 Protective indices of sterols during heating of trioleylglycerol (triolein) at frying temperatures (e.g. 180°C) (from Quaglia *et al.*, 1998)

Additive	% Compound	Concentration PI after heating for 24 hr	48 hr	72 hr
α-tocopherol	0.02	0.97	1.01	1.01
BHA	0.02	0.96	1.04	1.00
Cholesterol	0.1	0.96	1.12	1.12
Stigmasterol	0.1	0.95	1.11	1.09
Fucosterol	0.1	1.52	1.71	1.66
Δ^5-avenasterol	0.1	1.78	1.69	1.59
Δ^5-avenasterol	0.05	1.16	1.19	1.09
Δ^5-avenasterol	0.01	1.10	1.09	1.07

Table 12.15 Efficacy of α-tocopherol in various virgin olive oils

Oil	α-tocopherol (mg/kg)	% overall antioxidant effect
1	236	12.9
2	130	21.1
3	297	79.0
4	138	24.9
5	172	63.9
6	182	25.7
7	172	27.6

because the presence of antioxidants affects the rate of lipid oxidation and consequently may play a role in flavour production.

The effect of the tocols (i.e. tocopherols and tocotrienols) as free radical quenching agents and, therefore, antioxidants is well-known and documented (Hamilton, 1983). It has also been established that delta-5-avenasterol (i.e. Δ^5-avenasterol) and fucosterol have antioxidant properties as do chlorophyll, pheophytin and the polyphenols. However, other sterols, such as cholesterol and stigmasterol, are ineffective (Gordon and Magos, 1983).

Quaglia *et al.* (1998) describes the protection index (protection index = PI). This provides a guide to antioxidant activity with those compounds having a PI greater than 1.00 considered to be useful antioxidants. From Table 12.14, it will be seen that Δ^5-avenasterol and fucosterol demonstrate significant antioxidant activity. It is suggested that the mode of action of Δ^5-avenasterol involves the sterol reacting with the lipid radical to produce a sterol free radical which is comparatively stable and does not initiate further oxidation reactions.

Table 12.15 lists the α-tocopherol content of a series of virgin olive oils and the antioxidant effect. It will be seen that sample 3 contains a high tocopherol content and has a significant antioxidant activity. In contrast, sample 1 which has a similar tocopherol content has an antioxidant activity one-sixth that of sample 3. It was suggested by Quaglia *et al.* (1998), that this reflects the effect of minor components which might have antioxidant properties, and might be important in the assessment of the oxidative stability of oil.

Man and Tan (1999) investigated the effects of four antioxidants (BHA, BHT, oleoresin rosemary, and sage extract) when added at 200 mg/kg to RBD palm olein. Results were compared with a control sample (i.e. palm olein with no added antioxidant). The frying trials involved cooking potato chips for 3.5 hours per day for seven consecutive days. The oil was analysed chemically to determine a number of oxidative changes including PV, TBA, IV, FFA, polymer content and colour. A sensory assessment of the chips was also conducted immediately the chips were produced and over a subsequent 14 week storage period at ambient. TBA values were also determined on the chips.

The results indicated that the inclusion of oleoresin rosemary in the frying oils led to the lowest rate of increase of oxidation indicators, indicating that this had the greatest antioxidant potential at the added concentration. The order of efficacy was described as being (at 95% confidence):

oleoresin rosemary > BHA > sage extract > BHT > control

The chips were subjected to sensory evaluation before storage and it was found that the flavour, odour, texture and overall acceptability of the product was independent of added antioxidant. During storage over 14 weeks, there was no observed difference in the sensory evaluation of the chips from any of the oil-antioxidant systems. It was, however, noted that the chips produced from oils and antioxidant had lower TBA values; the antioxidant efficacy was of the same order as shown directly above.

Gordon and Kourimska (1995) studied the effects of antioxidants on rapeseed oil during heating at 80°C and during deep-fat frying of potato chips (French fries) at approximately 162°C. As with previous reports, the oil degradation during frying was assessed by determining: (i) the formation of polymers; (ii) peroxide value; (iii) the increase in mass of the oil during heating (i.e. uptake of oxygen); (iv) changes in the concentration of the tocopherols.

The order of antioxidant activity varied slightly being dependent on method of assessment. However, during heating it was:

TBHQ > lecithin > ascorbyl palmitate > rosemary extract > BHT > BHA > D-δ-tocopherol.

It should be noted that the level of addition of the above to oil was 1g/kg for lecithin and rosemary extract and 0.2g/kg for the others. The difference in the concentrations was to reflect the concentration of active components within each antioxidant.

Rosemary extract and ascorbyl palmitate reduced the formation of both dimers and PV throughout the deep-fat frying trials and also retarded the losses of natural tocopherols. For example, after frying operations, oils containing rosemary extract, ascorbyl palmitate and no added antioxidant had α-tocopherol contents of 164, 139 and 25.5 mg/kg, respectively. The initial concentration of α-tocopherol was 215 mg/kg. The dimer content of the frying oil and the oils extracted from the potato chips were similar (26g/kg and 23 g/kg, respectively after 12 fryings) which indicated that the potatoes were not extracting polymers preferentially.

Lai *et al.* (1991) examined the effects of the addition of sodium tripolyphosphate (STPP), oleoresin rosemary (OR) and TBHQ on the stability of lipids in restructured chicken pieces during refrigerated and frozen storage. This was achieved by measuring TBA values, and conducting sensory and chromatographic examination. It was shown that a mixture of STPP/OR was equivalent to STPP/TBHQ in preventing lipid oxidation in the above chicken pieces; this was supported by sensory studies. It appeared that OR and STPP acted synergistically because individually these compounds were less effective at reducing oxidation. However, the oxidative stability of the chicken nuggets could not be improved by adding OR to the frying oil.

Table 12.16 Composition of olive oil deodoriser distillate and its unsaponifiable fraction (from Abdalla, 1999)

Unsaponifiable fractions	Olive oil deodoriser distillate (g/kg)	Unsaponifiable matter (g/kg)
Squalene	284	774
Total sterols, of which:	48.4	130.8
campesterol	1.2	2.8
stigmasterol	0.5	1.4
β-sitosterol	42.7	116.4
avenasterol	3.8	10.4
Total tocopherols, of which	18.1	49.4
α-tocopherol	17.1	46.5
γ-tocopherol	1.1	3.1

Sharma *et al.* (1997) produced potato chips, banana chips and fried Bengalgram (*Cicer arietinum*) by cooking the respective products in sunflower oil at an initial temperature of 200°C which fell to 150–180°C on addition of the product. The antioxidants (BHA, BHT or TBHQ) were added to the fried product at approximately 2% and stored at 37°C in polypropylene packs. Assessment of the PV, total steam volatile carbonyls, and sensory evaluation indicated that protection of the oil in the product was provided by the above compounds, the greatest by TBHQ the least by BHA.

The effect of the addition of olive oil deodoriser distillate on frying oil and potato chip quality was examined by Abdalla (1999). In these studies, olive oil deodoriser distillate (OODD) was collected and refluxed with alkali to saponify the glyceride material; the unsaponifiable matter extracted with ether and appropriate clean-up techniques were deployed. The composition of the OODD was FFA (as oleic acid) 30.4%, neutral glycerides 32.8%, and unsaponifiable matter 36.7%. The composition of the unsaponifiable fractions is presented in Table 12.16. The antioxidant properties of OODD were attributed to its content of tocopherols, sterols (e.g. Δ-5-avenasterol), and squalene.

The unsaponifiable matter extracted from OODD was added to the frying medium (sunflowerseed oil) at concentrations of 0.2%, 0.5% and 1.0%. Potato slices were fried at 180°C for seven minutes, with eight batches being cooked each hour. Chip production took place over ten days. Chip samples were taken and stored in the dark at 22±3°C for three months. The results indicated that the addition of 1% unsaponifiable matter to sunflower oil (SFO) had the greatest effect in retarding oxidation of oil during frying. The protective nature of the addition of OODD unsaponifiable matter to the frying medium is illustrated in Table 12.17.

After ten days frying the concentrations of the squalene, avenasterol and tocopherols had decreased significantly, suggesting that these components possess antioxidant properties, a point borne out by the observation that the concentration of other components (e.g. campesterol, stigmasterol, β-sitosterol) did not decrease (see Table 12.18).

Table 12.17 Effect of various levels of addition of unsaponifiable matter (UM) on the oxidative stability of sunflowerseed oil frying medium (abridged from Abdalla, 1999)

Treatment/ Frying day		Colour at 400 nm	Iodine value (Wijs)	PV ($meqO_2/kg$)	FFA (% as oleic)	*p*-anisidine	% Total polar compounds
SFO - Control	Day 0	0.03	128.5	1.2	0.11	1.4	4.9
SFO - Control	Day 10	0.18	108.3	22.8	1.55	46.4	43.6
SFO + 0.2%UM	Day 0	0.03	128.4	1.6	0.14	1.8	4.9
SFO + 0.2%UM	Day 10	0.10	116.2	22.5	1.20	27.2	39.7
SFO + 0.5%UM	Day 0	0.03	128.4	1.9	0.16	1.8	4.9
SFO + 0.5%UM	Day 10	0.07	121.7	13.3	0.90	21.8	23.9
SFO + 1.0%UM	Day 0	0.04	128.4	1.7	0.15	1.8	4.9
SFO + 1.0%UM	Day 10	0.06	124.4	9.9	0.80	18.3	17.2

Table 12.18 The percentage reduction in concentration of components of the unsaponifiable matter of the olive oil deodoriser distillate after ten days frying at 180°C (abridged from Abdalla, 1999)

Treatment	Squalene (%)	Total tocopherols (%)	Campesterol (%)	Stigmasterol (%)	β-sitosterol (%	Δ-avenasterol (%)
SFO – Control	78.7	68.6	1.9	1.8	1.3	57.8
SFO + 0.2%UM	70.2	57.1	0.7	1.1	1.1	69.4
SFO + 0.5%UM	69.2	44.3	1.4	1.4	0.9	54.4
SFO + 1.0%UM	64.6	42.4	1.4	1.0	1.0	53.3

Table 12.19 The protective effect of OODD on oil extracted from potato chips before and after storage for three months at 22±3°C (from Abdalla, 1999)

Treatment and frying day		Before/ after storage	Oil absorbed (%)	IV (Wijs)	PV (meq O_2 /kg)	FFA (% as oleic)	Total tocopherol (mg/kg oil)	Total sterols (mg/kg oil)	Squalene (mg/kg oil)
SFO – Control	Day 1	Before	39.3	121.5	12.3	0.3	59.5	29.1	10.1
SFO – Control	Day 10	Before	38.7	96.4	30.5	2.6	0	0	0
SFO – Control	Day 1	After	39.0	119.1	19.8	0.26	41.6	21.2	0
SFO – Control	Day 10	After	38.5	83.1	39.7	2.89	0	0	0
SFO + 1.0%UM	Day 1	Before	38.3	127.2	5.3	0.18	96.4	40.2	536.4
SFO + 1.0%UM	Day 10	Before	38.6	122.4	10.7	1.25	0	0	0
SFO + 1.0%UM	Day 1	After	38.3	125.8	6.5	0.20	66.3	28.5	285.3
SFO + 1.0%UM	Day 10	After	38.1	121.8	13.5	1.34	0	0	0

The frying medium was extracted from batches of potato chips before and after 3 months storage at 22ºC. The amount of oil absorbed by the potatoes varied slightly during the trial and it was also observed that the unsaponifiable matter absorbed by the chips played a role in protecting the product from deterioration during the storage period, an effect that was dose related (i.e. the greater the concentration of squalene, avenasterol and tocopherol the greater the protection) (see Table 12.19).

Those chips produced from oil containing 1% OODD were shown to have pleasant characteristics such as good flavour, colour and acceptable flavour. These observations were supported by chemical analyses of oil extracted from chips.

A study was conducted to establish the efficacy of nitrogen and carbon dioxide flushing in reducing canola (rapeseed) oil oxidation during heating at 195±5ºC. It was found that flushing with either gas at a slow rate tended to increase oxidation rather than stopping it. It was considered that, because of its greater solubility in oil and higher density, carbon dioxide provided greater protection than nitrogen. The authors (Przybylski and Eskin, 1988) suggested that to prevent oil oxidation during heating at frying temperatures, the following should be considered:

- use carbon dioxide rather than nitrogen
- the oil should be flushed with nitrogen for 15 minutes or carbon dioxide (5 minutes) to eliminate any dissolved oxygen
- the linear flow of gas in the containers should be 50 cm/minute
- the ratio of oil height to diameter of the vessel should be a minimum of three
- the vessel should not be filled to more than 70% of its height.

12.8 Oil uptake by fried food

12.8.1 Introduction

Oil uptake during frying is considered because the fat content of a product will affect its flavour/odour and general organoleptic properties. The frying oil not only acts as a heat transfer medium. As the oil is heated to conventional frying temperatures of approximately 170–180ºC it will start to degrade through hydrolysis and oxidation of the fatty acids. The breakdown products can themselves give rise to flavour and can react with carbohydrates, proteins and their decomposition products to produce taste traditionally associated with fried foods.

Blumenthal (1996) reviewed the work by Lyderson (1997) and Keller and Eschel (1989) whereby the frying process is compared to that of a pump. In this analogy, the water migrates from the central parts of the food being fried to the edges to replace that already lost by evaporation. In terms of heat transfer, the hot oil causes the water in the food to turn to steam and to leave the system and, providing that steam continues to leave the food, then its temperature will remain approximately 100ºC and prevent the product from charring (Blumenthal, 1996).

12.8.2 Surfactant theory

The oil transfers heat initially by surface contact and then by capillary action into the product, a process that is governed by oil viscosity and surface tension. As the oil degrades from fresh, the contact time between it and the food will increase. This process is described by Blumenthal in terms of surfactants. In a triglyceride oil and a watery food (e.g. raw potato) there is little to bring the immiscible foods together. However, as the oil degrades by oxidation at the oil-air surface, or the food-oil interface, the oil changes and the surface tension between the food and oil decreases, probably as result of oxidised surface active agents. There is a balance. If the oil becomes highly degraded the oil-food contact will increase. It has been suggested, therefore, that a new hot oil's contact time with food is only about 10% of the immersion time of the food in the oil. As the oil degrades, this increases to an optimum of about 50% thereby increasing heat transfer and flavour production (Blumenthal, 1996).

Pinthus and Saguy (1994) developed an equation in which oil uptake was found to be a power function of the interfacial tension. Therefore, an abused oil with a high concentration of surfactants would result in low interfacial tension, high contact time and high oil absorption, a factor that would be compounded by an abused oil's higher viscosity. It has been observed that food will tend to preferentially extract polymers into its surface layers (Pokorny, 1980). This contrasts with the work of Gordon and Kourimska (1995) who found that dimers were not preferentially extracted by potatoes (see page 297).

12.8.3 Surface area

The surface area of the sliced potato plays an important part in oil uptake. For a given length of potato section, a wavy slice will have a greater surface area than a flat slice of the same thickness. It is reported (Gould *et al.* 1989), however, that in commercial chip production wavy chips are more likely to be thicker than flat chips to prevent breakage. Consequently, comparisons between the two types of chips are difficult. The effects of raw potato slice thickness on oil content of the finished chip are shown in Table 12.20. The data show that for all frying temperatures examined, thinner slices (i.e. those that have a greater surface area per volume) adsorb more oil. At lower frying oil temperatures, for the same slice thickness, the oil content increases. This is probably a direct result of the raw potato requiring a greater frying time at a lower temperature.

Frying regular and wavy chips of any given thickness at low temperatures (325°F/162.8°C) resulted in the latter having a greater oil content in comparison with the regular chip again perhaps due to the increased surface area. However, when both styles were fried at higher temperatures (375°F/190.5°C) the wavy chip was found to adsorb less oil. When fried at 350°F/176.6°C there was little difference in oil adsorption between the two styles.

Equations relating chip oil content to slice thickness and frying oil temperature were developed (Gould *et al.*, 1989).

Table 12.20 Effect of slice thickness on oil content of potato chips fried at various temperatures to a moisture content of 1.5% for regular style slices (from Gould *et al.*, 1989)

Slice thickness	Oil content (%) for frying temperatures of		
in inches	375°F (190°C)	350°F (177°C)	325°F (163°C)
0.050	41.3	42.1	42.9
0.060	36.3	37.7	39.2
0.070	33.8	35.9	38.0

It is suggested that there is an interaction between slice thickness and fryer temperature and that these variables can account for approximately 43% of the variation in the oil content of the potato chips.

12.8.4 Alternative and additional processing techniques

It has been suggested that approximately 17% of the variation in oil content of potato chips is due to the following factors:

- variations in the finished moisture content of the chip
- methods of washing the raw slice
- sampling practices of the raw or finished product.

The oil content of a chip could be controlled complete or partial replacement of frying with an alternative drying operation.

As frying involves dehydration, drying chips instead of frying them might decrease the oil content of the finished product as it would decrease the time chips are able to absorb oil. For example, partial frying may be followed by infrared or microwave energy to remove excess moisture. Modified frying vessels have been used successfully to reduce the oil content of potato chips. For example, the oil content of vacuum fried chips was more than 6% lower than those fried in a conventional process. (Smith, 1968).

The following are patented methods for reducing the oil content of chips; it is suggested, however, that while they may be of research interest few are used commercially. The text is taken more or less verbatim from Gould *et al.* (1989).

- Soaking raw potato slices in hydrogenated vegetable oil at 100°F (37.7°C) for 5, 10 and 15 minutes, followed by drying in air at 315°F (157.2°C) or 325°F (162.8°C) for 10 minutes. This produced potato chips lower in oil owing to a shorter soaking period; the results are shown below (from Smith, O., 1968):

Time soaked (mins.)	Oil content (%)
5	23.3
10	24.9
15	27.3

Table 12.21 Effect of drying processes to 50% of original moisture on oil content of potato chips using the variety 'superior' tubers (from Gould *et al.*, 1989)

	% oil content of chips for the two specific gravity lots	
Tuber specific gravity	1.070	1.080
Conventional frying	43.0	40.1
Microwave	31.1	27.9
Vacuum oven drying	30.8	29.0
Tray drying	28.2	28.5

- Using a fluidised bed of hot sodium chloride particles to cook the raw potato slices to their final moisture content. Oil could then be applied by spraying, permitting control of the amount of oil in the final product (15 to 20%). Heating the oil-sprayed chips at 200°F (93.3°C) for five minutes alternatively gives 'the desired potato chip flavor' (Talburt and Smith, 1975).
- Using supercritical carbon dioxide to extract chips fried by conventional processes removes 50% of the oil in the chip but maintains 'texture and flavor characteristics that were very similar to the original product.' Liquid carbon dioxide near its critical point will dissolve fats and oils without removing proteins or carbohydrates.
- The use of superheated steam has been proposed for 'washing' the fat or oil from chips.

In a 1978 experiment, Baroudi of Ohio State University (Baroudi, 1978) demonstrated that a 13.6% difference in oil content between two portions of potato chips (cv Norchip) could be achieved. One portion of slices was fried at 375°F (190°C) for about 12 seconds and dried on a continuous air belt oven for 3.5 to 4 minutes. The temperature in the dryer was originally 220°F (104.4°C) falling to 180°F (82.2°C) after 27 seconds, and rising linearly to 310°F (154.4°C) after 180 seconds. The chips were then fried at 290°F±5°F (143.3°C±2.7°C) for approximately 65 seconds. This produced potato chips with an oil content 33% lower than the control group which had an oil content of 46.6%.

In a second experiment, Baroudi compared vacuum oven and freeze-drying methods. Norchip tubers which had been stored at 40°F (4.5°C) were sliced and fried at approximately 330°F (165.5°) for 55±5 seconds. The potato slices were subjected to either drying under vacuum or freeze-drying. Following frying, chips from the former had an oil content of 29.8% and from the latter 21.3%. Both of these results were lower than the mean oil content for the control sample, which was 42.55%.

Plimpton (1985) from OSU used 'Superior' tubers separated into specific gravity lots of 1.070 and 1.080. The tubers were sliced to 0.060 inches and dried by: (i) tray dehydration; (ii) microwave; (iii) and vacuum oven to 50% of their initial moisture content, and then fried in oil at 375°F (190.6°C) to a 1.25% moisture content (Table 12.21).

The average oil contents of the two specific gravity lots that were tray dried were not significantly different; however, there was a significantly lower content for microwave lots with the higher specific gravity lot (3% lower). It was reported that the experimental drying techniques resulted in the finished products having oil contents 26 to 34% lower than the control samples. Taste panel tests at OSU indicated that pre-dried chips were equal to or better than the same slices made under conventional techniques.

12.8.5 Pre-fry and post-fry procedures

It has been suggested that oil uptake by food is a surface effect involving equilibrium between adhesion and oil drainage as the product is removed from the fryer. It has been found (Moreira *et al.*, 1997) that during frying tortilla chips absorbed approximately 20% of the product's final oil content, the remainder being on the surface. As the food cooled, it absorbed approximately two-thirds of its final oil content.

Pre-fry

The observation by Moreira *et al.* (1997) is interesting and has led to changes in commercial practice which might be attributable to the above. It has been reported (Rossell, 2000) that potatoes chips (crisps) can be made with 25% less fat and produce a lighter, drier and crisper product. This process involved part-frying the potato slices to a moisture content of 5–10%. The oil content of the product was reduced to approximately 2% by exposure to steam and then dried with the final food having a moisture content of 1%. Table 12.22 provides a summary of the composition of the conventional product and the low-fat version produced using the above novel technology.

Selman (1989) also described a series of pre- and post-fry treatments that could be used to alter oil content. The first system involved drying. By reducing the initial moisture content followed by post-drying (i.e. microwave or superheated steam) the oil content of chips/crisps may be controlled to an extent. Regression equations linking the initial moisture content and final oil level were derived as shown below for microwave pre-fry dried and air pre-fry dried potato slices. The frying times were controlled to either: (i) produce an acceptable food; or (ii) be limited to a set time (120 seconds).

Table 12.22 Summary of approximate compositional information of novel and conventional chips described above (Rossell, 2000)

Component	Conventional product	Low-fat product
Energy	2300 kJ	2100 kJ
Protein	6g	7g
Carbohydrate	46g	56g
Fat	38g	28g

Microwave pre-fry dried (variable frying time) $y = 25.8 + 0.238x$ ($r^2 = 86.4$)
Microwave pre-fry dried (constant frying time) $y = 32.4 + 0.123x$ ($r^2 = 80.8$)
Air pre-fry dried (variable frying time) $y = 25.2 + 0.241x$ ($r^2 = 91.0$)
Air pre-fry dried (constant frying time) $y = 25.1 + 0.211x$ ($r^2 = 83.4$)
where x = pre-fry moisture content and y = final oil content.

At the end of frying, a potato slice has steam-filled gelatinised starch in the heated potato cells. It was suggested by Selmen that if oil uptake were a process that occurred only at the end of frying, the oil would be drawn into the product in a consistent manner regardless of pre-fry conditions and processing time and temperature. This is not seen in experimental conditions and, furthermore, oil content was observed to be a function of drying time suggesting that it was influential in determining oil uptake.

Microwave drying has different effects and causes the potato slice to dry heterogeneously giving rise to selective loss from specific sites formed as a result of structural weaknesses in the potato. Consequently, the raw microwaved slice will have areas of high and low moisture and, following frying, the chip will have variable oil content throughout its structure. In general, air drying will be more homogeneous.

Blanching causes starch to gelatinise and this leads to increased oil uptake. Stutz and Buriss (1948) reported that this effect could be reversed by blanching in an ionic salt solution. It was found that the final oil content was related to the blanch temperature and time, and the concentration of the ionic solution. In addition, the cation affected the oil absorption, with calcium being the most effective under the conditions used. In comparison with pure water, it was reported that blanching record potato slices at 70ºC for 60 seconds in sodium chloride (2M) reduced the final oil content by 10%.

Post-fry solutions

As stated above, oil is clearly absorbed during the frying process. Nonetheless, oil is also taken up after the process has finished and tends to be found in the surface of the chips. The objective of post-fry treatments is to reduce surface oil. Selmen (1989) treated chips in two ways at the conclusion to frying at 180ºC. Method 1 involved allowing the excess surface oil to drain naturally during post-fry cooling at ambient. In method 2 the chip was transferred directly to an oven at 160ºC for 3 minutes, followed by blotting to remove excess oil. The moisture content of the chips was approximately 2% after 125 seconds. A comparison of methods 1 and 2 suggested that after frying at 180ºC for 200 seconds, chips from methods 1 and 2 had oil contents of approximately 38% and 30% respectively. Using steam jets and water washing along with blotting to remove excess surface oil and holding the French fries at 160ºC led to a reduction in oil content of approximately 60% in comparison with French fries that were allowed to drain naturally during post-fry cooling.

There are other ways by which the fat content of fried food can be reduced. For example, the ability of the dry ingredients curdlan and cellulose derivatives to lower oil uptake and moisture loss in doughnut production was evaluated during deep-fat frying. The addition of curdlan showed a linear effect on reducing oil uptake and moisture loss over the application range 0–0.5%. The researchers (Funami *et al.*, 1999) suggested that this observation was a result of curdlan's thermal gelling property, and the heat-induced gel formed during frying probably acted as an oil absorption barrier and prevented moisture escaping. It was reported that cellulose derivatives were less effective in these respects.

12.8.6 Moisture content of food

As moisture is removed during frying the oil uptake increases and many studies suggest that foods with higher initial water content lead to a higher oil content (Gamble and Rice, 1988 and Lulai and Orr, 1979). Although mathematical models have been developed to link initial moisture content and oil uptake (Moreira *et al.*, 1995) a variety of studies have not been able to establish a relationship (Ufheil and Escher, 1996).

It has been found that raw potatoes with high specific gravity (SG), high solids content and starch levels, tend to produce chips that have a lower oil content (Lulai and Orr, 1979). In this study Norchip potatoes with SG from 1.060 to 1.110 were divided into 49 samples. It was found that the oil content of the chips produced from these correlated with the SG, such that:

Oil content (%) = 329.11 − 266.10 (specific gravity)

for specific gravities between 1.060 and 1.110; (i.e. SG = 1.06, oil content = 47.04%; and SG = 1.110, oil content = 33.74%).

The above was supported by Selman (Selman, LFRA Proceedings, 1989) where it was reported that the dry matter of potatoes was dependent on variety, growing and storage conditions and that these might affect the oil content of fried potato products. For example, it was suggested that as potato starch is heated, amylose is lost from the cell and may contribute to intercellular adhesion. It was also reported that Racenis (1959) found that as the amylopectin content of the potato increased the oil content did the same. As will be seen from Table 12.23, fat contents exceeding 40% were obtained if the SG was lower than 1.085.

Lisinska (1989) studied the relationship between oil content of chips and other components of potato tubers and found a negative link existed between oil uptake and dry matter and starch content (see Table 12.24).

12.8.7 Frying time and temperature

It is generally agreed that as frying time increases the oil content of the product does likewise. Gamble *et al.* (1987) conducted batch-frying experiments using Record potatoes and found that the resulting oil content of the product was proportional to the square root of frying time. As an approximation, it was suggested that the oil content was twice the square root of frying time in

Table 12.23 Influence of specific gravity on oil content of chips (from Lisinska, 1989)

Specific gravity of potatoes (gcm^{-3})	Approximate oil content of chips (%)
1.060	47.0
1.065	45.7
1.070	44.4
1.075	43.1
1.080	41.7
1.085	40.4
1.090	39.1
1.095	37.7
1.100	36.4
1.105	35.1
1.110	33.7

seconds. It was also observed that as frying temperature increased from 145, 165 to 185°C. the oil content of the product was reduced for a particular frying time. However, the same researchers found that oil content was not closely related to frying temperature but was directly related to the remaining moisture present.

Mohamed *et al.* (1998) investigated the effect of certain food ingredients on oil absorption and the crispness of fried batter based on rice flours. It was found that crispness was positively correlated with amylose content whilst oil absorption was negatively related. The addition of pregelatinised starch improved crispness, but also increased oil absorption which was considered to be a result of increased porosity of the product. The inclusion of different protein sources (egg yolk, gluten, skimmed milk ovalbumin and whey) was investigated; of these ovalbumin reduced oil uptake. As observed by Ng *et al.* (1957), the addition of calcium reduced uptake. However, in both cases it was important that the inclusion of calcium and ovalbumin did not exceed optimum concentrations since this would reverse the trend and increase oil absorption. The use of modified tapioca starch and diglyceride emulsifiers increased oil content of the product. The best recipe for a crisp batter with low oil content included: 30 g/kg ovalbumin; 20 g/kg oil; 20 g/kg emulsifier and a water/flour ratio of 2:1; 850 g/kg pregelatinised rice flour; 150 g/kg tapioca starch; with an amylose/amylopectin ratio of 18:67.

Table 12.24 Effect of dry matter and starch content in potato tubers on oil content of chips (Lisinska, 1989)

Dry matter (%)	Starch (%)	Approximate oil content of chips (%)
18.95	12.75	43.1
19.83	14.37	42.5
21.24	16.31	41.6
23.30	17.44	36.1
27.14	20.81	34.0

Through the development of some innovative processing, Lujan-Acosta and Moreira (1997) made tortilla chips with lower oil content as a result of impingement drying. They found that air impingement-dried tortillas lost moisture significantly more quickly with increasing drying air temperature. It was concluded that the drying rate increased by increasing: (i) the temperature of the drying air, and (ii) increasing the convective heat transfer coefficient. The texture of the product was influenced greatly by the drying air temperature, which was crisper at higher temperatures (e.g. 177°C). The low-fat product (14.3%) (traditional chips = 22–28% fat), was produced by first baking the tortillas to lower the water content from about 55% to 40%. This was followed by application of air impingement drying at 177°C which reduced the water content to 10–15%. The partially cooked product was finally fried at 200°C for half a minute. The low-fat tortilla chips were considered to have a good flavour and texture.

Tseng *et al.* (1996) found that the quality of the frying medium (fresh or abused) played an important role in the finished product. Although the oil content of the chips was not influenced by the oil quality, it was observed that the oil adhering to the surface of the tortilla was significantly higher in the tortilla chips fried in degraded oil than in fresh oil.

The absorption of oil by food during frying is complex and a variety of variable findings have been reported in the literature. What follows, however, is an attempt to summarise some of the findings from the literature (Gebhardt, 1996).

Frying oil temperature: increasing the frying oil temperature tends to decrease oil uptake because the product spends less time in the fryer. It might be that this process is aided by the formation of a crust which acts a barrier to further oil uptake. In addition, it might prevent water from leaving the food to an extent and consequently hinder the ingress of oil. However, it is important to find the optimum frying temperature to prevent a semi-raw and oily product as a result of too low a cooking temperature and a burnt and only partially cooked product from too high a frying temperature. (Note that the crust will form as a result of dehydration and reactions between amino acids, carbohydrates, lipids and their breakdown products.)

Specific gravity (SG) of potatoes: High specific gravity is equivalent to high solids and low moisture. The higher the solids, the lower the oil content of the product and the higher the chip yield. This is the position that potato processors want to achieve. However, if the SG of potatoes is low (high moisture and low solids) will lead to higher oil uptake with a lower yield.

Frying time: For all products there is an optimum cooking time for an optimum cooking temperature. If the temperature is constant, but the frying time exceeds the optimum, then the product will tend to have a higher than desirable fat content. Conversely, if the product is not given sufficient time to cook, it will not release the retained moisture and the starch will not develop as required thereby yielding a semi-raw food.

Frying medium quality: If the frying medium is highly oxidised and polymersied a number of changes will have occurred within the oil. The polymer and polar compound content of the product will be high. This results in an increase in the time required for the oil temperature to recover following immersion of the raw product. Two compounding issues arise: (i) firstly, the cooking product will tend to be exposed to a lower temperature as the oil struggles to recover its initial temperature; and (ii) the more viscous medium will tend not to drain from the product as rapidly as a fresh oil. These two aspects mean that frying in an old abused oil will tend to increase the fat content of the product.

Potato slicing: finally, if the blades used to cut through a potato, for example, are over-used and blunt, they are likely to tear the product rupturing more cells, producing an uneven surface. This will lead to a more porous surface and a greater surface area which will in turn lead to an increase in oil content, a relationship easily overlooked.

12.9 Effect of frying techniques, frying regime, cooking method, and additives on flavour of fried food

Yang *et al.* (1994) studied the sensory and nutritive qualities of pork strips, which contained little attached fat following cooking, using: (i) microwave oven; (ii) stir-frying; and (iii) broiling. The tests were conducted in triplicate and the sensory evaluation conducted by a trained panel of 14 men and women. Those pork strips that were stir-fried were more brown and tender than those microwaved or broiled. Overall the sensory characteristics of those slices stir-fried had a more appealing flavour. While this observation is not greatly surprising, it was interesting to note that greater quantities of vitamin B_6, thiamin, iron, magnesium and zinc were retained in the products that were fried.

John *et al.* (1992) examined the sensory qualities of raw jack fruit following steam cooking and frying at 150±30°C for 5 minutes and 180±20°C for five minutes, respectively. Sensory evaluation revealed that the texture and flavour of the samples were more acceptable than those of the steam-cooked samples. Not surprisingly, the fat content of the fried jack fruit (8.2%) was significantly greater than that of the steam-cooked version (0.6%).

Waimaleongora-ek and Chen (1986) studied the effect of frying shortening quality and subsequent holding conditions on selected flavour volatiles in deep-fat fried chicken parts. It was established that holding fried chicken pieces at: (i) 60°C for 3 hours; (ii) 2–4°C for 24 hours; and (iii) −18°C for 24 hours did not influence the flavour volatiles examined. Reheating refrigerated and frozen chicken parts also had no effect on the carbonyl and hydrogen sulphide contents except in the skin and coating portions of the frozen parts. The methanthiol and free ammonia contents of the refrigerated skin and coating portions decreased during the reheating process.

The relationship between the intensity of warmed over flavour in cooked stored pork prepared by different methods of cooking (i.e. microwaving, boiling in water, pan-frying and contact grilling) at varying temperatures, with different energy inputs and with different times of heating was examined by Satyanarayan and Honikel (1992). Contact grilling produced a product that did not give rise to off-odours. In contrast, microwave cooking produced significant undesirable aromas which were detectable immediately following cooking and increased on storage. This was not observed with the other forms of cooking, in which a few hours storage were required before unpleasant odours were released. In conventional cooking with water and in well roasted samples from pan-grilling, no detectable odours were seen for over 2 days. It was noted that the temperature required to induce surface browning of meat was more important than the time of exposure. This also affected the detection of odours following storage. It was found that grilling at lower temperatures and for shorter times produced higher warmed over flavour when compared with higher temperatures. In grilling and pan-frying, fried flavour notes dominated.

Use of meat minced immediately before cooking, small variations in pH and total fat, did not influence odour formation. In contrast, however, it was found the use of stored minced meat adversely affected odour after cooking and storage in comparison with non-minced meat. It was found difficult to relate the thiobarbituric acid (TBA) values with the sensory evaluations conducted. In the opinions of Satyanarayan and Honikel (1992), microwaving was not a good means of preparing or reheating meat owing to the high susceptibility of it producing unpleasant flavours.

Higgins *et al.* (1999) examined the effect of feeding tocopherol to turkeys. These were divided into two groups and fed: (i) either 20 or 600 mg *all-rac*-α-tocopherol acetate per kg food for 21 weeks prior to slaughter. Breast and leg meat was removed and four batches of patties were produced from each. Two batches were from birds fed 20 mg tocopherol/kg food, one of which contained 1% salt. Two similar batches were produced containing meat from birds fed 600 mg/kg food, again one batch had 1% salt added. The patties were fried, cooled and wrapped in high-oxygen-permeable film. The patties were displayed under fluorescent light (616 lux) at 4ºC. Lipid oxidation was assessed by measuring thiobarbituric acid reacting substances over a ten-day period. Taste panels assessed the warmed over flavour (WOF). The patties from birds fed 600 mg tocopherol/kg food were the most stable to lipid oxidation. Salt promoted oxidation. The patties produced from birds that had been fed tocopherol produced less WOF than the control patties and those to which salt had been added. A linear relationship was found between thiobarbituric acid reacting substances (TBARS) and WOF for all batches tested.

Concern has been expressed at the potential for food-borne micro-organisms from chickens to cross-contaminate food preparation surfaces or to induce food-borne illness directly. Hathcox *et al.* (1995) treated 180 whole raw chickens with: (i) tap water-control; (ii) 12% trisodium phosphate; (iii) lactic acid-sodium benzoate (0.5%/0.05%), in an attempt to reduce the bacteriological load of the birds. Panelists assessed raw, treated whole chickens and fried breasts and thigh pieces. The results indicated that the application of trisodium phosphate, or the lactic acid/sodium benzoate solutions did not affect external colour, texture, flavour or overall acceptability of fried chicken. Nor did it appear to influence consumer purchase intent. However, when compared with controls, raw chickens 90 minutes after treatment with lactic acid/sodium benzoate or after 7 days storage at 1°C were judged to be of poorer quality. The sensory quality of the chickens treated with trisodium phosphate was not adversely affected. The authors concluded that the treatment of the raw chickens with 12% trisodium phosphate or lactic acid/sodium benzoate were suitable for the treatment and did not cause adverse organoleptic degradation of the product.

12.10 The influence of the food being fried

The following is a very brief review of a handful of reports in the literature regarding some fried foods. The food being fried will release a variety of compounds into the oil during the frying process. These can adversely affect the oxidative stability of the oil and therefore products fried in it. For example, Pokorny (1998) states that during the frying of cabbage and brassica vegetables, glucosinolates are released and decompose to produce nitriles, indolyl derivatives or vinyl oxazolidinethione. Chloropyhlls and their decomposition products the pheophytins pass from food into the frying medium darkening the oil. Vitamins, such as ascorbic acid, pyridoxin, riboflavin and thiamine, are broken down on heating and, if oil soluble, decomposition products can leach from the food to the oil giving rise to increased odour and flavour. This can result in an increasing rate of oil oxidation (Pokorny, 1998).

It is also true that the reverse can happen. The oxidation of frying oils may be reduced or inhibited by antioxidants leached from the food being cooked, providing that they are not volatile at frying temperatures. Such antioxidants include sulphur compounds, ascorbic acid and phenolic substances. Spices such as ginger oleoresin have been found to have antioxidant activity in soyabean frying oil (Kim and Ahn, 1993). It has also been found that rosemary and sage oleoresin are active under frying conditions and are not volatilised at frying temperatures. It was found that acetone and ethyl acetate extracts inhibited the formation of polymers in oil during French fried potato production (Reblova *et al.*, 1997). Porkorny (1980) found that oil-soluble material from carrots, potatoes or oat flakes reduced the rate of lipid oxidation in both frying oil and hydrogenated frying oil. It has also been observed that phospholipids reduce oil oxidation (Kourimska *et al.*, 1994) and protected tocopherols (Kajimoto *et al.*, 1987).

While benefits can be derived by antioxidant material leaching from the food being fried into the frying oil and thereby adding protection, it should be noted that more often, the reverse occurs. For example, oil leaching from the food can introduce emulsifiers, FFA, alkaline reacting material and trace metals (Rossell, 1989). The topic of potato product frying and the influence of oil quality has been reviewed by Smith (1987).

As Rossell (1989) points out, the introduction of foreign oils is of greatest concern when fatty fish is fried. This arises because of the highly unsaturated nature of certain fish oils which can oxidise rapidly. For this reason, Rossell recommended that fish products should be battered before frying to inhibit the unsaturated lipid from leaching into the frying oil. It is also reported that dairy fats should be avoided because they contain short-chain fatty acids which are more polar and volatile and prone to hydrolysis and, therefore, reduce the smoke point of the oil and increase the likelihood of excessive foaming. The accidental presence of emulsifiers from food in a frying oil is also likely to cause foaming. However, particular care must be exercised if products are likely to introduce copper or iron to the oil as these metal ions, particularly the former, are lipid pro-oxidants.

12.11 Sensory issues

The text that follows gives a brief insight into the biology of taste and odour perception, and describes some of the vocabulary and techniques available to assess the sensory characteristics of fried food.

12.11.1 The biology of flavour and smell

Highly sensitive and specialised sense organs are found in the tongue and mouth and contain the receptors for the taste sensation. These receptors are found in clusters of approximately 50 cells in a layered ball called a taste bud. These cells are more like skin (e.g. epithelial) than nerve and exist for a few days. Cells differentiate from the surrounding tissue and link into the taste bud structure and there make contact with the sensory nerves. The upper part of the taste bud makes contact with the saliva in the mouth and the molecules giving rise to flavour are considered to bind to the cilia. The taste cells connect with the primary taste nerves. A series of neurotransmitter molecules are released into the synaptic gap to stimulate the primary taste nerves and the sensation of taste is sent to the brain. (Lawless and Heymann, 1998).

12.11.2 The biology of odour

Although it is not necessarily well known, it is considered that the largest contribution to the flavour and taste actually arises from volatile compounds detected by the olfactory receptors. To illustrate this the following example has been taken from Lawless and Heymann (1998). The organoleptic characteristics of lemons are not derived from the taste associated with lemons (which can be broken down into sour, sweet and bitter) but from the aroma compounds terpene (particularly limonene) that volatilise in the mouth and flow up the nasal cavity from the rear direction (retronasally). In effect, this is the opposite to breathing in aroma compounds. Consequently, the organoleptic properties of a food may be detected nasally from afar and, once consumed, retronasally.

The two olfactory receptors are small (approximately 1–2 cm^2) portions of epithelium located high in the nasal cavity. The location of these sensitive areas might be protective but it does mean that only a small fraction of the volatile compounds in the air will reach the odour-perceiving sites. However, to overcome this there are several million receptors on either side of the nose. These contain many cilia which are short hair-like structures approximately 10 μm long. Their purpose is probably to increase the surface area onto which the odour inducing volatiles can react. Furthermore, the millions of receptors send signals to approximately a thousand glomerular structures in the olfactory bulb. This is an area of dense synaptic contact of the olfactory pathway. Although there are many branches from one nerve to another, with the potential to dilute the signal, there are also many opportunities for the initially weak senses to be concentrated onto the next level of nerve cells in the pathway to higher neural structures of the brain. It has been suggested that many of these olfactory routes link closely with emotion and memory centres in the brain.

12.11.3 Why conduct sensory evaluation?

The above briefly describes the workings of the human sense flavour and odour organs and indicates that it is a complex and interrelated system and that, to an extent, flavour and odour cannot be separated completely. Chemical methods of analysis such as Rancimat analysis can provide an indication of the induction period of an oil, and PV, TBA and *p*-anisidine values can give an indication of the oxidative state of the oil in a food. Chromatographic systems now exist that can separate and detect a very large number of volatile flavour/aroma compounds. However, the issue is complex because as human beings we tend to perceive flavours and aromas as a whole rather than reducing them to individual components. Consequently, the use of sensory evaluation is a very useful adjunct to many analytical techniques. We should always remember that customers rely on the odour, texture and taste of food in deciding whether to repeat purchase.

12.11.4 Fried food flavour

The flavour of fried food can be determined immediately following cooking or after ageing to mimic storage condition and hence give an indication of shelf-life. While there is no reason not to taste and test the flavour quality of any fried product (e.g. chicken), sensory evaluation of fried potato products is often conducted because, in comparison with meat and fish, it does not provide strongly competitive tastes and aromas nor do potatoes contribute much lipid, unlike meat products. Frankel *et al.* (1985) developed a system for assessing the immediate and longer term stability of fried food. In this work squares of white bread were used for frying and storage tests. The resulting fried bread was tasted immediately after production and after storage at ambient or higher temperatures (accelerated testing).

Although the flavour components produced during frying are numerous and dependent on the oil, the food being fried and the conditions of use, the consumer is very sensitive to the aroma of the frying medium during cooking and will reject it if the oil gives rise to unpleasant odours. Methods were developed thirty years ago (Evans *et al.*, 1972) to assess the room odour of heated oils and this will be discussed further.

12.11.5 Sensory panels

Kathleen Warner (1996) has produced an excellent review of flavours and sensory evaluation of fats and fatty foods and it is suggested that readers might wish to peruse this for detailed information. What follows here is a summary of the points contained within the above with reference to other works as appropriate.

There are two main types of panels, with the main difference being the degree to which the panelists are trained. Consumer panels can be used as a means of determining market acceptability of a product and are particularly useful because the panel is often reflective of the purchasing public. In general, consumer panels would assess prepared foods. The design of sensory panels can vary and with increasing complexity, the four main types are described very briefly below.

Basic panel: this aims to answer fairly straightforward questions such as whether a recently arrived oil supply has the same taste as a previous batch. The room in which the tasting takes place should be odour-free and quiet. The sensory analysts should be assessed before the trial begins to ensure that they are capable of detecting simple differences. The panelists can be trained further to not only identify differences but also to provide information as to why the differences arise. This might involve the panelists being trained in scalar scoring and descriptive analysis.

Intermediate panel: the objective of the panel is to identify the type of difference and also to quantify this; evaluation is often conducted to assess effects of processing or ingredient changes. The panelists should be trained to use a scoring scale (e.g. 0–2 very poor; 6 fair; 10 excellent) and to rate the quality of samples. It is important that the assessors are separated from one another while conducting analysis to avoid one panelist inadvertently influencing another. The data from the analysis are frequently subjected to analysis of variance (ANOVA).

Upgraded intermediate panel: this is usually implemented in product development and also where it is important to relate sensory and instrumental analysis results. The panel should be trained in descriptive analysis to describe fully the odours and flavours detected in products. The environment in which the evaluation takes place should have an area for the panelists but, as before, with dividers between each to prevent discussion during assessment. The evaluation should let onto an area in which the products are produced. Clearly, this preparation room should not be seen by the analysts. Lighting should be used to ensure that differences in product colour or texture are concealed if this is appropriate.

Sophisticated panel: a company conducting such analyses would have a high throughput of samples and might have a number of different sensory panels running each day. In such a system it is likely that communication between the sensory scientists running the trial and the panelists will be electronic. This is usually accomplished by computer input/output of data; panelists should be separated. In effect a series of questions regarding the sample may be posed to the panelists via a computer screen and the analysts respond accordingly.

12.11.6 Assessment of frying oils and fried food

Evans *et al.* (1972) devised a system whereby a pan of oil was heated in an empty room. Panelists entered the room and assessed the oil odour and characteristics. This procedure was refined by the introduction of rooms of specified size and controlled airflow and temperature (Mounts, 1979). Warner *et al.* (1985) established that high-oleic sunflowerseed oil and low-linolenic acid soyabean oil gave low values when heated at 190ºC under these conditions. In contrast, low-erucic rapeseed oil (LEAR) and olive oil gave high values (7–8) indicating a strong odour. In the case of LEAR, this may be a result of the linolenic acid oxidation, and, in the case of olive oil, the volatile flavours and odours associated with oils that have not been fully refined.

Fried food can be assessed either immediately after cooking or after storage. It is suggested that the panelists assess the overall flavour of the fried food and whether there are additional flavours from the oil and, if so, whether these are objectionable or pleasant. A quality scale is developed for assessment of fried food (i.e. 2 weak; 5 moderate; and 8 strong). This scale should be used to assess the following criteria of fried food: stale; fishy; hydrogenated; waxy; rancid; and painty. With the vast bulk of sensory panels, it is important to ensure that adequate control samples are provided to act as a reference for the sensory scientists. It is possible to link the sensory analysis to complex chemical analysis (e.g. GC, GC-MS with dynamic or static headspace).

12.12 Application of flavours

This section will address:

- flavour definitions
- flavour carriers
- ingredients available (e.g. herbs and spices, oleoresins and essential oils, etc.)

- free flow agents
- flavour enhancement
- the importance of fat in flavour issues and fat reduction
- salt
- flavour modification
- flavour compounding
- formulation and production considerations (e.g. factors affecting the application techniques)
- application techniques paying particular attention to fried foods. Legal issues are not discussed.

12.12.1 Definition of a flavouring

A flavouring may be defined as an additive consisting of material used or intended for use in or on food to impart odour, taste or both, provided that such a material does not consist entirely of:

- any edible substance (including herbs and spices) or product, intended for human consumption as such with or without reconstitution, or
- any substance which has exclusively a sweet, sour or salt taste, and the components of which include at least one of the following:
  - natural flavouring – a single flavour chemical or concentrate derived by physical methods (i.e. grinding or drying but not chemical or biochemical methodologies) from a natural source;
  - nature identical flavouring – a single chemical derived by chemical synthesis or process that is chemically identical to a natural flavouring substance;
  - artificial flavourings – a single chemical derived by chemical synthesis not chemically identical to a natural flavouring substance (comparatively few of these substances are now used);
  - smoke flavourings – smoke extracts used in traditional smoking processes. These might for example include distillation of smoke and cover discrete categories which are tightly governed through legislation.
  - processed flavourings (including enzymically modified cheeses (EMC) and enzymically modified dairy products (EMDI)) are formed by the reaction between proteins and carbohydrates including the lipid fraction, with the selection of the starting material greatly affecting the flavour of the finished product. EMCs may be added to biscuit dough containing cheese powder, this will reduce the amount of cheese powder required to give the same cheesy flavour and thereby reduce the cost of the biscuit.
  - natural flavouring preparations are formed by multiple microbial and enzymatic processes, and, as before, selection of the starting materials modifies the finished product flavour.

Table 12.25

	Natural	Nature identical	Artificial
Advantages	Authentic pack claims.	Economic, can use in concentrated form, authentic.	Economic, can use in concentrated form.
Disadvantages	Can provide a 'weak' taste	May lack some flavour notes	Can be perceived as unhealthy

There are advantages and disadvantages of using natural, nature identical and artificial flavourings as shown in Table 12.25.

12.12.2 Flavour carriers

These can act as a means of protecting the flavour during its application to food as well as a means of taste delivery to the consumer. The main systems include:

- adsorption – involving dispersion of the flavour (e.g. an oleoresin) on to a cheap powder carrier (e.g. salt);
- spray drying – the flavour is mixed with a maltodextrin (partially hydrolysed starch) or gum prior to spray drying;
- spray chilling – involves dissolving or dispersing the flavour into a high melting point fat (e.g. palm oil or partially hydrogenated palm oil). This is followed by spray drying in a cold chamber – the incorporation of the flavour into a fat that melts at approximately 32–34°C. These flavours provide: (i) a pleasant mouthfeel; and (ii) a rapid release of flavour as the fat melts and consequent cooling sensation.
- encapsulation – the flavour is covered with a high melting point fat or gelatine.

On going from adsorption to spray drying to spray chilling and encapsulation, the degree to which the flavour is protected from loss and damage increases; however, so does the cost.

The flavour carrier systems above can undergo one of three further processes:

1. agglomeration – in which the flavour particle is increased in size to ease blending;
2. freeze drying – is applicable only when the water content is high but then provides a rapid reduction in water content;
3. roller drying – a traditional technique in which the liquor is passed between hot rollers and then ground.

12.12.3 Ingredients: herbs and spices

The trade in herbs and spices dates from Chinese and Egyptian use for medicinal and culinary purposes with records dating back to around 3000BC. Herbs may

Table 12.26

Herb	Active ingredient	Spices	Description/active ingredient/use
Bay	Cineole/eucalyptol	Pepper/Ginger	Pungent spices
Thyme, Oregano	Thymol/carvacrol	Nutmeg/Mace	Aromatic fruit
Basil	Sweet alcohol	Cumin/Anise	Umbeliferous fruits
Sage	Thujone	Cassia	Aromatic barks
Mint	Menthol	Clove	Phenolic/Eugenol
		Paprika/Saffron	Colour

be described as soft stemmed plants that die back after flowering and have culinary or medicinal properties. They are often grown in temperate climates such as the Mediterranean; for example:

- Rosemary from Spain and France
- Thyme from Morocco
- Oregano from Turkey and Greece.

Spices are vegetables that provide aromatic or pungent odours and flavours. Such products are grown in hot climates within a few degrees of the Equator. For example:

- Chilli/Capsicum from Mexico and India;
- Ginger from Africa and the Caribbean;
- Nutmeg from the East and West Indies;
- Cinnamon from Sri Lanka and Seychelles.

Herbs and spices may be classified as shown in Table 12.26. It is easier to be more precise regarding the active ingredient in herbs than in spices.

There are two common reasons for applying herbs and spices: improving the product's (i) appearance (parsley, chilli and paprika); and (ii) flavour (basil, coriander, cayenne, chilli and cumin). The flavours are often added as either essential oils or as oleo resins. Essential oils are the principal but not sole flavouring component of herbs and spices and may be dry distilled, vacuumed distilled or dry pressed, and tend to be comparatively pure. Oleo resins are substances that contain a variety of components including: essential oils, resins and certain non-volatile fatty acids. The addition of flavours derived from herbs and spices can be made in a variety of ways each of which have benefits and drawbacks, as shown in Table 12.27.

12.12.4 Free-flow agents and anti-caking agents

To ensure that a powdered flavour is added in the required and consistent manner, it is important to include a free-flow agent in the flavour's composition. The seasoning mixture is likely to contain a complex mixture of particles. These might be described as rough, smooth and cubic (e.g. sodium chloride), spherical,

Table 12.27

	Dried	Essential oil	Oleo resin
Benefits	Slow flavour release; easy to handle; few labelling issues; good appearance	Standard flavour; no colour; good stability; less storage space required	Standard flavour flavour similar to dry product; good shelf-life; less storage space required.
Drawbacks	Bulk quantity; flavour variation; adulteration and contamination possible.	Oil and flavour can be lost at high temperatures; different flavour profile and release; ease of oxidation; difficult to disperse.	Highly concentrated and viscous, difficulty in measuring correct dose.

irregular, elastic and adhesive. The addition of the free flow agent reduces the internal cohesive forces but in doing so it is necessary to strike a balance to ensure that the flavouring is added evenly throughout the product. For example, it is important to obtain a mixture that has physical characteristics somewhere between 'flooding' (where the seasoning over-runs in a manner similar to the way salt pours) and 'flushing' (analogous to the flow of flour).

One of the most commonly used free flow agents is silica (i.e. silicon dioxide SiO_2) a naturally occurring mineral. It has the ability to adhere to the surface of particles within the seasoning and thereby narrows the particle size distribution. It is important, however, to ensure that the silica is not over-mixed as this may drive the silica into the particles and reduce its efficacy as a free flow agent. In addition to free flow agents, anti-caking compounds are frequently added to control the moisture content of the seasoning. Perhaps the most commonly added anti-caking agent is tri-calcium phosphate [$Ca_3(PO_4)_2$]. Derived from a naturally occurring mineral it helps to prevent clumping. The addition of the free flow and anti-caking agents should be made at the end of the seasoning production and should involve gentle blending for as short a time as possible.

12.12.5 Flavour enhancement

Flavour enhancement has been linked to 'Umani' and some consider it to be the fifth element of taste perception, the others being salt, bitter, sweet and sour. Flavour enhancers are compounds that have the ability to heighten or improve the perceived flavour of food, without imparting a particular taste of its own. Sodium chloride (common salt) is the original flavour enhancer but owing to dietary health concerns associated with hypertension, its use is now more circumspect. Lower sodium products often involve complex blends with potassium chloride, magnesium sulphate and L-lysine hydrochloride. Other enhancers include: mono-sodium glutamate; ribosides such as salts of 5'-inosine monophosphate (IMP) and 5'-guanine monophosphate (GMP), proteinaceous substances, sugars, and herbs and spices.

The mechanism by which components enhance flavour is complex and not completely understood. However, it appears that 5'-nucleotides have the ability to bind on cells responsible for transmitting the taste sensation to the brain. This induces the exposure of more taste receptors and this leads to an enhancement of flavour perception. There are a number of alternative flavour enhancers. Yeast, yeast extracts and hydrolysed vegetable protein (HVP) impart enhancing effects due to the naturally occurring MSG together with complementary effects from protein hydrolysis.

Hydrolysed vegetable protein
HVP is predominantly produced by acid catalysed hydrolysis of plant protein, with the raw material generally being spent grains and nuts from vegetable oil production. HVP tends to impart flavour profiles associated with meat which may be amplified as a result of it containing ribotides and glutamates.

Sugars and artificial sweeteners are also employed as flavour enhancers. For example, maltol, and its ethyl ester, can improve flavour, by maximising sweetness and increasing creaminess; this masks bitterness. Onion powder is frequently used as a base in cooking as it provides a flavour enhancing effect in savoury products which is considered to arise as a result of its complex constituent sugars.

Acidic mixtures such as acetic acid/sodium diacetate and fruit acids (e.g. citric, malic and tartaric acids) are used to add flavour notes of their own but also to impart a succulence to snack foods. Modified starches (e.g. maltodextrins) may be adopted to provide adhesion for powdered flavours and increase fat-like mouthfeel.

12.12.6 The importance of fat and low-fat products

Reducing or eliminating the fat from foods causes a number of problems that need to be overcome. Fat provides many attributes. Firstly, it provides gloss, colour, and opacity. Secondly, it imparts a flavour of its own (and in the case of fried food, its oxidation products can impart additional flavour characteristics) and can also act as a carrier of other flavours. Furthermore, fats with the correct physical characteristics can release flavour quickly as they melt rapidly in the mouth, which also leads to a pleasant cooling sensation. Thirdly, fat can provide structure, texture, consistency, mouth coating, and lubrication and succulence. Finally, it has non-sensory effects the most notable being satiety.

Low-fat products tend to be baked rather than fried. The application of flavours involves the use of gums or low percentage dose rates of spraying oil onto the product in order to immobilise the flavour powder.

12.12.7 Salt

Like fat, salt is a component of fried snack foods that many wish to reduce or eliminate altogether. Processed foods are considered to contribute up to 80% of the salt in the diet, although it should be pointed out that chips, despite their comparatively high salt content, do not form a major part of an adult's salt intake. It is considered that the salt perception in the mouth changes if the levels of salt to which one is exposed are altered. Lower-sodium salt mixtures containing potassium chloride are used occasionally. Reducing salt concentrations can lead to a reduction in the attractive salty taste associated with certain fried snack products and in the case of sodium reduced 'salt' impart a non-characteristic flavour often identified as bitter or metallic.

12.12.8 Flavour modification

In developing any flavour, it is necessary to appreciate that the development chemist is building up a taste profile and that, on occasions, it is necessary for this to be modified, which may be achieved in a variety of ways. For example, harsh acidity introduced by the addition of an acid can be reduced by adding sodium citrate, sodium acetate, or maltol/ethyl maltol. Sweetness and bitterness can be moderated by the inclusion of maltol/ethyl maltol.

Some herbs and spices can modify the perception of flavours. For example, Szechwan pepper contains hydroxy-α-sanahool, which stimulates both pain, cold, and touch receptors in the mouth. Gymnema sylvestre contains gymnemic acid, which is able to numb sweet receptors, and synsepalum dulcilicum contains taste-modifying enzymes that can alter the perception of sour to sweet.

The stereochemistry of compounds can have an effect. Sugars for example are generally considered to be sweet, however, L-glucose is slightly salty, while the naturally occurring version, D-glucose, is sweet as would be expected. In a similar vein, D-carvone from caraway seeds is spicy, while L-carvone in Spearmint provides a sweet mint flavour.

12.12.9 Flavour compounding

The flavour cocktail can be viewed as a pyramid and its development takes place in three phases. The first involves base flavours (e.g. yeast extracts, salt, MSG, and HVP). The second group involves specific notes from process flavourings and preparations (e.g. from modified cheese). Finally, top notes are added consisting of compound flavourings, natural extracts and herbs and spices (e.g. cheese flavour and onion oil).

12.12.10 Formulation and production considerations

With regard to particle characterisation, the size distribution should be narrow, all particles should have similar diameters (i.e. 150–350 μm). Such product characteristics allow the particles to move in a similar manner and to flow appropriately. Usually, the flavour will be dampened with oil to hold the cocktail together preventing separation.

In terms of handling the flavour mixture, it is important not to over-mix as this can reduce the efficacy of the free flow agent. Equally important is the need to ensure that the optimum amount of free flow agent is added to provide the correct flow characteristics. When applying the cocktail in the production hall, it is necessary to consider the level of free flow. Too much can lead to high levels of dust in the atmosphere and poor application, the cocktail will quite literally free flow freely off of the substrate. Too little and it will not flow evenly giving rise to patchy application.

12.12.11 Application techniques

The substrate or base is obviously very important. A potato crisp is a coarse surface while an extruded snack would be comparatively smooth. If the flavour cocktail were to be dusted on to an extruded base the flavour would not adhere well to the surface. The method of application varies, generally being dependent on the substrate. For example, in potato chips the flavour is dusted on as the product's surface is rough and it has a high oil content. This standard technique may be augmented by using electrostatic equipment to help in the transfer of the flavour on to the base. In many lower-fat snacks, it can be added by dusting on to an oiled or pre-gummed base, or as a slurry either using oil or gum solutions. In baked products (e.g. biscuits) the flavours can be baked in with an additional topical seasoning if desired. The factors that affect the adhesion of the flavour cocktail include seasoning flow properties, humidity during storage and application, application rate, drop points on the application line, and the quality of the raw materials.

There are three types of dust on units: a simple static type that is vibrated flat or inclined; one that is shear edged which provides a greater area of application; and a combination of weir and shear edge. The latter is perhaps the most effective and delivers the least pulsing of flavour thereby ensuring a greater degree uniformity of application. An electrostatic charge is applied to some seasonings or slurries, which is then attracted to the product. This assists in the overall dispersion. On occasions, flavours are added in an oil slurry. In a batch process a known concentration of flavour in oil is prepared. In a continuous process, a known quantity of flavour and oil are metered, mixed and sprayed onto the product.

12.13 The future

There is little doubt that sales of snack foods and particularly fried food will continue to grow. This arises because of their pleasant flavours, textures and convenience. There are, however, concerns over dietary fat and this may lead to an expansion in the low-fat snack and fried food markets. It is also possible that use of synthetic fat replacers will increase. Perhaps the most well-known of these is olestra, a generic name given to sucrose polyesters produced by Procter and Gamble (IFST, 1999). At the molecular level this group of compounds is composed of sucrose esterified to 6 to 8 fatty acid residues. In effect, the sucrose molecule acts as the backbone to the fatty product in the same way glycerol does in conventional triglycerides. Olestra can be composed of any fatty acid of choice. For example, it is possible to esterify the fatty acids from a conventional oil (e.g. rapeseed or palm) to sucrose by a variety of processes and thereby produce a rapeseed or palm oil olestra.

The physical and chemical properties of the resultant product depend greatly on the fatty acids esterified to the sucrose backbone. Taking the above as examples, the rapeseed oil olestra would be liquid and prone to the flavour developments associated with conventional rapeseed oil, and the palm oil olestra would be semi-solid. The importance of olestra comes from its indigestibility and consequently from its zero calorific value. This arises because the 6–8 fatty acids that surround (all of which should contain 8 or more carbon atoms) prevent lipolytic enzymes from hydrolysing fatty acids from the sucrose and, consequently, the olestra is not absorbed and does not provide any dietary energy. It does, however, provide the taste, mouthfeel and aspects of flavour development associated with conventional oils.

Olestra has been granted approval by the US Food and Drug Administration for use as a partial fat replacer in certain snack foods, subject to specified US labelling conditions and fortification with fat-soluble vitamins. Olestra is not yet approved in the UK and there is no application currently pending.

The use of genetic modification offers the potential to produce oils with fatty acid compositions that are resistant to excessive oxidation while maintaining the production of certain desirable flavour compounds. For example, Mounts *et al.* (1994b) extracted oil from three genetically modified soyabeans. The oils contained 1.7, 1.9 and 2.5% linolenic acid and were compared with oil extracted from a commercial grade of soyabean (Hardin) which had a linolenic acid content of 6.5%. It was found that the low linolenic oils were more resistant to oxidation and had greater flavour stability in accelerated storage tests. It was also observed that the low linolenic oils when stored at 190ºC for 1 hour or 5 hours, had lower fishy and acrid/pungent odour, respectively, than those from conventional soyabean oils.

It was also noted that the overall flavour quality of potatoes fried in the modified oils was good and significantly better than that produced in the high linolenic acid oil. Furthermore, potatoes fried in the modified oils did not give rise to fishy taints. It was concluded that by lowering the linolenic acid content the oil, quality increased both in terms of oxidative stability and flavour production. In addition, the flavour of food fried in the modified oils was better than that from conventional soyabean oil.

It has also been observed (Warner and Knowlton, 1997) that genetically modified corn oils containing 65% oleic acid had significantly lower total polar compound levels after 20 hours heating and frying at 190ºC in comparison with conventional corn (maize) oil. It was also explained that high oleic corn oil had a better flavour and oxidative stability in comparison with conventional corn oil after ageing at 60ºC and after high temperature frying.

It was also noted that potato chips produced from genetically modified low-linolenic acid canola oil (C18:3 3.7%) had a slightly greater stability than those fried in conventional canola oil (C18:3 10.8%) (Petukhov *et al.*, 1999). The above is a small sample of the literature available on the potential benefits of modifying oil composition and this may be the subject of significant further study during the forthcoming years.

12.14 Acknowledgement

The author would like to thank Simon Hewlett of Griffith Laboratories, Derbyshire, UK for his help in the production of this chapter.

12.15 References

ABDALLA, A.E.M. (1999) Antioxidative effect of olive oil deodorizer distillate on frying oil and quality of potato chips. *Fett/Lipid* 101, 2, 57–63.

BAROUDI, A. (1978) Evaluation of factors affecting colour and acceptance of potato chips. Ph.D. Dissertation, The Ohio State University, Columbus, Ohio, USA.

BELITZ, H.D. and GROSCH, W. (1999) *Food Chemistry*, 2nd edition, Springer-Verlag, Berlin, p. 211.

BLUMENTHAL, M.M. (1996) Frying Technology in Y.H. Hui Editors of Bailey's Industrial Oil and Fat Products. Volume 3: Edible oil and Fat Products: Products and Application Technology (5th edn) John Wiley & Sons, Inc. 605 Third Avenue, New York, NY, USA.

BREWER, M.S., VEGA, J.D. and PERKINS, E.G. (1999) Volatile compounds and sensory characteristics of frying fats. *Journal of Food Lipids* 6, 47–61.

BUTTERY, R.G. (1973) *J. Agric. Food Chem.* 21, 31.

BUTTERY, R.G., SEIFERT, R.M., GUADAGNI, D.G. and LING, L.C. (1971) *J. Agric. Food Chem.*19, 969.

BUTTERY, R.G. and LING, L.C. (1972) *J. Agric. Food Chem.*, 20. 698.

BUTTERY, R.G. and LING, L.C. (1973) *J. Agric. Food Chem.* 21, 745.

CARLIN, J.T., JIN, Q.Z., HUANG, T.C., HO, C.T. and CHANG, S.S. (1986) Identification of alkyloxazoles in the volatile compounds from French-fried potatoes. *J. Agric. Food Chem.* 34, 621.

CARLIN, J.T. HO, C-T., CHANG, S.S., VELLUZ, A. and PICKENHAGAN, W. (1990) Analysis of French fried potato flavor: identification of 3-(methlythio) alkanals. *Lebensm. Wiss. u. Technol.*, 23, 276.

CERNY, C. and GROSCH, W. (1993) Quantification of character-impact odour compounds of roasted beef. *Z. Lebensm. Unters. Forsch.*, 196, 4170–22.

CHAN, H.W.S. LEVETT, G. and GRIFFITHS, N.M. (1978) *J. Sci. Food Agric.* 29, 1055.

CHIU, E.M., KUO, M.C., BRUECHERT, L.J. and HO, C.T. (1990) Substitution of pyrazines by aldehydes in model systems. *J. Agric. Food Chem.* 38, 58.

CHUN, H-K. and HO, C-T. (1997) Volatile nitrogen-containing compounds generated from Maillard reactions under simulated deep-fat frying conditions. *Journal of Food Lipids*, 4, 239–44.

DECK, R.E. and CHANG, S.S. (1965) *Chemical Industry* p. 1343.

DECK, R.E., POKORNY, J. and CHANG, S.S. (1973), *Journal of Food Science*, 38, 345.

DORNSEIFER, T.P. and POWERS, J.J. (1963) *Food Technology*, 17, (10), 118.

DORNSEIFER, T.P. and POWERS, J.J. (1965) *Food Technology*, 19, (5), 195.

EVANS C.D. and SHAW, R. (1968) Proceedings of Third Annual Sunflower Conference, Crookston, MN p. 101.

EVANS, C.D., LIST, G.R., HOFFMAN, R.L. and MOSER, H.A. (1969) *J. Am. Oil Chem. Soc.* 46, 501.

EVANS, C.D., WARNER, K., LIST, G.R. and COWAN, J.C. (1972) *J. Am. Oil Chem. Soc.*, 49, 578.

FISCHER, J. and MULLER, K. (1991), *Potato Research*, 34, 159.

FRANKEL, E.N., WARNER, K. and MOULTON, K.J. (1985) *J. Am. Oil Chem. Soc.*, 62, 1352.

FUNAMI, T., FUNAMI, M., TAWADA, T. and NAKAO, Y. (1999) Decreasing oil uptake of doughnuts during deep-fat frying using curdlan. *Journal of Food Science*, 64, 883–8.

GAMBLE, M.H., RICE, P. and SELMAN, J.D. (1987) Distribution and morphology of oil deposits in some deep fried products. *J. Fd Sci.*, 52 (6), 1743-3 + 1745.

GAMBLE, M.H. and RICE, P. (1988) The effect of slice thickness on potato crisp yield and composition. *Journal of Food Engineering*, 8, 31–46.

GEBHARDT, B. (1996) Oils and Fats in Snack Foods in Y.H. Hui Editors of Bailey's Industrial Oil and Fat Products. Volume 3: Edible oil and Fat Products: Products and Application Technology (5th edn) John Wiley & Sons, Inc. 605 Third Avenue, New York, NY, USA.

GOBURDHUN, D. and JHURREE, B. (1995) Effect of deep-fat frying on fat oxidation in soybean oil. *International Journal of Food Sciences and Nutrition*, 46, 363–71.

GORDON, M.H. and MAGOS, P. (1983) The effects of sterols on the oxidation of edible oils. Fd Chem., 10, 141.

GORDON, M.H. and KOURIMSKA, L. (1995) The effects of antioxidants on changes in oils during heating and deep frying. *J. Sci. Food Agric.* 68, 347–53.

GOULD, W.A., SOWOKINOS, J.R., BANTTARI, E., ORR, P.H. and PRESTON, D.A. (1989) *Chipping Potato Handbook*, The Snack Food Association, 1711, King Street, Alexandria, VA 22314, (703) 836–4500, USA.

GUADAGNI, D.G., BUTTERY, R.G. and TURNBAUGH, J.G. (1972) *J. Sci. Food Agric.* 23, 1435.

GUSTAFSSON, I-B., HAGLUND, A. and JOHANSSEN, L. (1993) The taste of dietary fat based on rapeseed oil was superior to that based on sunflower oil when used for frying and baking. *J. Sci. Food Agric.* 62, 273–81.

HA, J. K. and LINDSAY, R.C. (1991) *J. Am. Oil Chem. Soc.*, 68, 294.

HAMILTON, R.J. (1983) 'The Chemistry of Rancidity in Foods' in ALLEN, J.C. and HAMILTON, R.J. *Rancidity in Foods*, Barking, Essex, England, Applied

Science Publishers Ltd., 1–21.

HARTMAN, G.J., JIN, Q.Z., COLLINS, G.J., LEE, K.N., HO, C-T. and CHANG, S.S. (1983) Nitrogen-containing heterocyclic compounds identified in the volatile flavor constituents of roasted beef. *J. Agric. Food Chem.*, 31, 1030–3.

HATHCOX, A.K. HWANG, C.A., RESURECCION, A.V.A. and BEUCHAT, L.R. (1995) Consumer evaluation of raw and fried chicken after washing trisodium phosphate or lactic acid/sodium bezoate solutions. *Journal of Food Science*, 60, 3, 604–5 and 610.

HAWRYSH, Z.J., ERIN, M.K., KIM, S.S. and HARDIN, R.T. (1995) Sensory and Chemical Stability of Tortilla Chips Fried in Canola Oil, Corn Oil, and Partially Hydrogenated Soybean Oil. *J. Am. Oil Chem. Soc.* 72, 10, 1123–30.

HAWRYSH, Z.J., ERIN, M.K., KIM, S.S. and HARDIN, R.T. (1996) Quality and stability of potato chips fried in canola, partially hydrogenated canola, soybean and cottonseed oils. *Journal of Food Quality*, 19, 107–20.

HIGGINS, F.M., KERRY, J.P., BUCKLEY, D.J. and MORRISSEY, P.A. (1999) Effects of α-tocopherol acetate supplementation and salt addititon on the oxidative stability (TBARS) and warmed–over flavour (WOF) of cooked turkey meat. *British Poultry Science*, 40, 59–64.

HO, C-T., JIN, Q.Z., LEE, M.H. and CHANG, S-S. (1983) Positive identification of new alkyloxazoles, alkylthiazoles and piperidine in roasted peanut flavor. *J. Agric. Food Chem.*, 31, 1384–6.

HO, C.T., BRUECHERT, L.T., ZHANG, Y. and CHIO, E.M. (1989) Contribution of lipids to the formation of heterocyclic compounds in model systems p. 105, in *Thermal Generation of Aromas*. Parliament, T.H. MCGORRIN, R.J. and HO, C.T. Eds., American Chemical Society, Washington DC.

HORVAT, R. J. LANE, W.G., NG, N. and SHEPARD, A.D. (1964) *Nature* 203, 523.

IFST (1999) Institute of Food Science and Technology website IFST: Current Hot Topics dated 23 June 1999 updating 14 September 1996. Address http://www.ifst.org/hottop13.htm

KIM, E-J. and AHN, M-S. (1993) Antioxidative effect of ginger extract. *J. Korean Soc. Food Sci.* 9, 37–42.

JARVI, P.K., LEE, G.D., ERICKSON, D.R. and BUTKUS, E.A. (1971) *J. Am. Oil Chem. Soc.* 48, 121.

JEON, I.J. and BASSETTE, R. (1984) *Journal of Food Quality*, 7, 97.

JOHN, P.J., SARVAMANGALA, G.K. and NARASIMHAM, P. (1992) Textural and sensory attributes of curried raw jack fruit, fried and pressure cooked. *Journal of Food Quality*, 15, 295–302.

JOSEPHSON, D.B. and LINDSAY, R.C. (1987) *Journal of Food Science*, 52, 328.

KAJIMOTO, G., ONISHI, H., YOSHIDA, H. and SHIBIHARA, A. (1987) Effects of phospholipid on the decomposition of tocopherol in heated vegetable oil. *Nippon Nogeikagaku Kaishi*, 61, 191–8.

KELLER, C. and ESCHER, F. (1989) Presented at International Congress of Engineering and Food V, Cologne, Germany.

KIRYUKHIN, V.P. and GUROV, V.A. (1980) *Kartofel Ovoshchi*, 5,10.

KOEHLER, P.E., MASON, M.E. and ODELL, G.V. (1971) *Journal of Food Science*, 36, 816.

KOURIMSKA, L., POKORNY, J. and REBLOVA, Z. (1994) Phospholipids as inhibitors of oxidation during food storage and frying. *Prehr. Technol. Biotechnol. Rev*. 32, 91–4.

LAI, S-M., GRAY, J.I., SMITH, D.M., BOOREN, A.M., CRACKEL, R.L. and BUCKLEY, D.J. (1991) Effects of Oleoresin rosemary, tertiary butylhydroquinone and sodium tripolyphosphate on the development of oxidative rancidity in restructured chicken nuggets. *Journal of Food Science*, 56, 3, 616–20.

LAWLESS, H.T. and HEYMANN, H. (1998) *Sensory Evaluation of Food*. Chapman and Hall, 115 Fifth Avenue, New York, NY 10005, USA.

LEDAHUDEC, J. and POKORNY, J. (1991) Effect of free fatty acids on the flavour of frying oil. *Die Nahrung* 10, 1071–5.

LEE, K.N., HO, C-T., GIORLANDO, C.S., PETERSON, R.J. and CHANG, S.S. (1981a) Methyl-3, 4-dimethyl-5, 6-dihydro-α-pyran-6-carboxylate in roast beef volatiles: identification and synthesis. *J. Agric. Food Chem.*, 29, 834–6.

LEE, M-H., HO, C-T. and CHANG, S.S. (1981b) Thiazoles, oxazoles and oxazolines identified in the volatile flavor of roasted peanuts. *J. Agric. Food Chem.*, 29, 684–6.

LESZCZYNSKI W. Potato tubers as a raw material for processing and nutrition. In LISINSKA, G. and LESZCZYNSKI W. (1989) *Potato Science and Technology*, Elsevier Applied Science, London.

LISINSKA, G. (1989) Chapter 4 Manufacture of potato chips and French fries. In LISINSKA, G. and LESZCZYNSKI W. (1989) *Potato Science and Technology*, Elsevier Applied Science, London.

LIU, H-R and WHITE, P.J. (1992) High-temperature stability of soybean oils with altered fatty acid compositions. *J. Am. Oil Chem. Soc.* 69, 6, 533–7.

LULAI, E.C. and ORR, P.H. (1979) Influences of specific gravity on yield and oil content of chips. *American Potato Journal*, 56, 7, 379–90.

LUJAN-ACOSTA, J. and MOREIRA, R.G. (1997) Reduction of oil in tortilla chips using impingement drying. *Lebensm. Wiss. u. Technol.*, 30, 834–40.

LYDERSON, A.L. (1997) *Mass Transfer in Engineering Practice*, John Wiley & Sons, New York.

MAGA, J.A. (1973) *J. Food Sci.*, 38, 1251.

MAGA, J.A. (1994) Potato Flavor, *Food Reviews International*, 10 (1), 1–48.

MAGA, J.A. and SIZER, C.A. (1978) *Lebensm. Wiss. u. Technol.* 11, 181.

MALCOLMSON, L.J., VAISEY-GENSER, M., PRZYBYLSKI, R. and ESKIN, N.A.M. (1994) Sensory stability of canola oil: present status of shelf-life studies. *J. Am. Oil Chem. Soc.* 71, 4, 435–40.

MAN, Y.B.C. and HUSSIN, W.R.W. (1998) Comparison of frying performance of refined, bleached and deodorized palm olein and coconut oil. *Journal of Food Lipids* 5, 197–210.

MAN, Y.B.C. and TAN, C.P. (1999) Effects of natural and synthetic antioxidants on changes in refined, bleached and deodorised palm olein during deep-fat frying of potato chips. *J. Am. Oil Chem. Soc.* 76, 3, 331–9.

MANCUSO, J. and CAPOSSELA, A. 1968. Low-fat potato chips. General Foods Corp. US Patent 3,205,074, September 1965.

MANDIN, O., DUCKHAM, S.C. and AMES, J.M. (1999) Volatile compounds from potato-like model systems. *J. Agric. Food Chem.*, 47, 2355–9.

MAY,W.A., PETERSON, R.J. and CHANG, S.S. (1978) Synthesis of some unsaturated lactones and their relationship to deep-fat flavor. *J. Food Sci.*, 43, 1248–52.

MAY, W.A., PETERSON, R.J. and CHANG, S.S. (1983) Chemical reactions involved in the deep-fat frying of foods: IX. identification of the volatile decomposition products of triolein. *J. Am. Oil Chem. Soc.*, 60,5, 990–5.

MEIGH, F., FILMER, A.A.E. and SELF, R. (1973) *Phytochemistry*, 12, 987.

MOHAMED, S., HAMID, N.A. and HAMID, M.A. (1998) Food components affecting oil absorption and crispness of fried batter. *J. Sci. Food Agric.* 78, 39–45.

MOOKHERJEE, B.D., DECK, R.C. and CHANG, S.S. (1965) *J. Agric. Food. Chem.* 13, 131.

MOREIRA, R.G. PALAU, J.E. and SUN X.Z. (1995) Deep-fat frying of tortilla chips: an engineering approach. *Food Technology*, 49, 146–50.

MOREIRA, R.G., SUN, X.Z., CHEN, Y.H. (1997) Factors affecting oil uptake in tortilla chips in deep-fat frying. *Journal of Food Engineering*. 31, 485–98.

MOUNTS, T.L. (1979) *J. Am. Oil Chem. Soc.*, 56, 659.

MOUNTS, T.L., WARNER, K., LIST, G.R., NEFF, W.E. and WILSON, R.F. (1994a) Low-linolenic acid soybean oil – alternatives to frying oils. *J. Am. Oil Chem. Soc.* 71, 5, 495–9.

MOUNTS, T.L., WARNER, K. and LIST, G.R. (1994b) Performance evaluation of hexane-extracted oils from genetically modified soybeans. *J. Am. Oil Chem. Soc.* 71, 2, 157–61.

MOUNTS, T.L., WARNER, K., LIST, G.R., KLEIMAN, R., FEHR, W.R., HAMMOND, E.G. and WILCOX, J.R. (1988) Effect of altered fatty acid composition on soybean oil stability. *J. Am. Oil Chem. Soc.* 65, 4, 624–8.

MURRAY, K.E. and WHITFIELD, F.B. (1975) *J. Sci. Food Agric.* 26, 973.

NAFISI, K. and MARKAKIS, P. (1983) *J. Agric. Food Chem.* 31, 1115.

NG, K.C., BROWN, H.D., BLACKMORE, R.H. and BUSHNELL, J. (1957) The relation of the Ca content of potato tubers to the quality of potato chips. *Food Technol.* 11, 118–22.

OHNISHI, S. and SHIBAMOTA, T. (1984) Volatile compounds from heated beef fat and beef fat with glycine. *J. Agric. Food Chem.* 32, 987.

PETUKHOV, I., MALCOLMSON, L.J., PRZYBYLISKI, R. and ARMSTRONG, L. (1999) Storage stability of potato chips fried in genetically modified canola oils. *J. Am. Oil Chem. Soc.*, 76, 8, 889–96.

PFNUER, P., MATSUI, T., GROSCH, W., GUTH, H., HOFMANN, T. and SCHIEBERLE, P. (1999) Development of stable isotope dilution assay for the quantification of 5-methyl-(E)-2-hepten-4-one: application to hazelnut oils and hazelnuts. *J. Agric. Food Chem.*, 47, 2044–7.

PINTHUS, E.J. and SAGUY, I.S. (1994) Initial interfacial tension and oil uptake by deep-fat fried foods. *J. Food Sci.* 59, 804–7.

PLIMPTON, S. (1985) Factors affecting the oil content of potato chips and the application of near-red reflectance to on-line moisture on oil contents measurements. Ph.D. Dissertation. The Ohio State University, Columbus, Ohio, USA.

POKORNY, J. (1980) Effect of substrates on changes of fats and oils during frying. *La Revista Italiana delle Sostanze Grasse*, 57, 222–5.

POKORNY, J. (1998) Substrate influence on frying process. *Grasas y Aceites*, 49, 3–4, 265–70.

PRZYBYLISKI, R. and ESKIN, N.A.M. (1988) A comparative study on the effectiveness of nitrogen or carbon dioxide flushing in preventing oxidation during the heating of oil. *J. Am. Oil Chem. Soc.* 65, 4, 629–33.

PRZYBYLISKI, R. and HOUGEN, F.W. (1989) *JAOCS*, 66, 1465.

PRZYBYLISKI, R., MALCOLMSON, L.J., ESKIN, N.A.M, DURANCE-TOD, S., MICKLE, J. and CARR, R. (1993) Stability of low-linolenic acid canola oil to accelerated storage at 60ºC. *Lebensm. Wiss. u. Technol.* 26, 205–9.

QUAGLIA, G., COMENDADOR, J. and FINOTTI, E. (1998) Optimization of frying process in food safety. *Grasas y Aceites*, 49, 3–4, 275–81.

RACENIS, V. (1959) Factors affecting browning of potato chips – new types of model systems for studies for browning. *Dissertation Abstracts*, 20(1), 268–9.

REBLOVA, Z., KUDRNOVA, J., TROJAKOVA, L., VALENTOVA, H., POKORNY, J. and ZAVODSKA, H. (1997) Application antioxidants for stabilisation of fats and oils used for intermittent deep frying. *Proc. Meeting Technol. Anal. Fats Oils* 35, 155–7.

ROBERTSON, J.A., MORRISON, W.H., BURDICK, D. and SHAW, R. (1972) *American Potato Journal* 49, 444.

ROBERTSON, J.A. MORRISON, B.G. LYON, D. and SHAW, R. (1978) *Journal of Food Science* 43, 420.

ROSSELL, J.B. (1989) Factors affecting the quality of frying oils. In Symposium Proceedings Number 25: March 1989 Frying: Proceedings of a Symposium held at Leatherhead, 25 February 1988. The British Food Manufacturing Industries' Research Association, Leatherhead, UK.

ROSSELL, J.B. (1994) Measurement of rancidity. In ALLEN, J.C. and HAMILTON, R.J. (eds) *Rancidity in Foods*. Blackie Academic and Professional, London and New York.

ROSSELL, J.B. (1998) Development of purity criteria for edible vegetable oils. In HAMILTON, R.J. (ed.) *Lipid Analysis in Oils and Fats*. Blackie Academic and Professional, London and New York.

ROSSELL, J.B. (2000) Personal communication.

SATYANARYAN, V.T. and HONIKEL, K-O. (1992) Effects of different cooking methods on warmed-over flavour development in pork. *Zeitschrift für Lebensmitteluntersuchung und Forschung*, 194, 5, 422–5.

SELKE, E., MOSER, H.A. and ROHWEDDER (1970) *J. Am. Oil Chem. Soc.* 47, 393.

SELMAN, J. (1989) Oil uptake in fried food products. In Symposium Proceedings Number 25: March 1989 Frying: Proceedings of a Symposium held at Leatherhead, 25 February 1988. The British Food Manufacturing

Industries' Research Association, Leatherhead, UK.

SEVERIN, T. and LEDL, F. (1972) Thermische zersetzung von cystein in tributyrin. *Chem. Mikrobiol. Technol. Lebensum.*, 1, 135.

SHARMA, G.K., SEMWEL, A.D., NARASIMHA, MURTHY, M.C. and ARYA, S.S. (1997) Suitability of antioxygenic salts for stabilization of fried snacks. *Food Chemistry*, 60, 1 19–24.

SHEN, N., FEHR, W. and WHITE, P. (1997) High-temperature stabilities of oils from soybeans that lack lipoxygenases. *J. Am. Oil Chem. Soc.* 74, 4, 381–5.

SHIMODA, M., NAKADA, Y., NAKASHIMA, M. and OSAJIMA, Y. (1997) Quantitative comparison of volatile flavor compounds in deep-roasted and light-roasted sesame seed oil. *J. Agric. Food Chem.*, 45, 3193–6.

SMITH, O. (1968) *Potatoes, Production, Storing Processing*. The AVI Publishing Co. Inc., Westport, Connecticut, USA.

SMITH, O. (1987) Potato chips. In TALBURT, W.F. and SMITH, O. (eds) *Potato Processing*, 4th edn. AVI Westport, Connecticut.

STUTZ, R.A. and BURISS, R.H. (1948) Factors influencing oil content of chips. *Food Industry*, 20, 97–197.

SUYAMA, K. and ADACHI, S. (1980) Origin of alkyl-substituted pyridines in food flavour: formation of the pyridines from the reaction of alkanals with amino acids. *J. Agric. Food Chem.* 28, 546.

TALBURT, W.F. and SMITH, O. (1975) *Potato Processing*. 3rd edn. The AVI Publishing Co. Inc., Westport, Connecticut, USA.

TALLEY, E.A. and EPPLEY, G.A. (1985) *Lebensm. Wiss. u. Technol.* 18, 281.

TANG, J., JIN, Q.Z., SHEN, G-H., HO, C-T. and CHANG, S.S. (1983) Isolation and identification of volatile compounds from fried chicken. *J. Agric. Food Chem.*, 31, 1267–92.

TSENG, Y-C., MOREIRA, R. and SUN, X. (1996) Total frying-use time on soybean-oil deterioration and on tortilla chip quality. *International Journal of Food Science and Technology*, 31, 287–94.

UFHEIL, G. and ESCHER, F. (1996) Dynamics of oil uptake during deep-fat frying of potato slices. *Lebensm. Wiss. u. Technol.*, 29, 640–4.

VAN GEMERT, L.J. (1996) Sensory properties during storage of crisps and French fries prepared with sunflower oil and high oleic sunflower oil. *Grasas y Aceites* 47, 1–2, 75–80.

VERNIN, G. and PARKANYI, C. (1982) Mechanisms of formation of heterocyclic compounds in Maillard and pyrolysis reactions. In VERNIN, G. (ed.) *Chemistry of Heterocyclic Compounds in Flavours and Aromas*. Ellis Horwood, Chichester, p. 151.

WAGNER, R. and GROSCH, W. (1997) Evaluation of potent odorants of French fries. *Lebensm. Wiss. u. Technol.*, 30, 164–9.

WAIMALEONGARA-EK, C. and CHEN, T.C. (1986) Effects of frying shortening quality, holding conditions, and reheating on selected flavor volatiles of deep-fat fried chicken parts. *Poultry Science*, 65, 2043–50.

WARNER, K. (1996) Flavors and Sensory Evaluation in Y.H. Hui Editors of Bailey's Industrial Oil and Fat Products. Volume 1: Edible oil and Fat

Products: General Applications. (5th edn) John Wiley & Sons, Inc. 605 Third Avenue, New York, NY, USA.

WARNER, K., EVANS, C.D., LIST, G.R., BOUNDY, B.K. and KWOLEK, W.F. (1974) *Journal of Food Science*, 39, 761.

WARNER, K., MOUNTS, T.L. and KWOLEK, W.F. (1985) *J. Am. Oil Chem. Soc.*, 62, 1483.

WARNER, K. and KNOWLTON, S. (1997) Frying quality and oxidative stability of high-oleic corn oils. *J. Am. Oil Chem. Soc.* 74, 10, 1317–22.

WARNER, K. and MOUNTS, T.L. (1993) Frying Stability of Soybean and Canola Oils with Modified Fatty Acid Compositions. *J. Am. Oil Chem. Soc.* 70, 10, 983–88.

WARNER, K., ORR, P., PARROTT, L. and GLYNN, M. (1994) Effects of frying composition on potato chip stability. *J. Am. Oil Chem. Soc.* 71, 10, 1117–21.

WARNER, K., ORR, P. and GLYNN, M. (1997) Effect of fatty acid composition on flavor and stability of fried foods. *J. Am. Oil Chem. Soc.* 74, 4, 347–56.

WHITFIELD, F.B. (1992) Volatiles from Interactions of Maillard Reactions and Lipids. *Critical Reviews in Food Science and Nutrition*, 31, 1–58.

XU, X-Q, TRAN, H., PALMER, M., WHITE, K. and SALISBURY, P. (1999a). Monitoring deep frying performance of Monola and Monosun oils. *Food Australia*, 51, 4, 157–63.

XU, X-Q, TRAN, H., PALMER, M., WHITE, K. and SALISBURY, P. (1999b). Chemical and physical analyses and sensory evaluation of six deep-frying oils. *J. Am. Oil Chem. Soc.* 76, 9, 1091–9.

YAMATO, T., KURATA., KATO, H. and FUJIMAKI, M. (1970) Volatile carbonyl compounds from heated beef fat. *Agric. Biol. Chem.* 34, 88.

YANG, J., SULAEMAN, A., SETIAWAN, B., ATUGHONU, A., GIRAUD, D., HAMOUZ, F.I. and DRISKELL, J.A. (1994) *Journal of Food Quality* 17, 33–40.

ZHANG, Y. and HO, C.T. (1989) Volatile compounds from thermal interaction of 2,4-decadienal with cysteine and glutathione. *J. Agric. Food Chem.* 37, 1016.

Appendix: Flavours and aromas derived from lipid oxidation

Flavour	Compounds
Aromatic	2-methylbutyl propanoate, propionic acid
Artichoke, green, flowers	*trans*-3-hexenal
Banana	3-methylbutyl acetate, *cis*-3-hexen-1-ol, *cis*-2-penten-1-ol
Beany	alkanals, non-2-enol
Bitter, almonds, green	*trans*-2-hexenal
Brown beans	oct-2-enal
Burnt	Guaiacol
Cardboard/tallowy	n-octanol; n-alkanals (C_9-C_{11}); alk-2-enal (C_8-C_9); 2,4-dionals (C_7-C_{10}); nona-2-*trans*-dienal
Creamy	4-*cis*-heptenal
Cucumber	nona-2-*trans*-6-*cis*-dienal
Deep-fried fat	2-*trans*-4-*trans*-decadienal, *trans*-2,4-nonadienal, 2,4-nonadienal
Ethyl acetate-like	2-methyl propan-1-ol
Fatty	heptanal, *trans*-2-octenal, *cis*-3-nonenal, *cis*-2-nonenal, 2-decenal, *trans*-2-decenal
Fatty, tallowy	*trans*-2-nonenal
Fishy	n-alkanals (C_5-C_{10}); alk-2-enals (C_5-C_{10}); 2,4-dienals (C_7); oct-1-en-3-one; deca-2-*trans*-4-*cis*-7-*trans*-trienol, 2-methylbutan-1-ol,
Fragrant	butan-2-one
Fresh	methyl decanoate
Fruity	n-alkanals (C_5, C_6, C_8, C_{10}); aliphatic esters; isobutyric acids, ethyl-2-methylpropanoate, ethyl-2-butanoate, *cis*-3-hexenyl acetate, ethyl cyclohexanoate, heptan-2-one, aldehyde C_6 branched, methyl nonanoate, 6-methyl-5-hepten-2-one, nonan-2-one
Fruity, aromatic	hexan-1-ol
Fruity, mushroom-like	C_8 ketone
Fruity, soap	2-octenal
Greasy	2-*trans*-hexenal; nona-2,6-dienal.
Green banana, fruity	3-hexenyl acetate
Green beany	3-*cis*-hexenal
Green leaves, grassy	*cis*-3-hexenal, *trans*-2-hexen-1-ol
Green, apple-like	*cis*-3-hexenal, *cis*-2-pentenal, hexanal,
Green, grassy	hexanal
Green, pleasant	*cis*-2-pentenal
Hardened, hydrogenated	6-*trans*-nonenal
Malty	3-methylbutanal, 3-methylbutanol,
Melons	nona-3-*cis*-6-*cis*-dienals, non-6-*cis*-enal
Metallic	pent-1-en-3-one; oct-1-en-3-one, *trans*-4,5-epoxy-*trans*-2-decenal
Mild, pine-like	3-*trans*-hexenal
Moldy	3-(4-methyl-3-pentenyl) furan
Mushroom	oct-1-en-3-ol
Nutmeg	octadienal

Oily	n-alkanals (C_5-C_7); hex-2-enal; 2,4-dienals (C_5-C_{10})
Oxidised	oct-1-ene-3-one; octanol; hept-2-enal; 2,4-heptadienal; n-alkanols
Painty	n-alkanals, (C_5-C_{10}); alk-2-enals(C_5-C_{10}); 2,4-dienals (C_7); 2-alkanone (C_7); pent-2-enal
Potatoes	penta-2,4-dienal
Pungent	propionic acid, acetic acid, pentan-1-ol
Putty-like, unpleasant	3-methyl-2-butenyl acetate
Rancid	2-*trans*-nonenal, volatile fatty acids (C_4-C_{10})
Rancid hazelnut	2-*trans*-4-*trans*-heptadienal
Rotten apple	2-*trans*-4-*cis*-heptadienal
Sharp citrus	octanal
Sharp, pungent	1-penten-3-one
Soapy	fatty acid soaps, free capric, lauric and myristic acids
Soapy, citrus-like	decanal, nonanal, octanal
Solvent-like	octene, 2-methylbut-2-enal, methyl benzene
Strong	ethyl benzene
Sweet	ethylfuran, pentan-3-one, 4-methylpentan-2-one, alcohol C_6 branched, tridecene
Sweet aldehyde	2-*trans*-4-*cis*-decadienal
Sweet, aromatic	ethyl acetate
Sweet, fruity	3-methylbutanal, hexyl acetate,
Sweet, honey-like	2-phenylethanol, phenylacetaldehyde
Sweet, strawberry, apple	pent-1-en-3-one, ethyl propanoate
Vanilla	Vanillin
Wet earth	1-penten-3-ol

Adapted from: Morales, M.T., Rios, J.J. and Aparicio (1997) *J. Agric. Food Chem.*, 45, 2666–73; Kochhar, S.P. and Meara, M.L. (1976) A Survey of the Literature on Oxidative Reactions in Edible Oils as it Applies to the Problem of Off-flavours in Foodstuffs. *Scientific and Technical Surveys* No. 87, November 1975; Boskou, D. ed. (1996) *Oliver Oil Chemistry and Technology*, AOCS Press, Champaign, IL 1996 (pp. 78–79).

13

Improving the texture and colour of fried products

C-S. Chen, C-Y. Chang and C-J. Hsieh, Da-Yeh University, Taiwan

13.1 Instrumentation for measuring the texture and colour of fried products

This chapter looks first at the instrumentation available for measuring texture and colour. It then considers some of the main influences on the texture and colour of fried products. Given the number of such influences and their complex interactions, a key issue is an appropriate modelling tool able to identify the significance of any one variable on texture and colour. Response Surface Methodology (RSM) provides just such a technique. The chapter concludes with a case study looking at the use of RSM in optimising textural and colour quality in the production of gluten balls.

Consumer preferences are product and habit oriented. The colour of a fried food, for example, can be seen as one of a range of input signals perceived by consumers, rather than just as a physical characteristic of the food. The linkage between colour and consumer perceptions of quality is often psychological. For example, depending on experience, golden yellow fried chicken pieces may indicate to consumers the use of fresh raw materials and fresh oil together with the right frying technique. A dark brown colour might be seen as suggesting poorer quality, for example poorer raw materials, prolonged frying or a re-used frying oil. Alternatively, it might indicate a salty and heavy taste, if soy sauce were used in preparation. Consumer preferences need to be tracked by the use of appropriate sensory evaluation techniques. Instrumental measures are best used as a complement to such sensory analysis, and can only be seen as reliable if instrumental results are validated against sensory measurements. However, within this context, instrumental measurement of texture and colour can offer a quantified basis for manipulating processing variables for quality improvement.

13.1.1 Measuring texture

The terms 'rheological' and 'mechanical' properties are frequently used interchangeably in describing the texture of solid foods. However, although all rheological properties are mechanical properties (both study the behaviour of materials under applied physical forces), some mechanical properties do not involve deformation and should not be regarded as rheological properties (Szczesniak, 1983). Rheological properties of solid foods have been analysed by observing the deformation (strain) of a specimen under applied stress (force per unit area). Typical methods include compression (direction of force points toward the specimen centre), tension (force that pulls the specimen apart), and shear (force parallel to the plane of action).

Compression testing is both relatively simple and popular as a way of obtaining rheological information about solid foods in the food industry. The test commonly involves pressing a food specimen against a probe (i.e. a plunger). The specimen platform, driven by a motor, travels at a preset speed toward a stationary probe connected to a force transducer. An alternative arrangement is for a stationary specimen platform and a mobile probe. Rheological information is obtained by analysing the force (sensed by the probe) against the distance (deformation) curve that describes the stress–strain relationship. Generally, as a solid food is being compressed, the resistance to deformation initially increases linearly with distance. This initial slope of the stress–strain curve is referred to as the 'elastic modulus' (Szczesniak, 1983) or the 'modulus of deformability' (Mohsenin and Mittal, 1977) and may be considered to be a measure of firmness. As the sample is further compressed, the resistance (sensed by the probe) starts to increase nonlinearly with the strain until a point where some structural elements begin to fail and the resistance starts to decline. This point is called the bioyield point. Since the rupture of structural elements is local, the resistance may increase (in some cases, remain virtually constant) upon further compression until a point, the rupture point, where massive failure occurs, is reached. After the point of massive failure, the resistance declines quickly (Szczesniak, 1983).

During mastication, the wedging action of teeth imposes tensile stresses on foods (Voisey and deMan, 1976). Tensile measurement is important in such areas as quantifying dough strength and meat quality. The measurement has been used by a number of workers to study mechanical properties of raw and cooked muscle fibres (Bouton and Harris, 1972; Bouton *et al.*, 1975; Stanley, 1976). Shear tests were initially designed for muscle mechanical property measurements and are widely used in meat research, for example. The widely used Warner–Bratzler shear apparatus, for example, is operated by applying force across muscle fibres via a blunt edge. Voisey and Larmond (1974) pointed out that during the shear test, compressive and tensile stresses are also involved in the deformation. A similar conclusion that pure shear forces are rarely encountered in food products has also been reported by other workers (Szczesniak, 1983; Peleg, 1987). Szczesniak and co-workers (1970) used the

Kramer shear press on a variety of foods and published shear–deformation curves that are characteristic of individual foods.

Currently, many manufacturers offer computerised, multifunctional measuring rheometers equipped with different types of probes for different tests. They all share a number of common elements: a probe, a driving system, a sensing unit, and a readout system. Types of probe include: flat plunger, plate, piercing rod, penetrating cone, cutting blade, shearing jaws, or cutting wires (Szczesniak, 1983). A variable drive electric motor, weight/pulley arrangement or hydraulic system can be used as the driving mechanism. A sensing element, such as a simple spring, strain gauge, load cell, or force transducer, is used to transform forces sensed by the probe into signals that can be read or processed. For computerised models, force signals are transformed into analog electrical signals (voltage or current) which can be recognised by a recorder, or converted into a digital signal through an analog to digital (AD) interface card plugged into a personal computer. In more advanced models, the testing machine can be operated, through an AD/DA interface card, by a software program residing in a personal computer.

13.1.2 Measuring colour

The simplest way to designate the colours of foods is by visual comparison to colour standards of painted paper, plastic or glass. The Munsell system offers a wide range of colour standards, which had been adopted in some official USDA grading systems, for example (Francis and Clydesdale, 1975). Although easy and straightforward, the method has its shortcomings: colour standards may decay with use, visual judgements are subjective, and the colour of the food may fall between existing standards. Instrumental methods, on the other hand, are more flexible, providing more sensitive and quantitative results.

The basic theory behind spectrophotometric measurement of colour is that one can match just about any spectral colour (obtained by using a prism or grating) by adjusting the amount of red, green and blue light (obtained by filtering 3 separate light sources). Extending this basic principle, the CIE (the initials stand for Commission Internationale d'Eclairage) XYZ system was developed using mathematical models coupled with the use of x y z standard observer curves (response to wavelength curves) that were standardised in 1932 (Francis and Clydesdale, 1975). Anther system is the Judd–Hunter L a b solid, in which, L is lightness or darkness, a is greenness, $+a$ is redness, b is blueness and $+b$ is yellowness. A colorimeter using either system needs only three basic elements: a light source, three glass filters with transmittance spectra that duplicate the X, Y, Z curves, and a photocell that senses reflected or transmitted light waves. With the aid of a digital computer, instrumental measurement of colour is now relatively easy.

13.2 Influences on the texture and colour of fried products

There is a wide range of factors influencing texture and colour. This section picks out a few which might form the basis of experimentation and analysis using techniques such as RSM.

13.2.1 Raw materials

Water is one of the most important constituents determining the texture of fried foods. As a result of water evaporation, heat transferred from oil is carried off as latent heat of vaporisation. Blumenthal (1991) has pointed out that, during deep frying, this process of energy removal maintains the temperature of the food/oil interface virtually at 100°C, preventing charring or burning of the food. As water content decreases, the amount of thermal energy carried off as latent heat of vaporisation starts to decline and the food/oil interface temperature begins to increase, leading potentially to charring or burning of the food.

In the process of water evaporation during frying, one important phenomenon is the rapid increase of the molecular volume of water during the phase change from liquid to gas. This increase often leads to volume expansion of the fried object if the vapour does not have a clear passage to the food/oil interface, particularly with products wrapped in a thick outer coating.

In the process of vaporisation, volume expansion of water, from liquid to vapour phase, also leads to the porous structure of the crust (particularly for battered fried products), while the rate of dehydration determines the pore size. The volume expansion of the fried product also depends on the relative ease of migration of water through the surface matrix, which depends on the strength of the walls (or membrane) surrounding each chamber. For finished fried products, the final water content of the food is often related to its perceived tenderness and/or juiciness. For example, a crispy exterior with a juicy interior is normally perceived as an indictor of quality for fried chicken. The water-holding capacity of the food, its initial water content and remaining water after frying are all of great importance in controlling the texture of both the interior and exterior of a fried food.

The reactions between various food constituents at elevated temperature during frying include both physical (i.e. phase and volume changes) and chemical (i.e. chemical bond destructions and formations) changes. Protein denaturation and starch gelatinisation are typical phenomena of the combined effect of multiple-order chemical reactions. They involve breaking of hydrogen bondings, formation of covalent bondings between amino acids, rearrangements of three-dimensional structures, hydration and dehydration. The resulting protein–starch network is important to the texture of the battering and/or the breading, as well as the texture of the interior food. Adding protein and starch can also affect water-holding capacity, and consequently influence the amount of water loss through dehydration during frying. Kadan *et al.* (1997) clearly showed that protein/starch ratio is a factor in optimising the textures of both the interior portion and the crust.

Protein is not only an essential component in the structure of many fried foods, but also participates in the Maillard browning reaction with reducing sugars (i.e. glucose, fructose) to form flavours and brown pigments. A variety of proteins have been used in modern batters – cheese powder, egg albumen, whey proteins, gluten, soy protein, some of them prehydrolysed. Added protein can serve as an emulsifying agent in some cases and film-forming agent in others. Soy protein is sometimes added to meat products to improve water-holding capacity, flavour, and cohesiveness (Kotula, 1976; Brewer *et al.*, 1992). Another good example of protein manipulation for improving water-holding capacity is the manufacturing of fish balls. Fish meat is first fine ground to paste-like meal, so that the fibrous protein structure is destroyed, and the meal is then shaped and cooked or fried. Unlike fibrous protein in fish fillet which tends to lose water and became tough and dry after prolonged cooking or frying, smaller and less organised protein molecules are less likely to form aggregates during cooking or frying, and are more capable of retaining water.

The presence of phosphate has also been shown to improve protein's water-holding capacity in meat (Moore *et al.*, 1976; Neer and Mandigo 1977; Whiting, 1984). Lin and Kuo (1994) studied low-temperature (0ºC and −20ºC) stored chicken breast coated with batters and breadings, and compared the effect of injection of phosphate solution, soy protein and oil emulsions on the texture and palatability of the fried chicken breast pieces. They found, by panel studies, that for chicken breast stored at 0ºC, injection of phosphate significantly improved tenderness and juiciness of the meat. Lowering the storage temperature to −20ºC, together with a slow freeze–thaw process to separate water from the fibrous muscle protein, and injection of olive oil emulsion, was also found to provide superior improvements to the fried product.

Starch is the major component in many commercial premixed battering powders, which are responsible for the body of the crust of batter fried products. Starch gelatinisation is crucial in frying: it holds water and provides volume expansion. Carbohydrates are added in new formulation of batters for various purposes and in many different forms: gums, pregelatinised starch, modified starch, high amylose starch and dietary fibres. Kadan *et al.* (1997) studied the effect of amylose (starch with less branched structure) and protein on the texture of extruded rice-based fries. They proposed that in high protein content rice-based fries, protein molecules tended to form a barrier around starch granules and retarded water uptake during starch gelatinisation. Consequently, the water-holding capacity was reduced, and caused more water to be lost as steam during extrusion (at 90ºC). This, after frying, resulted in fries that were hard and tough in texture. This example shows that the types, state, interactions and the ratio of the two macromolecules, protein and starch (originally present or added) exert an important influence on the texture of both the interior and exterior portions of the fried food.

Fat is particularly important in the mouthfeel of a fried product, for example in improving the tenderness of turkey breast (Moran and Larmond, 1981; Larmond and Moran, 1983; Moran, 1992). Other additives such as salt, chemical leavening agent and stabilisers, although minor in quantities, are also significant. Leavening agents, added to batters, provides volume expansion during frying, and affects the texture of the crust. The presence of salt increases the water boiling point and influences the rate of heating.

The influence of frying oil on product quality is discussed elsewhere in this book. One example of its importance is the 'Surfactant Theory of Frying' (Ohlson, 1983; Blumenthal and Stockler, 1986; Blumenthal, 1991). The theory states that as the frying oil degrades, more surfactants (i.e. metal salts of fatty acids) are formed, which increase contact between frying oil and (water-based) food. As a result, the heat transfer rate to the surface is increased, leading to darkening and drying of the food surface. The quality of the frying oil (amount of polar, or free fatty acid released) is therefore an important variable.

13.2.2 Frying conditions

Frying temperature is a major influence on product quality. With too high a frying temperature, the Maillard browning reaction would proceed to an unacceptable level. The result would be to burn or char the product, with an accompanying dark colour, bitter taste and unpleasant mouthfeel, while the centre portion remained undercooked. For low-temperature frying, although charring could be prevented, prolonged frying tends to pump more water to the surface and leave the product with a dry interior. Since the heat transfer rate is governed by the temperature gradient and surface area of heat transfer, and the time required for energy to be transferred from surface to centre is distance dependent, product shape governing surface area to volume ratio is also important, as is the initial temperature of the food. In deep frying, a large volume of frying oil exerts a damping effect in reducing temperature fluctuations once fried foods are added at low temperature. However, the cost of using a large volume of oil is high. The optimum ratio of oil volume to quantity of fried food is also important in keeping temperature fluctuation in an acceptable range.

The time of immersion in frying oil at fixed temperature, not only affects the degree of browning and flavour development, but also influences the texture of both the outer crust and the inner food body. Initially, drastic temperature change causes a high rate of water evaporation and migration towards the surface, resulting in volume expansion. As the dehydration process goes on, the outer crust is simultaneously formed and its structure becomes stronger. In this middle phase of frying, water in the interior portion of the food continues to evaporate and escape through the porous matrix of the crust. Meanwhile, the crust together with the partially dehydrated and cooked food (e.g. swelled starch) begins to form barriers for water transportation, and increases the inner vapour pressure, therefore increasing the boiling point of water. Finally, the crust turns crispy upon continued dehydration. Continued dehydration, if not controlled, may lead to a dry interior and muscle protein would soon become fibrous and tough. For fried tofu, for example, extensive dehydration, either by long immersion time or elevated temperature, would result in a product with elastic, rubbery and brown skin, while in the interior, protein aggregation due to dehydration, would lead to a less tender texture.

Pressure is another variable. Frying under positive pressure will increase water boiling point, and raise the cooking temperature, which will impart a softer texture to the breading as compared to a crispy texture when frying under atmospheric pressure. Rao and Delaney (1995) studied deep frying of breaded chicken pieces under positive pressure. They compared the densities and moisture content of the breadings after pressure- and atmospheric-frying, and found the breadings using atmospheric-frying had lower density and moisture content. Scanning electron micrographs also showed that breading from chicken pieces fried under atmospheric pressure had a more porous protein–starch network (Rao and Delaney, 1995). Low density and moisture content, together with high porosity, led to a crispy texture of the breading.

13.3 Using response surface methodology (RSM)

The previous section illustrates both the range of influences on product quality and their complex interactions in determining the texture and colour of a fried product. Analysing this complex picture requires appropriate modelling tools. RSM is particularly useful for optimisation of sophisticated multifactor systems when the quantitative relationship between key variables is not always clear, as is the case with a complex operation such as frying. RSM can simultaneously consider several factors at many different levels, and the corresponding interactions between these factors, using a relatively small number of observations.

13.3.1 The principles of RSM

For a single variable function of the form: $y = f(x)$, the dependent variable y changes with the variation of the independent variable x according to the pattern defined by f. On x–y plane, this relationship, in a narrow interval of concern, often exhibits a curve for quadratic functions of the form: $f(x) = a^2 + bx + c$. Finding the optimum point on a quadratic curve can be achieved by simply locating the highest (or lowest) point on the curve, or by differentiation and equating the first derivative of the function to zero ($f'(x) = 0$), then solving for the root which gives the exact location of the optimum point (calculus reference). The method of finding the optimum point by differentiation is based on the fact that at the extrema (highest or lowest point), the slope or the rate of change of $f(x)$ is zero.

When it comes to functions of two independent variables ($z = f(x, y)$, we are looking at a surface in three-dimensional space, on which z, the dependent variable, changes with the variations of x and y according to the relationship defined by f. Standing at the highest point of a surface, calculus still tells us that the first partial derivatives of the function with respect to the two independent variables are zero. But the reverse is not necessarily true (calculus reference).

There are times that one can find the optimum point by changing one independent variable at a time while fixing other variable(s): the one-factor-at-a-time technique. This process is equivalent to solving $f_x = 0$ for x while holding y constant, and solving $f_y = 0$ for y while holding x constant (f_x and f_y represent partial differentiation of f with respect to x and y respectively). Although the one-factor-at-a-time technique is strategically simple, the method does not take into account the potential interactions between independent variables. In cases of significant interaction effect(s) among factors, the optimum point obtained by the one-factor-at-a-time technique could be significantly different from the true optimum point.

Frequently, for real applications, the function f that defines the relationship between the dependent and the independent variable(s) is not known. One can only know the approximate shape of f by doing experiments. Theoretically, having experimental results converted into graphical form, one can visually locate the optimum point. However, besides the difficulty in drawing a surface from data points in three-dimensional space, drawing curves connecting data points on the x–y plane is prone to error, not to mention that statistical questions such as the validity of the data remain unanswered. In addition, since there are many possible lines passing through the neighbourhood of the three points that are not exactly on the same straight line, given the nature of error inherent in experimental data, deciding which line best represents given data by eye is not easy. It is even more confusing when one is drawing a curve out of a set of scattered experimental data. Regression is a mathematical tool that helps us to decide which line, or curve, or surface best fit (represent) given experimental data by minimising the sum of squares (sum of the square of the distances between the line, curve, or surface and the data points).

A mathematical model (the function f that describes the relationship between the dependent variable and the independent variable(s)) is necessary before regression is possible. Unfortunately, such mathematical models seldom exist. Based on current understanding of the subject and available mathematical tools, the wish to deduce a mathematical model that is able to express, for example, the quality of a fried food product as a function of oil temperature, food composition and geometry in a wide range of intervals would be very difficult if not impossible. One way of bypassing this problem is by assuming that in the neighbourhood of an optimum point, concavity (whether concave up or down) of an arbitrary surface makes it reasonable to use a quadratic function to approximate the surface (in the neighbourhood of the optimum point). Although this assumption holds only in the area near the optimum point, most of the time this simplification is good enough for practical purposes.

The concept of optimisation using RSM was first proposed by Box and Wilson in 1951 (Box and Wilson, 1951). RSM uses quadratic function of the form (two-factor model): $Y = a_0 + a_1X_1 + a_2X_2 + a_3X_1X_2 + a_4X_1^2 + a_5X_2^2$, for approximation of the arbitrary surface where, Y is the dependent variable (e.g. expansion volume of the fried food product), X_1 and X_2 are the independent variables (e.g. frying temperature and water content), and a_0 to a_5 are constants to be determined by regression using experimental data. The shape of the response surface is determined by experiments that record how the dependent variable responded to the variations of the independent variables near the optimum point.

Theoretically, manipulating six parameters (a_0 to a_5) should be capable of adjusting the shape of the surface to fit any set of data that are smooth and quadratic in nature (in the vicinity of the optimum point). However, in practice, it is quite likely that one or more of the following situations might occur:

- experimental data tend to fluctuate, and sometimes, the true object function is not quadratic even in the neighbourhood of the optimum point (poor degree of fitness for quadratic model)
- independent variables do not fall in the vicinity of the optimum point
- one or more of the chosen factors might not affect the dependent variable to a significant level to be included in the model.

Statistical analysis is therefore necessary to verify the data (see below). Key questions to be answered in applying RSM are how many experiments will be sufficient to achieve conclusive results and how should values of the independent variables (factors) be assigned? On one hand, there is pressure to keep the number of experiments to a practical minimum. On the other hand, more data means statistically more reliable results. Experiment design is a technique developed to establish an optimal number of experiments (Mason *et al.*, 1989; Montgomery, 1984; Thomson, 1982). The basic theory and sources of various experiment design will be given in the next section. The place of experimental design in RSM is illustrated in Fig. 13.1.

13.3.2 Applying the principles of RSM

An experiment design for one-factor-at-a-time optimisation is shown in Table 13.1, for a two-factor-five-level design, where a coded form of independent variables was used so that the design could be applied on any type of system. In the coded form, one unit could represent 10°C difference of oil temperature, or 5% water content. From run number 1 to 5, X_2 was held constant at 0 (the centre), and $X_1 = -1$ was found to produce highest response (Y_2). From run number 6 to 9, X_1 was fixed at -1, while X_2 varied from -2 to $+2$, and $X_2 = +1$ was found to be optimum. The combination of $X_1 = -1$ and $X_2 = 0$ had been carried out in run number 2. As mentioned above, the true optimum point could be somewhere else, should interaction effects be significant.

Central Composite Design (Mason *et al.*, 1989; Montgomery, 1984; Thomson, 1982) is commonly employed for systems of potential interaction effect(s) between factors. For n-factor-five-level design, five levels ($-d$, -1, 0, $+1$, $+d$) were assigned to each independent variable, where d is the extended

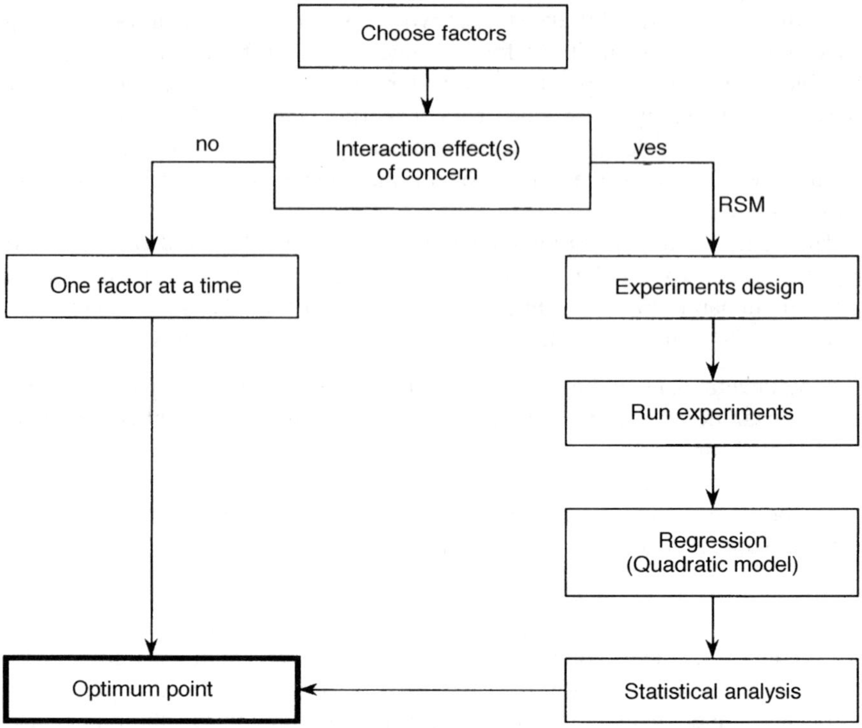

Fig. 13.1 Flow chart of optimisation procedure when the search domain lies in the vicinity of the optimum point.

Table 13.1 Experiment design for one-factor-at-a-time optimisation (two-factor five-level)

Run number	Independent variables		Dependent variable
	X_1	X_2	Y
1	−2	0	Y_1
2	−1	0	Y_2
3	0	0	Y_3
4	+1	0	Y_4
5	+2	0	Y_5
6	−1	−2	Y_6
7	−1	−1	Y_7
8	−1	+1	Y_8
9	−1	+2	Y_9
Optimum point	−1	+1	$Y_{optimum} = Y_8$

Table 13.2 Central composite design (two-factor five-level)

Run number	Independent variables		Dependent variable
	X_1	X_2	Y
1	1	1	Y_1
2	1	−1	Y_2
3	−1	1	Y_3
4	−1	−1	Y_4
5	0	−1.414	Y_5
6	0	1.414	Y_6
7	1.414	0	Y_7
8	−1.414	0	Y_8
9	0	0	Y_9
10	0	0	Y_{10}

level, and $d = (2)^{n/4}$. An example of two-factor five-level design is given in Table 13.2. The extended level, d, for 2-factor design, $d = (2)^{2/4} = 1.414$, for 3-factor design, $d = (2)^{3/4} = 1.682$, for 4-factor design $d = (2)^{4/4} = 2$, and so on. In Table 13.2, run numbers 1 to 4 are the 2-level factorial design, and run numbers 9 and 10 are duplicate experiments so that experimental error could be estimated by statistical diagnosis. In general, more replications at the 'centre' (level 0) will lead to better approximation of experimental error. Comparing Tables 13.1 and 13.2, it is clear that Central Composite Design provides more information for approximately the same number of experiments.

Statistical analysis is carried out by analysis of variance (ANOVA). One of the most important terms in ANOVA in applying RSM is R^2, which reflects the goodness of fit of the mathematical model. R^2 is determined by calculating the ratio of regression sum of squares (SS_R) to the total sum of squares (SS_T): $R^2 = SS_R/SS_T$. Since SS_T is the sum of SS_R and SS_E (the error sum of squares, which comes from lack of fit and pure experimental error), the closer R^2 is to 1 (or 100 on a percentage basis), indicating smaller SS_E, therefore, the more intimate relationship there is between the model predictions and the true responses. Another quantity, the P-value, is also important in determining whether each term (e.g. X_1, X_1^2, X_1X_2) in the model is significant or not. The P-value obtained from ANOVA states that for the term X_i in the model, the probability that X_i is not significant to the response is P_i. For example, $P_2 = 0.05$ means the probability that X_2 is not significant to the response is 0.05, and it is said that X_2 is significant at 5% level.

Recently, advances in statistical applications in engineering and sciences have opened the market for computer software packages that merge computer graphics, experiment design, regression, statistical analysis, worksheet and documentation. SAS (SAS, 1989), the Statistical Analysis System is an example. Design-Expert (1996) by Stat-Ease is very user friendly. The software has a powerful tutorial system that will guide new users through the steps of RSM optimisation. Similar software like STATISTICA (2000), SPSS (2000) are also popular on the market.

RSM optimisation has wide applications in the field of food sciences. Many successful examples can be found in the literature. As an example, Jungqua *et al.* (1997) maximised the production of microbial transglutaminase, an important enzyme in protein-related foods, using RSM. They determined the experimental domain by the one-factor-at-a-time technique and used the obtained optimum point as the centre (level 0) of central composite design in subsequent RSM optimisation. A three-fold increase of transglutaminase production was achieved using RSM. Mahoney *et al.* (1974) optimised lactase production by a conventional one-factor-at-a-time technique. Building on this work Chen *et al.* (1992) used central composite design in RSM optimisation to further increase lactase production by 60%. Chen *et al.* (1998) have also studied fried gluten balls using RSM, and concluded that better quality fried gluten balls could be obtained by simply adjusting oil temperature.

13.4 A case study: fried gluten balls

The value of RSM can be demonstrated by its application to optimising texture and colour in the manufacture of gluten balls. Fried gluten balls are popular in the Chinese community. The manufacturing process is as follows:

- wheat flour is washed with water to separate gluten from the starch
- the wet gluten is immersed in water for 30 minutes
- it is then removed and cut and shaped into wet gluten balls, which are then deep fried using three or four deep frying pans at differing temperatures
- in the first and second deep frying pans, water evaporation expands the gluten balls, establishing their basic volume, shape, texture and colour
- frying in the final pans completes the ageing of the balls

Due to the high water content of the wet gluten balls, a high rate of water evaporation leads to significant volume expansion (as much as 15–20 times the original size) during the first stages of frying, producing highly porous, dried and crispy gluten balls. These early stages in frying are the most important in determining the final quality of the gluten balls. In the experiment described here, the main variable analysed was the optimum frying temperature in the first and second deep frying pans (Chen *et al.*, 1998).

13.4.2 The RSM design

The study used the rotatable central composite design (Mason *et al.*, 1989), consisting of a two-factor-five-level pattern with 10 design points (8 combinations with 2 replications of the centre point). The two factors and the coded values of the 5 levels of each factor are shown in Table 13.3, and the experimental design of RSM for the frying temperature of the gluten balls is

Table 13.3 Coded values and corresponding real values of independent variables

Independent variables	Coded levels				
	−1.414	−1	0	1	1.414
$X_1(T_1)$	126	130	140	150	154
$X_2(T_1)$	151	155	165	175	179

(Chen *et al.* 1998, reproduced with permission from John Wiley & Sons)

shown in Table 13.4. The data from these experiments were analysed by the SAS's RSREG (response surface regression) procedure (SAS, 1989). In visualising the relationship between the response and the experimental levels of each factor, the response surfaces were generated from the fitted quadratic polynomial equation obtained from the RSREG analysis.

13.4.2 The experimental design

The flour used in this study was an untreated flour, milled commercially from a mixed grist of hard red wheat from America. Flour protein content, on a 13.5% moisture basis, was 13.96%, and ash content was 0.53%. 100 grams of wet gluten balls were fried continuously in three consecutive frying pans. Each pan (of identical dimensions) contained about 10 litres of soybean oil. The frying time in each pan was 120, 90 and 70 seconds respectively. The temperature of the third pan was fixed at 195 ± degrees C. The temperatures of the first and second pans were varied in line with the principal objective of the experiment.

13.4.3 Texture and colour measurement

Sensory evaluation was conducted by presenting samples of the fried gluten balls to a panel of 30. Panelists scored the samples in 3 ways: appearance score (AS), texture score (TS) and total acceptance score (TAS). AS and TAS were analysed by a hedonistic test, and TS analysed by a comparison test using three samples.

Instrumental measurements of texture and colour were correlated to panel studies to ensure the validity of the quantitative measurements (Chen *et al.*, 1997). Texture was measured by measurement of the peak force (PF) and brittleness breakdown (BB) using a rheometer, and quantitative results were correlated to consumer preferences. PF is a measure of hardness of the specimen and is defined as the maximum force at 75% compression during the first bite. BB is a measure of crispiness and is defined as the first major peak or force at failure before the maximum force is obtained during the first bite. Although PF and BB represent different textural qualities (hardness or softness, and crispiness), they show, in many instances, a high degree of correlation. For fried gluten balls, high PF reflects tough texture which often corresponds to a

Table 13.4 Quadratic model coefficients[b] and R^2 values for the response surfaces of different quality indices[a]

Coeff[b]	EV	ER	PF	BB	HB	AS	TS	TAS
a_0	−222.32	−17 666	1543.74	1737.92	157.88	−189.95	−73.94	−145.57
a_1	3.39	271.22	−236.28	−243.53	−0.36	2.98	1.78	2.92
a_2	0.058	4.38	183.79	187.35	−1.57	−0.082	−0.46	−0.57
a_3	−0.0079	−0.64	0.83	0.85	−0.0018	−0.009	0.0065	−0.0092
a_4	0.0034	0.28	−0.62	−0.64	0.0025	0.0018	0.0017	0.0029
a_5	−0.0079	−0.63	0.13	0.14	0.0056	−0.0035	−0.0006	−0.0028
R^2	82.57	82.55	92.89	93.29	91.02	89.74	93.72	90.22

[a] EV, expansion volume; ER, expansion ratio, PF, peak force; BB, brittleness breakdown; HB, Hunter *b* value; *AS, appearance score; TT, texture source; TAS, total acceptance score.*

[b] $Y = a_0 + a_1X_1 + a_2X_2 + a_3X_1^2 + a_4X_2^2 + a_5X_1X_2$, where X_1 is T_1 (the temperature of the first deep frying pan); X_2 is T_2 (the temperature of the second deep frying pan).

(Chen *et al.*, 1998, reproduced with permission from John Wiley & Sons.)

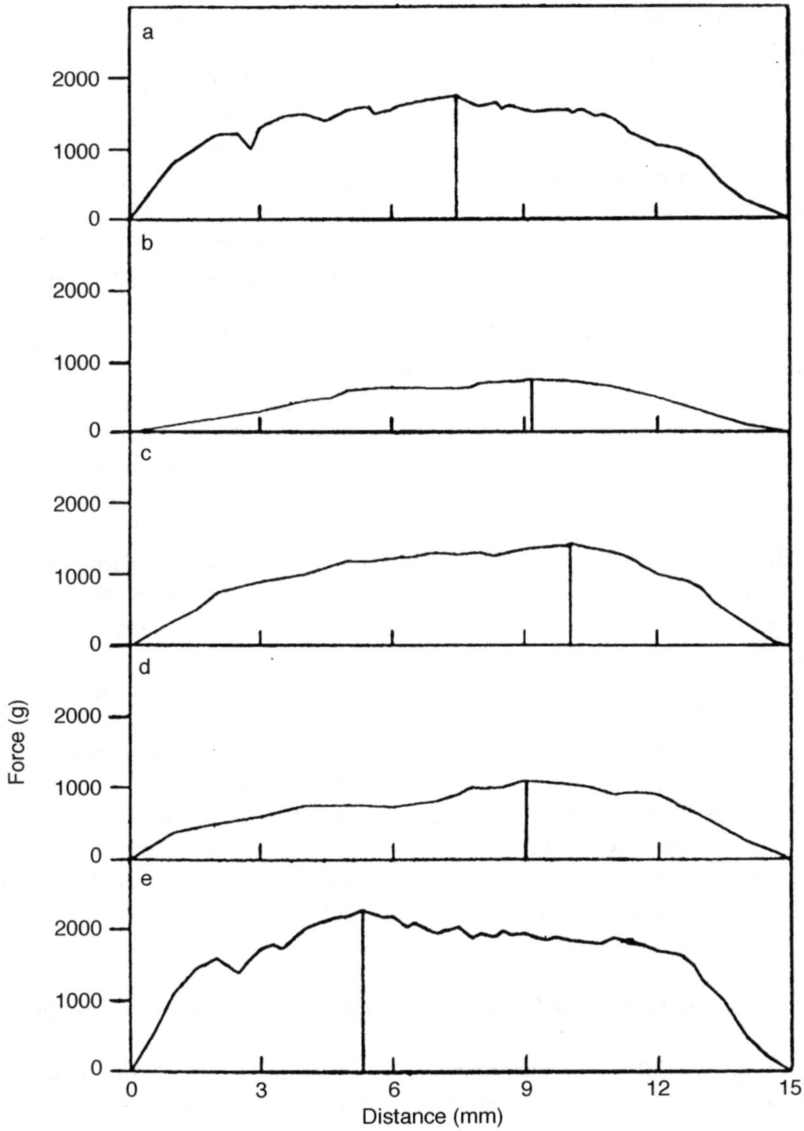

Fig. 13.2 Force-distance curves used to determine peak force (PF) and brittleness breakdown (BB) of fried gluten balls.

high peak or force at failure (BB). Experimental results showed that frequently, for fried gluten balls, numeric values of PF were very close to BB and sometimes coincided with each other. Generally speaking, consumers prefer puffy, yet soft fried gluten balls (low PF and BB). After sensory evaluation, PF and BB were negatively correlated to consumer preferences.

A Sun rheometer (Sun CR 200D, Sun Scientific Co Ltd, Japan), mounted with a plunger (adapter No. 14) was used to measure PF and BB. The setup included a microcomputer for continuous monitoring and recording of force variation sensed by the plunger during the travel of the platform on which the sample was placed. The sample platform travelled upward to the plunger at a speed of 60 mm/min, and the compression distance of the plunger was 12 mm. The measured values of 30 grains from the sample of fried gluten balls were averaged.

The obtained force-distance curves as shown in Fig. 13.2 were analysed by computer software to determine PF and BB. The force-distance curves also offer information about texture other than PF and BB. Volume expansion is responsible for the porous network structure, while for the same mass of gluten ball, larger size indicates larger void volume and thinner membranes dividing these void cells. A larger volume for the same weight of gluten ball indicates a softer texture preferred by panelists in sensory evaluation. Therefore, the expansion volume (EV) and expansion ratio (ER) were also considered relevant to the overall quality as dependent variables for this study. Higher EV/ER correlated to higher appearance score (AS) and higher total acceptance score (TAS).

The colour of 30 sample grains of the fried gluten balls was measured using a colorimeter (Color Analyzer, Color Mate OEM, Milton Roy Co., USA). Three determinations were conducted randomly on the surface of each fried gluten ball. The measured values of the 30 samples were averaged. Of the 3 parameters available, the Hunter b value (HB) was found to correlate best with AS, TS (texture score) and TAS. Panelists reported that a light yellow colour suggested to them that the frying oil was fresh and the gluten ball not over fried. HB was, therefore, negatively correlated to AS, TS and TAS.

13.4.4 Analysing the results: the optimum frying temperature for gluten balls

By using SAS's RSREG procedure (SAS, 1989), the quadratic regression equations for a range of quality indices in relation to the temperatures of the first and the second deep frying pans were obtained (shown in Table 13.3). From the satisfactory values of R^2 it is clear that these quality indices are significantly related to the frying temperatures of the first and the second frying pans.

Figure 13.3 shows the response surfaces of the quality indices of the fried gluten balls as functions of the frying temperatures of the first and the second deep frying pans. The response surfaces show that the expansion volumes, expansion ratios and sensory evaluation scores of the fried gluten balls increase

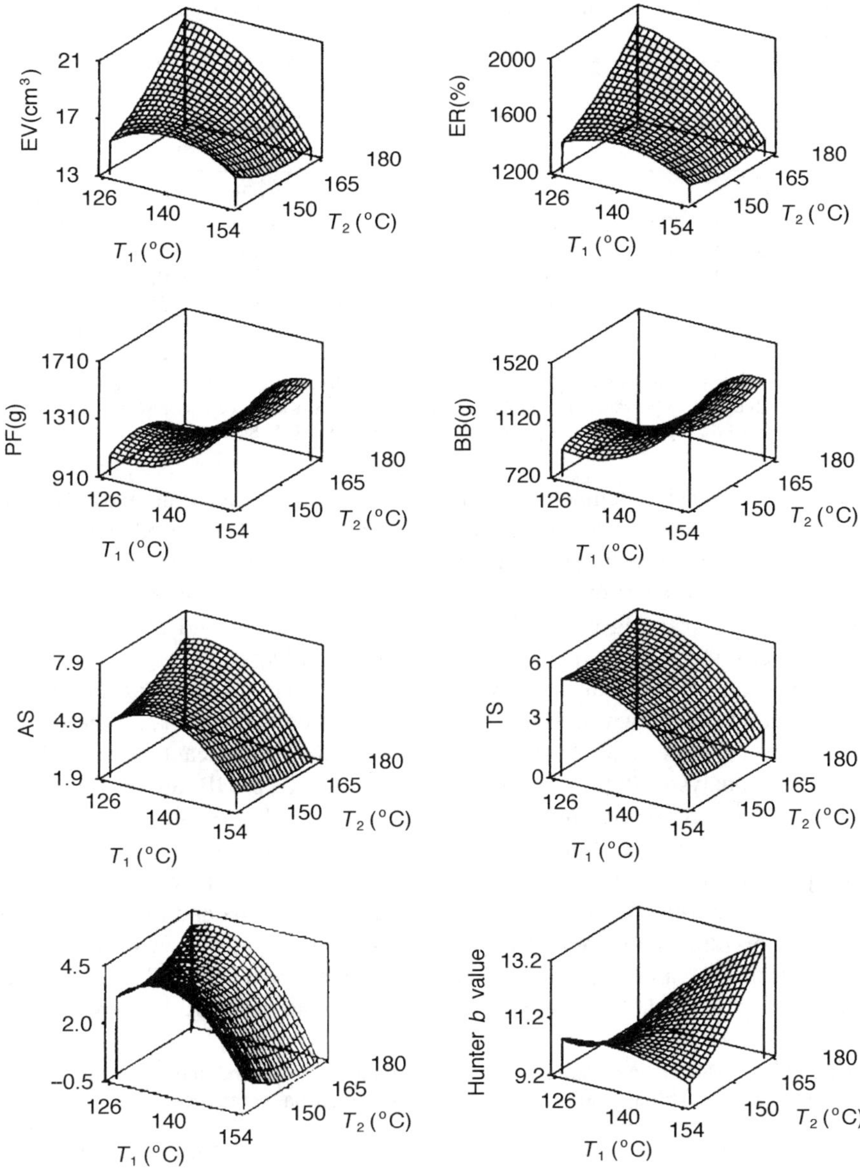

Fig. 13.3 The response surfaces of frying temperatures and quality indices of fried gluten balls. T_1, the temperature of the first deep frying pan; T_2, the temperature of the second deep frying pan; EV, expansion volume; ER, expansion ratio; PF, peak force; BB, brittleness breakdown; AS, appearance score; TS, texture score; TAS, total acceptance score.

(Chen *et al.*, 1998, reproduced with permission from John Wiley & Sons.)

Table 13.5 ANOVA for the frying temperatures vs the quality indices[a] of the fried gluten balls

Source	Response[a] (P values)							
	EV	ER	PF	BB	HB	AS	TS	TAS
Model	0.110	0.111	0.020	0.018	0.073	0.042	0.017	0.038
T_1	0.032	0.032	0.004	0.003	0.016	0.010	0.002	0.009
T_2	0.122	0.123	0.298	0.300	0.110	0.457	0.574	0.417
T_1^2	0.159	0.159	0.032	0.030	0.242	0.030	0.030	0.026
$>T_2^2$	0.601	0.600	0.176	0.167	0.462	0.618	0.515	0.427
$>T_1T_2$	0.291	0.292	0.772	0.741	0.228	0.38	0.82	0.478

[a] EV, expansion volume; ER, expansion ratio; PF, peak force; BB, brittleness breakdown; HB, Hunter *b* value; AS, appearance score; TS, texture score; TAS, total acceptance score.
(Chen *et al.*, 1998, reproduced with permission from John Wiley & Sons.)

with the decreasing temperature of the first deep frying pan and with the increasing temperature of the second deep frying pan. On the other hand, the peak force, brittleness breakdown, and Hunter colour *b* values increase with the increasing temperature of the first deep frying pan. Decreasing the temperature of the first deep frying pan and increasing that of the second deep frying pan should help improve the quality of the fried gluten balls.

The results of the analysis of variance for the quality indices against frying temperatures are shown in Table 13.5. The P values indicate that, although the expansion volume and expansion ratio are not significant for quadratic regression analysis ($P>0.1$), other quality indices, except HB, are all significant for quadratic regression analysis ($P<0.05$). The P value of the model for HB indicates a significant relationship. However, this only partially confirms that the quadratic regression analysis for the quality indices other than expansion volume and expansion ratio vs the temperatures of the first and the second deep frying pans are significant. Table 13.5 also lists P values against the two temperatures. It can be seen that T_1 is significant at 5% level to all the responses studied. T_1^2 is significant to PF, BB, AS, TS and TAS at 5% level. T_2 is

Table 13.6 The critical values of the frying temperature and the characteristics of stationary points for the quality indices[a] of fried gluten balls obtained using RSM

Process variables	PF	BB	HB	AS	TS	TAS
T_1(°C)	129.77	129.62	143.41	135.54	130.69	134.41
T_2(°C)	161.22	161.17	155.65	154.95	160.05	158.53
Stationary point	saddle	saddle	saddle	saddle	saddle	saddle

[a] PF, peak force; BB, brittleness breakdown; HB, Hunter b value; AS, appearance score; TS, texture score; TAS, total acceptance score.
(Chen *et al.* 1998, reproduced with permission from John Wiley & Sons).

significant to HB at 11% level but not to any of the other indices. This is in line with Fig 13.2 which shows that all the curves (of constant T_1) on the surfaces are relatively flat with respect to T_2.

The critical values of the frying temperature and the characteristics of stationary points of the response surfaces for the quality indices of the fried gluten balls obtained by using RSM are shown in Table 13.5. The optimal temperatures were picked by balancing between energy consumption (to achieve high temperature) and quality indices. Below $T_1 \sim 130°C$, PF and BB remained virtually constant (Fig. 13.2, constant T_2). AS and TAS were at their maximum points, with respect to T_1, when $T_1 \sim 135°C$ (Fig. 13.2, at fixed T_2). Although AS, TS and TAS showed further increases beyond $T_2 \sim 155°C$, the increases were minor. Besides, HB would increase (more browning) with $T_2 > 155°C$ (in the vicinity of $T_1 \sim 140°C$). 130–143°C and 155–161°C were chosen as the optimal ranges of temperatures for T_1 and T_2, respectively. The lack of significance of T_2 to those parameters discussed indicated that no serious precision control of the temperature of the second frying pan is needed.

In order to verify the optimal frying temperature of fried gluten balls obtained using RSM, 130±3°C and 155±3°C were selected as the frying temperatures of the first and the second deep frying pans, respectively. The temperature of the third deep frying pan was fixed at 195±3°C. The quality indices of the fried gluten balls produced by frying continuously through three consecutive deep frying pans are shown in Table 13.6. They show that the peak force and brittleness breakdown of the obtained fried gluten balls are lower. These data indicate that the fried gluten balls obtained in this test have a soft and elastic, not brittle and rigid, texture, which is more favoured by the customers. Moreover, the fried gluten balls obtained in this test have a bright yellow appearance (low Hunter *b* value), and their sensory evaluation scores, including appearance, texture and total acceptance scores, are all higher than the commercial standard scores of 4–5. The above results confirm the applicability of the frying temperatures, obtained by using RSM, for commercial production of fried gluten balls.

13.5 Conclusions

This case study illustrates the role of a technique such as RSM in analysing selected variables influencing the texture and colour of fried products. This approach can clearly be used to take in other variables. An obvious example would be to analyse differing types of flour or the process of gluten extraction. During the stage of gluten extraction, as the flour is washed with water, the networked structure of gluten is gradually forming (Huebner, 1977; Bietz and Wall, 1980). Upon hydration, glutenin becomes swollen, and at the same time, absorbs gliadin together with some of the albumin and globulin. The network structure of gluten is co-stabilised by disulfide bonds, hydrogen bonds, and hydrophobic interactions (Huebner, 1977). RSM could be used to analyse how the process of gluten extraction could be optimised for the amount and quality of gluten extracted per unit mass of flour processed. Further experiments might well establish different optimum temperatures for different quality flour/gluten. RSM is also clearly applicable to other fried products in isolating the mix of variables that will maximise colour, textural and other product qualities valued by the consumer.

13.6 References

BIETZ J A, WALL J S 1980 Identity of high molecular weight gliadin and ethanol-soluble glutenin subunits of wheat: Relation to gluten structure. *Cereal Chem* **57** 415–420.

BLUMENTHAL M M 1991 A new look at the chemistry and physics of deep-fat frying. *Food Technology* **45**(2) 68–71.

BLUMENTHAL M M, STOCKLER J R 1986 Isolation and detection of alkaline contaminant materials (ACM) in used frying oils. *J Am Oil Chem Soc* **63**(5)687–692.

BOUTON P E, HARRIS P V 1972 A comparison of some objective methods used to assess meat tenderness. *J Food Sci* **37** 218–221.

BOUTON P E, HARRIS P V, SHORTHOSE W R 1975 Possible relationships between shear, tensile and adhesion properties of meat and meat structure. *J Texture Stud* **6** 297–314.

BOX G E P, WILSON K B 1951 On the experimental attainment of optimum conditions. *J Royal Statist Soc* **B13** 1–45.

BREWER M S, MCKEITH F K, BRITT K 1992 Fat, soy and carrageenan effects on sensory and physical characteristics of ground beef patties. *J Food Sci* **57**(5)1051–1052.

CHEN C S, CHEN J J, WU T P, CHANG C Y 1998 Optimising the frying temperature of gluten balls using RSM. *J Sci Food Agric* **77**(1) 64–70.

CHEN J J, SHYONG P R, CHANG C Y 1997 Studies on the analytic method for the quality of oil-fried gluten ball. *J Chin Agr Chem Soc* **35**(1) 70–76.

CHEN K C, LEE T C, HOUNG J Y 1992 Search method for the optimal medium for the production of lactase by *Kluyveromyces fragilis*. *Enzyme Microb Technol* **14**(8) 659–664.

DESIGN-EXPERT 1996 Stat-Ease Corp., 2021 East Hennepin Avenue, Suite 191, Minneapolis, MN 55413, USA.

FRANCIS F J, CLYDESDALE F M 1975 Food Colorimetry: Theory and Applications. AVI Publishing Co., Westport, CT, USA.

HUEBNER F R 1977 Wheat flour proteins and their functionality in baking. *Baker's Dig* **51**(25) 154.

JUNQUA M, DURAN R, GANCET C, GOULAS P 1997 Optimization of microbial transglutaminase production using experimental designs. *Appl Microbiol Biotechnol* **48** 730–734.

KADAN R S, CHAMPAGNE E T, ZIEGLER G M, RICHARD O A 1997 Amylose and protein contents of rice cultivars as related to texture of rice-based fries. *J Food Sci* **62**(4) 701–703.

KOTULA A W 1976 Evaluation of beef patties containing soy protein during 12-month frozen storage. *J Food Sci* **41** 1142–1146.

LARMOND E, MORAN E T JR. 1983 Effect of finish grade and internal basting of the breast with oil on sensory evalution of small white toms. *Poultry Sci.* **62** 1110–1116.

LIN Y H, KUO J C C 1994 Palatability and storage stability of breaded chicken breast. *J Food Sci (Taiwan)* **21**(3) 216–227.

MAHONEY R R, NICHERSON T A, WHITAKER J R 1974 *J Dairy Sci* **58** 1620–1625.

MASON R L, GUNST R F, HESS J L 1989 *Statistical Design and Analysis of Experiments – With Application to Engineering and Science*. John Wiley & Sons, New York, USA.

MOHSENIN N N, MITTAL J P 1977 Use of rheological terms and correlation of compatible measurements in food texture research. *J Texture Stud.*, **8** 365–370.

MONTGOMERY D C 1984 *Design and Analysis of Experiments*. John Wiley & Sons, New York, USA.

MOORE S L, THENO D M, ANDERSON C R, SCHMIDT G R 1976 Effect of salt, phosphate and non-meat protein on binding strengths and cook yields of beef rolls. *J Food Sci.* **41** 424–429.

MORAN E T JR. 1992 Injecting fats into breast meat of turkey carcasses differing in finish and retention after cooking. *J Food Sci.* **57**(5) 1071–1076.

MORAN E T JR., LARMOND E 1981 Carcass finish and breast internal oil basting effects on oven and microwave prepared small toms: cooking characteristics, yields and compositional changes. *Poultry Sci.* **60** 1229–1236.

NEER K L, MANDIGO R W 1977 Effect of salt, sodium tripolyphosphate and frozen storage time on properties of a flaked cured pork product. *J Food Sci.* **42** 738–744.

OHLSON R 1983 Structure and physical properties of fats.

PELEG M 1987 The basics of solid foods rheology. In *Food Texture, Instrumental and Sensory Measurement* (MOSKOWITZ H R ed.), Marcel Dekker Inc., New York, NY, USA.

RAO V N M, DELANEY A M 1995 An engineering perspective on deep-fat frying of breaded chicken pieces. *Food Technology* April 1995 138–141.

SAS 1989 SAS/STAT User's Guide, Version 6 (4th edn, Vol 2). SAS Institute Inc., Cary, NC, USA.

SPSS 2000 SPSS Inc., 233 S. Wacker Drive, 11th floor, Chicago, Illinois 60606, USA.

STANLEY D W 1976 The texture of meat and its measurement. In *Rheology and Texture in Food Quality* (DEMAN J M, VOISEY P W, RASPER V F, STANLEY D W eds), AVI Publishing Co., Westport, CT, USA.

STATISTICA 2000 StatSoft Inc., 2300 E. 14th Street, Tulsa, Oklahoma 74104, USA.

SZCZESNIAK A S 1983 Physical properties of foods: what they are and their relation to other food properties. In *Physical Properties of Foods* (PELEG M, BAGLEY E B eds), AVI Publishing Company Inc., Westport, CT, USA pp 28–37.

SZCZESNIAK A S, HUMBAUGH P R, BLOCK H W 1970 Behavior of different foods in the standard shear compression cell of the shear and the effect of sample weight on peak area and maximum force. *J Texture Stud* **1** 356–378.

THOMPSON D 1982 Response surface experimentation. *J Food Proc Preserv* **6** 155–188.

VOISEY P W, DEMAN J M 1976 Application of instruments for measuring food texture. In *Rheology and Texture in Food Quality* (DEMAN J M, VOISEY P W, RASPER V F, STANLEY D W eds), AVI Publishing Co, Westport, CT, USA.

VOISEY P W, LARMOND E 1974 Examination of factors affecting performance of the Warner-Bratzler meat shear test. *Can Inst Food Sci Technol J* **7** 243–249.

WHITING R C 1984 Addition of phosphate, proteins and gums to reduced salt Frankfurter batters. *J Food Sci.* **49** 1355–1362.

Index

absorption
  digestion and absorption of lipids 62-4
  of oil in fried foods 115–16
    controlling 251–2
    flavour and aroma 303–12
AC-Check 188
acceptable daily intake (ADI) 24
accumulator 233
acidity *see* free fatty acids
ACM test (alkaline contaminant material) 187, 258
active filtration 153–4, 249
additives 21, 25–6, 202, 312–14
adsorption 320
adulteration 127–31, 177–8
  see also authenticity
aeration 151
agglomeration 321
aim value 222
air pre-fry drying 307–8
air radio frequent assisted (ARFA) dryers 203
aldehydes 278, 279
alkaline-reacting materials (ARM) 104, 144–5, 154
  ACM test 187, 258
alkylpyrazines 276
alkylpyrroles 275
alpha-tocopherol 297
Amadori products 273, 274
amino acids 272–6
analysis of variance (ANOVA) 347, 354–5
animal feeds 23, 24, 35, 177
animal tallows 88, 89, 118, 121, 129–30, 138
anisidine value 172, 173, 180
anti-caking agents 322
antioxidants
  leaching from food being cooked 314
  in oils
    antioxidant capacity 247–8, 256
    flavour and aroma 296–303
aroma 266–334
  aromas derived from lipid oxidation 335–6
  biology of odour 315–16
  degradation reactions 268–72
    Maillard and Strecker reactions 272–6
  effect of antioxidants in frying oils 296–303
  future development 325–7
  influence of the food 314–15
  oil uptake by fried food 303–12
  sensory evaluation 315–18
artificial flavourings 319–20
ascorbyl palmitate 99–100, 298
Austria 14
authenticity, oil 127–42
  criteria 132–4
    use of 135–40

authenticity, oil (*continued*)
current issues 127–31
measurement of quality and 165–93
potential future approaches 140–2
testing 131–2

bacterial proliferation 257–8
baseline data 167
basic sensory panels 317
batter coatings 204, 252
crispness 310
beef, roasted 284–5
beef tallow 88, 89, 118, 121, 129–30, 138
Belgium 14
belt blanchers 203
belt dryers 203
β-carotene 98–9, 104–5
biological oxygen demand (BOD) 34
biological risk 257–8
biotechnology 38
blanching 201–3, 308
brass 103
brassica vegetables 314
breading 210, 252
'break-in' oils 175–6
'breaking in' the oil 152
brittleness breakdown (BB) 349–55
browning 279–80
see also Maillard reaction
bulk road tankers 149–50, 151
butylated hydroxyanidsole (BHA) 95–6, 146
butylated hydroxytoluene (BHT) 95–6, 146

Campbell's 173
canola oil 108, 286–91, 293–4
see also rapeseed oil
carbohydrates 71
carbon dioxide flushing 303
carnosic acid 96, 97
carotenoids 98–9, 104–5, 146
central composite design 345–7, 348–9
certification of vendors 173–5
checking procedures 239
chemical oxygen demand (COD) 34
chemical risk 257
chemical tests 178–81
rapid
quick tests 182, 185–9
with instruments 182, 183–5
chicken 285, 298, 312, 313–14, 340, 341
chilled French fries 206, 211
chilling 206–7, 320
chips *see* French fries
chlorophyll 104
cholecystokinin (CCK) 63
cholesterol 60, 61–2, 64, 67, 70, 142–3
chylomicrons 65–6
citric acid 105, 146
city regulations 53–4
clarity test 182
cleaning 158, 219, 251
clear-coat batters 204
clusters 227
COAT (Cooking Oil Analysis Technique) 183–4
coating 204, 252
coconut oil 116–19, 295
Codex Alimentarius 20, 37
cold-pressed oils 130, 138
colipase 63
colour 280, 337–58
development in crisps 229
fried gluten balls 348–55
influences on 340–3
instrumentation for measuring 337, 339
oil colour 179–80, 182–3
RSM 343–8
colour compounds 104–5
composition of oils and fats 87–114
combined effects of natural products on stabilisation 105–8
and flavour production 285–96
future trends 108–9
minor components and stability 91–105
preventing degradation 247
and its relationship with suitability 116–27
modified oils 122–7
unmodified oils 116–22
types of oils and fats 88–91
compression testing 338
conjugated dienes and trienes 255
consumer panels 7–8, 317
consumer pressure 37–8
contaminants: safe levels 24
continuous filter 153
continuous fryer 2
control point management 222–3
convenience 9
cooking method 312–14
cooking temperature *see* temperature
cool zone 153, 248
copper 103, 150–1, 157–8, 248
coronary heart disease 70

corrective action 239
cottonseed oil 117, 121, 129, 136, 290–1, 294
crispness, batter 310
crisps 1–2, 107–8, 116, 215–35
 equipment 234
 flavour development 277–80
  fatty acid composition 285–91, 293, 294–6
 future trends 234–5
 managing the processing operation 222–32
  control point management 222–3
  finished inspection 231–2
  frying 228–30
  potato preparation 224–5
  process control 222
  processing objectives 222
  salting 230–1
  seasoning application 232
  slice washing 226–8
  slicing 225–6
 market 11–12
 oil and fat management 216–19
 oil uptake 304–7
  post-fry treatments 308–9
  pre-fry treatments 307–8
 origin of 1
 packaging 233–4
 process 216
  monitoring 234
 product 215–16, 234
 raw materials 234–5
  management 220–1
critical control points (CCPs) 238
 monitoring in frying process 252–9
 see also HACCP
critical limits 238
crude oils 171
Crum, George 1
crunchiness 72–3
crust 205, 252
curdlan 308–9
cut pre-fried potato products 198
cutting 200–1
cyclic monomers 29, 256

debris, proteinaceous 103–4, 144
deep-frying 236, 240
 role in fat intake 71–4
defect-cutter 201
degradation processes 175–6, 244–6, 268–72
 life of frying oils 28–34
 Maillard reaction 205–6, 272–6, 279–80, 282
 preventing 246–51
 Strecker reaction 272–6
'degrading' stage 176
delivery of oils and fats 149–52, 221
design of cooker/fryer 87
dietary lipids *see* lipids
digestion 62–4
dimer and polymer triglycerides (DPTG) 255
dimethyl-polysiloxane (DMPS) 91–2
directives 20
discarded frying oil (DFO) 75–7
distribution 210–11
 see also transport
draining 251
drum washer 227
drying 311, 320, 321
 crisps 305–7, 307–8
 French fries 203–4
due diligence 22, 23–4
durability 23
Durkex 500 123, 126
dusting on flavours 325

effluents, liquid 34–5
eicosanoids 68
elastic modulus 338
electronic nose 189, 258–9
encapsulation 320
environment 189–90
 protection and regulation 34–6
EPA 71
essential fatty acids 61
essential oils 321–2
ethylidene group-containing sterols 97–8
Europanel 7
European frozen food markets 12–15
European Union (EU) regulation 19–48
 basis of EU law 19–20
 environmental protection 34–6
 EU and national regulatory bodies 39–41
 future trends 36–8
 life of oils 28–34
 publishers of legislation 41–3
 sale of food 22–8
 sources of information 38–43, 44–7
 structure of frying industries 22
 supremacy of EU law 36
experiment design 345–7, 348–9
extrusion machines 210

fat 59, 60, 323
  intake 68–80
    health issues 68–71
    impact of repeated frying 74–5
    measuring the impact of frying 75–8
    role of deep-frying 71–4
  synthetic fat replacers 325–6
  see also lipids
fats for frying *see* oils and fats
fatty acids 60–1, 275–6
  composition of oils 88–90, 116–27, 135
    and flavour production 285–96
    modified oils 122–7
    prevention of degradation 247
    unmodified oils 116–22
  free *see* free fatty acids
  in French fries 90–1
  impact of deep-frying on concentration in foods 74
  oxidized (OFA) 169–71, 255
  quality control by analysis of 254
  release from adipose tissue 67–8
  at triglyceride 2-position 135
Federal Food, Drug and Cosmetic Act 49–50
filter systems 248–9
filtration 103–4, 153–4
fines removal box 227
finished inspection 231–2
finishing methods 199
Finland National Food Administration 56, 57
fire point 101
fish 10, 12–15, 74, 314–15
fish balls 341
fish oils 122
flash point 101, 149
flavour 147–8, 266–334
  application of flavours 318–25
    application techniques 325
    definition of flavourings 319–20
    formulation and production 324–5
  biology of 315
  degradation reactions 268–72
    Maillard and Strecker reactions 272–6
  development in foods 277–85
  effect of antioxidants in oils 296–303
  effect of frying techniques, cooking method and additives 312–14
  fatty acid composition and 285–96
  flavours derived from lipid oxidation 335–6
  future development 325–7
  influence of the food being fried 314–15
  oil uptake by fried food 303–12
  raw potatoes 267–8
  sensory evaluation 315–18
flavour carriers 320–1
flavour compounding 324
flavour enhancement 322–3
flavour modification 324
flow diagrams 239–41, 242–3
flow wheels 229, 230
foaming 100, 102–3, 143–4, 181, 227–8
Food Code 50
Food and Drug Administration (FDA) 49, 169
  regulations and guidelines 49–51
Food Oil Sensor (FOS) 185, 258
Food Safety and Inspection System (FSIS) 49
  guidelines and directives 52
FoodFuture campaign 37
force-distance curves 351, 352
formed pre-fried potato products 198–9
  key manufacturing processes 208–10
forming 210
France 12
fraud 127–31, 177–8
free fatty acids (FFA) 101–2, 177, 206, 217–18
  fraud 130–1
  testing 179
  used oil quality control 254–5
free-flow agents 322
freeze drying 321
freezing 206–7
French fries 2, 90–1, 106–7, 198
  flavour 280–3, 295–6
  flow diagram 241, 242–3
  key manufacturing processes 200–7
  key requirements 199–200
  oil degradation and 175–6
  quality 211–12
  storage and distribution 210–11
fresh oils
  characterization 167
  quality control 256–7
  quality limits 148–9
  regulation 29–30, 31
'fresh' stage 176
Fri-Check unit 183, 184
fried gluten balls 348–55
Fritest 185–6, 258

frozen foods 8
  French fries 206–7, 210–11
  market in other European countries 12–15
  UK market 10–11
fruit 312
frying
  crisps 228–30
  French fries 204–6
  process and oil quality 152–8
    evaluation during 158
    frying operation 153–8
    nature of food fried 152–3
  quality control during 175–7
  regime and technique 312–14
  temperature *see* temperature
  time 310–12, 342–3
frying equipment
  cleaning 158, 219, 251
  design and materials 87, 103
  preventing oil degradation 248–9
frying oil quality curve 175, 244–5
fume extraction systems 249

gas flushing 303
gas-packed French fries 207, 211
genetic modification (GM) 37–8, 140–1, 326–7
  GM standard 38
Germany 12
GiTIC (Gel-in-Tube Instant Chemistry) 189
global warming 9
gluten balls, fried 348–55
gluten extraction 355–6
glycerides, partial 102–3
'Good-Fry' oils 89, 90, 105–8, 109, 123, 126–7
good industrial practice 249–51
good manufacturing practice (GMP) 20, 26
grading 201, 207
grapeseed oil 117, 121
groundnut oil 117, 120, 127–8, 135–6, 139

HACCP (Hazard Analysis by Critical Control Points) 24–5, 189, 236–65
  approach 237–9
  FDA 50–1
  flow diagram examination 239–41
  hazard evaluation and preventive measures 241–52
  monitoring critical control points 252–8
hash-browns 208
hazard evaluation 241–52
hazelnuts 284
headspace sensors (electronic nose) 189, 258–9
health 38, 59–84
  consciousness 9
  dietary lipids 60–8
    digestion and absorption 62–4
    sources 61–2
    structure and function 60–1
    transport and metabolism 65–8
  fat intake 68–80
    health issues 68–71
    impact of repeated frying 74–5
    measuring impact of frying 75–8
    role of deep-frying 71–4
healthy stable frying oils 109
heat preservation 240–1
herbs 320, 321–2
Heynes products 273
high-density lipoproteins (HDL) 67
high-oleic sunflower seed oil (HOSO) 88–90, 122, 123, 294
high-speed air suction dryer 203–4
holding time 203
hot-pressed oil 130, 138
household size 9
hydrocarbons 141, 285
hydrogenated oils 88, 89, 123–6, 286–8
hydrolysed vegetable protein (HVP) 323
hydrolysis 100, 101–2
  rancidity 269–72
hydroperoxides 269

Iceland Environment and Food Agency guidelines 54–5
impingement drying 311
impulse savoury snacks 9, 11–12
inactivation of enzymes 202
incoming delivery checks 221
industrial frying oils 122–7
industrial structure 22
ingredients 26–8, 320, 321–2
inspection 231–2
inspection/trim table 224–5
interestification 171
intermediate sensory panels 317
iodine value 136, 181, 256
Ireland 15
Italy 13

jack fruit 312
Judd-Hunter system 339

Kaomel (Durkex 500) 123, 126
Kentucky Fried Chicken (KFC) 182
keto acids 270

labelling 26–8
lactase 348
lactic acid-sodium benzoate 313–14
lactones 270–1
lard 117, 121, 129–30, 138
  PCB contamination 23, 24
lauric acid 296
law
  areas covered by 20–1
  basis of EU law 19–20
  legal context of regulation 19–21
  publishers of legislation 41–3
  supremacy of EU law 36
  see also regulation
lecithins 102–3, 146
lid, floating 249
life of frying oils 28–34
  end of frying life 30–4
linoleic acid 288, 289, 294–5, 296
linolenic acid 286–91, 293, 296
lipids 60–8
  availability in world regions 69–70
  digestion and absorption 62–4
  oxidation *see* oxidation
  sources 61–2
  structure and function 60–1
  transport and metabolism 65–8
  see also fat
Lipofrac system 125–6
lipoprotein lipase (LPL) 65
lipoproteins 65–8
liquid effluents 34–5
loading the fryer 250–1
low-density lipoproteins (LDL) 66–7
low-fat products 323

Maillard reaction 205–6, 272–6, 279–80, 282
maize oil 117, 121, 129, 137
malabsorption of fat 64
margarines 291–2
market for fried food 7–18
  factors influencing 9–10
  frozen food market in other European countries 12–15
  future trends 15–16
  range of fried foods 8–9
  UK market 10–12
mashing 209
maximum recommended limit (MRL) 24
McDonald's 165
meat 11, 12–15, 313
  see also beef; chicken
Meat and Poultry Inspection Manual (USDA/FSIS) 49, 52
Mediterranean diet (MeD) 71–2
melting cycle 154
metabolism 65–8
metal sequestrants 99–100
metals
  surface metal contamination hazard 248
  trace metals 103, 144
microwave cooking 313
microwave drying 307–8
minced meat 313
minor components 142–8
  beneficial 92–100, 145–7
  detrimental 100–5, 142–5
  development of flavour during use 147–8
  and frying oil stability 91–105
MirOil Life Powder 91–2, 146–7
mixed micelles 63–4
mixing 210
modified oils 122–7, 286–8, 293
moisture content 205, 228–9, 250, 252, 255
  and oil uptake 305–7, 309–10
  and texture 340–1
monitoring
  critical control points 252–9
  establishing a monitoring system for each CCP 238–9
monounsaturated fatty acids (MUFA) 60–1, 71
moulding machines 210
Munsell system 339

national regulatory bodies 39–41
natural flavourings 319–20
natural products 105–8
nature identical flavourings 319–20
near infrared spectroscopy (NIR) 189, 258
neo-compound formation 241–6
Netherlands 13–14
neutral lipid exchange 67
nitrogen blanketing 150

nitrogen flushing 303
non-volatile substances of decomposition 253
Nu-Sun 88, 89
nutrition 26–8

oil colour 179–80, 182–3
oil turnover 157, 249–50
oils and fats 2–3
  absorption *see* absorption
  composition *see* composition of oils and fats
  degradation processes *see* degradation processes
  fat intake *see* fat
  and flavour
    effect of antioxidants 296–303
    fatty acid composition 285–96
    oil uptake 303–12
  genetic modification 37–8, 140–1, 326–7
  hazard evaluation and preventive measures 241–52
    controlling oil absorption 251–2
    preventing degradation 246–51
    thermal stress, neo-compound formation and toxicological hazard 241–6
  management 240
    crisp manufacture 216–19
  quality *see* quality of oils and fats
  quality control 252–7
  regulation and life of 28–34
    end of frying life 30–4
    fresh oils 29–30, 31
  safety and durability 23
  selection and storage 239–40
  sensory assessment 318
  stability 91–105
    combined effects of natural products on stabilisation 105–8
  testing *see* testing oils and fats
  types of 88–91
  waste oils 35
oleic acid 71, 269
oleo resins 321–2
olestra 325–6
olfactory bulb 316
olfactory receptors 315–16
olive oil 20–1, 118, 119, 178
olive oil deodoriser distillate (OODD) 299-303
one-factor-at-a-time optimisation 344–5, 346
  see also response surface methodology
optical scanners 234
optical sorters 201, 232, 234
'optimum' stage 176
organic standard 38
organoleptic property 254
origin of oils 131, 139–40
oryzanol 95
over-loading 250
oxazoles 281, 282
oxidation 2–3, 100, 268–9
  flavours and aromas derived from 335–6
  secondary oxidation products 269, 270, 275, 276
oxidized fatty acids (OFA) 169–71, 255
Oxifrit test 186, 258

packaging
  crisps 233–4
  declarations 26–8
  French fries 207
  regulation of materials 36
palm kernel oil 116–19
palm oil 88, 89, 117, 119, 128, 136
palm olein 88, 89, 107–8, 122–4, 128, 295–6, 297–8
palm stearin 128
partial glycerides 102–3
particles, potato 218–19
PCM (polar contaminant materials) test 187
peak force (PF) 349–55
peanuts 284
peeling 200, 224
peroxide value (PV) 180, 255–6
pesticides 177–8
phenolic compounds 95–7
phosphate 341
phospholipids 61, 62–3, 99
physical tests
  with instruments 182, 183
  without instruments 181–3
pizzas 11, 12–15
plant breeding 158–9
pneumatic salter 230–1
Polar Compound Tester (PCT 120) 185
polar compounds 29, 245–6, 250
  PCM test 187
  regulation 33–4, 171
  testing 178–9
  used oil quality control 253–4
polydimethyl siloxane (PDMS) 146, 147

polymerisation 100, 218
polymers 180–1
polyunsaturated fatty acids (PUFA) 60–1, 70–1
pork strips 312
Portugal 14–15
post-fry treatments 308–9
potatoes 1–2
  crisp production 234–5
    long-term storage 220–1
    preparation 224–5
    receiving deliveries 221
    requirements 220
  flavour of raw potatoes 267–8
  for French fries 199, 212
  slicing 312
  specific gravity 311
  see also crisps; French fries; pre-fried potato products
poultry 52
  see also chicken
pre-fried potato products 197–214
  future trends 212
  key requirements 199–200
  key manufacturing processes 200–10
    'formed' products 208–10
    French fries 200–7
  nature of 198
  quality of French fries 211–12
  range and use 198–9
  storage and distribution 210–11
pre-fry treatments 252, 307–8
preparation 240
pressure 343
preventive measures 241–52
process control 212, 222, 236–65
  continuous process monitoring 234
  flow diagrams 239–41, 242–3
  frying operation 155–8
  future trends 259
  HACCP approach 237–9
  hazard evaluation and preventive measures 241–52
    controlling oil absorption 251–2
    preventing oil degradation 246–51
    thermal stress and toxicological hazard 241–6
  monitoring critical control points 252–9
    fresh oil quality control 256–7
    fried food quality parameters 257–8
    quick frying process control tests 258–9
    used oil quality control 252–6
processing aids 21, 202
product specifications 166, 172, 173–5
protection index 296, 297
protein 340–1
proteinaceous residues/debris 103–4, 144
pseudo-legislative pressures 36–7
publishers of legislation 41–3
pyrazines 275, 278–9, 283
pyrolysis mass spectrometry 141–2

quality
  French fries 211–12
  fried food quality parameters 257–8
quality of oils and fats 115–64
  authenticity 127–42
  French fries 206
  frying process 152–8
  future trends 158–9
  measurement of authenticity and 165–93
    adulteration 177–8
    future for monitoring quality 189–90
    maintaining quality during frying 166–9
    purchasing specifications and vendor certification 173–5
    quality control during frying 175–7
    refining operations 171–2
    regulatory issues 169–71
    tests for fats and oils 178–89
  minor components 142–8
  monitoring in crisp production 219
  oil uptake and 311–12
  preventing degradation process 247–8
  properties and composition and relationship between composition and suitability 116–27
    modified oils 122–7
    unmodified oils 116–22
  quality control 252–7
    fresh oil 256–7
    unused oil 252–6
  quality limits for a fresh oil 148–9
  transport, delivery and storage 149–52
Quantitative Ingredient Declaration (QUID) 27
quantum satis (QS) 20, 26
quick tests 181–9, 258–9

rancidity, hydrolytic 269–72
random variation 222
range of fried foods 8–9

rapeseed oil 88, 89, 119–20, 137–8, 177
  canola oil 108, 286–91, 293–4
  and flavour 286–92, 298
  hydrogenated 88, 89, 123, 124–5
rapid tests 181–9, 258–9
raw materials 240
  crisp manufacturing 234–5
    management 220–1
  influences on texture and colour 340–2
  pre-fried potato products 199–200, 212
ready-cooked meals 2, 11, 12–15
reconditioning 221
records 239
refining operations 171–2
regulation 169–71
  EU 19–48
    environmental protection 34–6
    future trends 36–8
    life of frying oils 28–34
    legal context 19–21
    sale of food 22–8
    sources of information 38–47
    structure of frying industries 22
  US 49–58
    FDA 49–51
    state and city regulations 53–4
    USDA/FSIS 52
regulations 20
repairs 158
repeated frying 74–5
residues/debris 103–4, 144
response surface methodology (RSM) 337, 343–56
  case study of fried gluten balls 348–55
  principles 343–5
    applying the principles 345–8
restaurant frying 115
retail sale, frying for 115
rheological properties 338–9
rheometers 339
rice bran oil 105–6, 118, 120, 123, 126–7, 136–7
road tankers 149–50, 151
roasted products 284–5
roasted sesame oil 105
roasting 236
roller drying 321
rosemary 96–7, 105, 298
RULER (Remaining Useful Life Evaluation Routine) 183–4
'runaway' stage 176
safe levels of contaminants 24
safety 23
safflower seed oil 118
sage 96–7, 105
sale of food regulation 22–8
salt (sodium chloride) 322, 323–4
salting 230–1
sampling schedule 168
saturated fatty acids (SFA) 60, 62, 70
saturation, relative degree of 217
savoury snacks 9, 11–12
screw blanchers 202–3
seasoning application 232–3
secondary oxidation products 269, 270, 275, 276
secretin 63
Seed Crushers and Oil Processors Association (SCOPA) 150
sensory evaluation 315–18, 349, 350, 352–5
sensory panels 317–18
separating 209
sequestrants, metal 99–100
serrated roll salter 230–1
sesame seed oil 105–6, 118, 120–1, 123, 126–7, 137
sesame seeds 284
sesaminol isomers 92–5
sesamol 92–5
sesamolin 92–5
shallow frying 236
shear testing 338–9
Shortening Monitor 187–8
silica 322
Singapore Cocktail 128
sizing 231–2
slice thickness 304–5
slice washing 226–8
slicing 225–6, 312
smell, sense of 315–16
  see also aroma
smoke point 101, 149, 254
snacks, savoury 9, 11–12
soap 179
sodium tripolyphosphate (STPP) 298
sophisticated sensory panels 318
sorting 201
soya-bean oil 119–20, 127–8, 137–8
  and flavour 286, 291, 292–3, 295
  hydrogenated 124–5, 125–6
Spain 13, 72
specific gravity 309–10, 311
specifications 166, 172, 173–5

spectrophotometric colour measurement 339
speed washer 226–8
spices 320, 321–2
Spot Test 258
spray chilling 320
spray drying 320
squalene 98
stability of oils 91–105
  combined effects of natural products on stabilisation 105–8
stable carbon isotope ratio (SCIR) 137, 141–2
stack drip back 219
starch 309, 310, 340, 341
state regulations 53–4
steam peeling 200
sterols 61
  analysis 136, 141
  containing ethylidene group 97–8
storage
  fried foods 240–1
  long-term storage of potatoes 220–1
  of oil 149–52, 249
    in-plant 218
  pre-fried potato products 210–11
Strecker degradation 272–6
sugars 202, 276
sunflower seed oil 88–90, 118, 122, 127–8, 135–6
  and flavour 291–2, 295–6
  hydrogenated 123, 125
super palm olein 123, 124
supplier certification 173–5
supply chain 22
surface area 304–5
surface treatment 240, 251, 252
surfactant theory 175, 304, 342
Swedish National Food Administration 54, 55
Switzerland 14
synergists 105–8
synthetic fat replacers 325–6

tallow 88, 89, 118, 121, 129–30, 138
taste 202, 315
  see also flavour
temperature, frying 342
  batch frying 155
  control 218
  effect on crisp colour 229
  and oil uptake 310–12
  optimum frying temperature for fried gluten balls 352–5
temperature control apparatus 248
tensile measurement 338
tertiary butyl hydroquinone (TBHQ) 26, 96
testing oils and fats 178–89
  authenticity 131–42
  chemical methods 178–81
  rapid tests 181–9, 258–9
texture 202, 337–58
  case study of fried gluten balls 348–55
  influences on 340–3
  instrumentation for measuring 337, 338–9
  RSM 343–8
thermal stress 241–6
thiazoles 281, 282, 283
thiophenes 281, 283
Third International Symposium on Deep-fat Frying 36–7, 55–7, 91, 189–90
time, frying 310–12, 342–3
tocopherols 92, 93, 133, 135–6, 145–6, 297, 313
tocotrienols 92, 133
tolerable negative error (TNE) 28
tortilla chips 291, 311
toxicological hazard 241–6
total polar materials (TPM) 171, 178–9
trace metals 103, 144
trans-fatty acids 70
transglutaminase 348
transport
  mucosal transport of lipids 65–8
  oils and fats 22, 149–52, 172
tri-calcium phosphate 322
trisodium phosphate 313–14
triglycerides 60, 62–4
turnover of oil 157, 249–50

United Kingdom (UK) market 10–12
United States (US) regulation 49–58
  FDA 49–51
  state and city regulations 53–4
  USDA/FSIS 52
unsaturated fatty acids 60–1, 62
unused oils *see* fresh oils
upgraded intermediate sensory panels 317–18
US Department of Agriculture (USDA) 49, 169
  guidelines and directives 52
used oil quality control 252–6

Vanderbilt, Commodore Cornelius 1
vegetables 10–11, 12–15
vendor certification 173–5
Veri-test 258
very low-density lipoproteins (VLDL) 66
viscosity 254
  measuring 183
volatile substances of decomposition 253
volume expansion 340, 350, 352–5

warmed over flavour (WOF) 312–13
waste oils 35
water *see* moisture content
water knife cutting system 200–1
weights and measures 28
winterised oils 125–6
working women 9

XYZ system 339